Handbook of Materials

Handbook of Materials

Edited by
Reece Hughes

Willford Press
www.willfordpress.com

Published by Willford Press,
118-35 Queens Blvd., Suite 400,
Forest Hills, NY 11375, USA

ISBN: 978-1-68285-393-1

Cataloging-in-Publication Data

Handbook of materials / edited by Reece Hughes.
p. cm.
Includes bibliographical references and index.
ISBN 978-1-68285-393-1
1. Materials. 2. Materials science. I. Hughes, Reece.
TA403 .H36 2017
620.11--dc23

For information on all Willford Press publications
visit our website at www.willfordpress.com

WILLFORD PRESS

Printed in the United States of America.

Contents

Preface

This book on materials is an interdisciplinary approach to the design and manufacture of various materials. Engineering of materials takes into account certain facets of the material such as its characterization, physical and chemical structure, and their performance with consideration to the laws of thermodynamics and kinetics. The most common applications of materials engineering is nanomaterials, biomechanics and electronic engineering. Scientists and students actively engaged in this field will find this book full of crucial and unexplored concepts. For all readers who are interested in materials design and development, the case studies included in this book will serve as an excellent guide to develop a comprehensive understanding.

Various studies have approached the subject by analyzing it with a single perspective, but the present book provides diverse methodologies and techniques to address this field. This book contains theories and applications needed for understanding the subject from different perspectives. The aim is to keep the readers informed about the progress in the field; therefore, the contributions were carefully examined to compile novel researches by specialists from across the globe.

Indeed, the job of the editor is the most crucial and challenging in compiling all chapters into a single book. In the end, I would extend my sincere thanks to the chapter authors for their profound work. I am also thankful for the support provided by my family and colleagues during the compilation of this book.

Editor

1

Biomass Pyrolysis Solids as Reducing Agents: Comparison with Commercial Reducing Agents

Aitziber Adrados *, Isabel De Marco †, Alexander López-Urionabarrenechea †, Jon Solar †, Blanca M. Caballero † and Naia Gastelu †

Academic Editor: Rafael Luque

Department of Environmental and Chemical Engineering, University of the Basque Country UPV/EHU, Bilbao 48013, Spain; isabel.demarco@ehu.eus (I.D.M.); alex.lopez@ehu.eus (A.L.-U.); jon.solar@ehu.eus (J.S.); blancamaria.caballero@ehu.eus (B.M.C.); naia.gastelu@ehu.eus (N.G.)

* Correspondence: aitziber.adrados@ehu.eus

† These authors contributed equally to this work.

Abstract: Biomass is one of the most suitable options to be used as renewable energy source due to its extensive availability and its contribution to reduce greenhouse gas emissions. Pyrolysis of lignocellulosic biomass under appropriate conditions (slow heating rate and high temperatures) can produce a quality solid product, which could be applicable to several metallurgical processes as reducing agent (biocoke or bioreducer). Two woody biomass samples (olives and eucalyptus) were pyrolyzed to produce biocoke. These biocokes were characterized by means of proximate and ultimate analysis, real density, specific surface area, and porosity and were compared with three commercial reducing agents. Finally, reactivity tests were performed both with the biocokes and with the commercial reducing agents. Bioreducers have lower ash and sulfur contents than commercial reducers, higher surface area and porosity, and consequently, much higher reactivity. Bioreducers are not appropriate to be used as top burden in blast furnaces, but they can be used as fuel and reducing agent either tuyére injected at the lower part of the blast furnace or in non-ferrous metallurgical processes where no mechanical strength is needed as, for example, in rotary kilns.

Keywords: biomass; biocoke; bioreducer; slow pyrolysis

---

1. Introduction

The metallurgical industry is a great consumer of fossil fuels. It uses great amounts of coke as fuel and reducing agent [1–3], with the consequence of large emissions of carbon dioxide (CO_2), the primary greenhouse gas emitted by human activities [4].

International concerns about climate change are leading to a decrease in CO_2 emissions. Recently, the Intergovernmental Panel on Climate Change (IPCC) identified in the Fifth Assessment Report the world's carbon budget [5], *i.e.*, the amount of CO_2 the world can emit while still having a likely chance of limiting global temperature rise to 2 °C above pre-industrial levels. The 2 °C target was adopted by the countries within the United Nations Framework Convention on Climate Change (UNFCC) [6]. The international scientific community estimates this budget to be 1 trillion tons of carbon. In 2011, 52% of the budget had already been burned, and it is estimated that the budget will be exceeded before the end of 2045 if emissions rates continue with the same tendency [5]. To avoid this, new ideas and concepts must be developed by the industry.

The production of reducing agents (biocokes or bioreducers) by pyrolysis of biomass is one of these ideas and a step further in the reduction of the environmental impact of industry.

The key factor in the production of coke from biomass is the fact that biomass derived products and processes imply neutral CO_2 emissions. Biomass absorbs atmospheric CO_2 while it grows and returns it into the atmosphere when it is burned, all in a relatively short amount of time. Because of this, biomass utilization creates a closed loop carbon cycle. The substitution of fossil fuel coke by more environmentally friendly biocokes is of the most interest. For this reason, the objective of this paper is to compare the properties of biocokes obtained by pyrolysis of lignocellulosic biomass with those of typical commercial reducing agents used in non-ferrous processes, in order to evaluate the suitability of the biocokes as reducing agents.

The production of biocoke from biomass is not new; in fact it is the update and modernization of the traditional process of obtaining charcoal from wood. In recent years the effects of the type of biomass and of the operating parameters used on the properties of the obtained char have been investigated [7–12].

The production of a coke-like product from biomass requires more demanding process conditions than the traditional method, since traditional coke is a macroporous carbon material produced by carbonization of coals or coal blends of specific rank and characteristics [13]. In other words, to obtain a coke-like product from wood, the carbonization of wood followed by the carbonization of the resultant coal must occur in the same process. Griessacher and Antrekowitsch [14] and Griessacher, *et al.* [15] reported that a coke-like product can be obtained by slow pyrolysis of wood at high temperature (700–1000 °C).

One of the most important influencing parameter of the slow pyrolysis (carbonization) process is temperature. It is well known that by increasing temperature the amount of volatile components in the coal decreases and therefore the carbon content in the coal becomes higher reaching the carbon content of cokes [16]. Heating rate is also a very important parameter that influences the amount and characteristics of pyrolysis products. It has been reported [17] that a higher char yield is obtained with decreasing heating rate, since the slow emerging wood products have more time to react and carbonize. It may be expected that the lower the heating rate, the lower the char porosity and reactivity since the volatile products from the biomass emerge more slowly. The objective of the present work was to obtain solids from biomass useful as reducing agents for metallurgical processes (biocoke or bioreducers). Therefore, high temperatures and slow heating rates were used. On the other hand, the feedstock material is a key factor for the production of biocoke, and therefore, two woody biomass samples, olive tree cuttings, and eucalyptus, were used in this study. It is well known that woody biomass is the best biomass feedstock for coke production.

In order to evaluate the suitability of the pyrolysis solids as alternative reducing agents in metallurgy, it is important to characterize the biocokes, by determining their chemical and physical properties, and also their reactivity with CO_2. There are standardized reactivity measurement methods like the ASTM D 5341-99 standard [18,19] which was developed from the Nippon Steel Corporation (NSC) test procedure, and has been traditionally used for evaluating the quality of coal derived metallurgical cokes.

In this study, characterization and reactivity measurements of the biocokes obtained in pyrolysis experiments as well as of three commercial reducing agents used in metallurgical processes were carried out.

2. Materials and Methods

2.1. Materials

Two different biomass samples were pyrolyzed: an abundant waste woody biomass sample, olive tree cuttings, coming from the South of Spain and an extensive energy crop, eucalyptus (camaldulensis) wood, coming from South America.

The olive tree cuttings sample was selected since the major woody crop in Spain is olive. In Spain there are 2,605,252 ha of olive groves [20]. One hectare of olive groves generates about three tons of pruning residues [21] which means 7,815,756 tons of branches and leaves per year, most of which are illegally burnt or left on the ground partially for recirculation of nutrients.

On the other hand, the eucalyptus family was selected as an extensive energy crop because of its fast growth rate and abundance, especially in South America. It is worth mentioning that Eucalyptus occupies 70% of the 950,000 ha of the forest plantation in Uruguay and 60% of the 5,500,000 ha of forest plantation in Brazil.

The composition of both biomass samples is presented in Table 1.

Table 1. Proximate, ultimate and constituents analyses of the original pyrolyzed biomass samples.

(Weight %)		Olives	Eucalyptus
Proximate analysis (ar) [a]	Moisture	8.8	11.4
	Volatile matter	74.3	69.7
	Ash	2.1	2.1
	Fixed carbon [c]	14.8	16.8
Ultimate analysis (daf) [b]	C	49.4	52.7
	H	6.2	3.7
	N	0.3	0.1
	S	<0.1	<0.1
	Others (mainly O) [c]	44.1	43.5
HHV (ar) [a] ($MJ\cdot kg^{-1}$)		16.1	16.6
H/C atomic ratio		1.5	0.8
C/N weight ratio		146.7	455.0
Constituents (daf) [b]	Extractives	12.1	8.3
	Cellulose	32.9	38.5
	Hemicellulose	19.7	13.6
	Lignin	18.3	29.3
	Others [c]	15.7	10.3

[a] ar: as received basis; [b] daf: dry and ash free basis; [c] By difference.

It can be observed that the eucalyptus and the olive tree cuttings samples are quite different. The eucalyptus sample has more moisture, less volatiles and more fixed carbon than the olive sample. Concerning the ultimate analysis, the eucalyptus sample has more C%, less H% and less N% than the olives sample, and as a consequence has lower H/C and higher C/N ratios than the olives sample. The lower the C/N ratio, the higher the degradability of the organic matter, so the results indicate that eucalyptus is more resistant to biodegradation than the olives sample.

Concerning the constituents' composition, Table 1 shows that the eucalyptus sample has greater lignin content and lower hemicellulose content than the olives sample. With respect to extractives, it has been reported that they are more abundant in bark than in inner wood parts [22]. This justifies the fact that the eucalyptus sample, which has a lower bark/wood ratio than the olives sample, has less extractives.

Regarding the higher heating value of the biomass samples, there are no significant differences between them. Both values are around 16.4 $MJ\cdot kg^{-1}$, which are heating values comparable to those of low quality fuels like lignites ($\approx$16 $MJ\cdot kg^{-1}$).

In order to compare the properties of the pyrolitically produced biocokes, three commercial reducers provided by Befesa Zinc Aser S.A., an Electric Arc Furnace dust recycling company, located in Erandio (Biscay, Spain) were analyzed and tested: a metallurgical coke, a petroleum coke, and an anthracite. The characterization of these commercial reducers is presented in Table 5, and is discussed and compared to biocoke characteristics in Section 3.1.

2.2. Experimental Procedure

2.2.1. Pyrolysis Experiments

Pyrolysis experiments with biomass samples were carried out using two reactors connected in series: a first 3.5 L semi-batch non-stirred pyrolysis reactor where 100 g of biomass were pyrolyzed at 600 °C and 750 °C and with heating rates of 3, 15, and 20 $°C\cdot min^{-1}$, and a second tubular reactor where pyrolysis vapors were thermally upgraded at 900 °C. The vapors were swept with 1 L· min^{-1} of nitrogen (N_2) to a water cooled glass condensing system, where liquids were collected, followed by an activated carbon column (to retain the possible small particles that the gases could transport) before they were collected in plastic gas bags. The biocokes produced remained in the first reactor. The solid and liquid yields were determined by weight difference, while gas yield was calculated by difference to 100. Each yield presented in this paper is the mean value of the data obtained in two equivalent experiments which differed by less than three points. A flow sheet of the experimental setup used can be seen in Adrados, *et al.* [23].

2.2.2. Reactivity Experiments

The procedure used for the reactivity experiments is based on the ASTM D 5341-99 standard [18]. However, to meet the specifications of this standard, specific large equipment and a great amount of sample (250 g) is required. Since neither of them was available, an adaptation of the ASTM method had to be carried out trying to keep the different operational parameters as close as possible to the specifications of the standard. This ASTM standard is based on the Boudouard reaction ($C + CO_2 \leftrightarrow CO$), in which carbon reacts with CO_2 and forms carbon monoxide (CO). Depending on the reactivity of the material, the quantity of carbon consumed is variable.

Two parameters were determined in the reactivity tests: the so called coke reactivity index (CRI) and the *R* factor (1) and (2).

$$\text{CRI } (\%) = \frac{\text{initial sample weight} - \text{final weight}}{\text{initial sample weight}} \times 100 \quad (1)$$

$$R \text{ factor } (\%) = \frac{CO}{CO_2 + CO/2} \times 100 \quad (2)$$

The CRI is the mass weight loss after the treatment of the sample with CO_2 or the degree of conversion (wt %) and the R factor is the amount of CO in the gas stream after reaction with respect to the total amount of CO_2 used in the test (vol %). High values of both parameters indicate high reactivity of the sample.

The reactivity experiments with the biocokes and the commercial reducers were performed in the same plant used for pyrolysis experiments but adapted for the reactivity tests. In this case, the installation was composed of four units connected in series: (1) a first 3.5 L empty preheater, where the feed gases were pre-heated to 800 °C; (2) a second tubular reactor where a fixed bed of the biocoke was placed; (3) a dust trap (empty bubbler) where the small solid particles swept by the CO_2 flow were retained; and (4) Tedlar plastic bags where the gases of the whole test were collected. All the apparatus were connected with silicone tubing.

The tubular reactor used for the reactivity experiments was 2.54 cm in diameter and 60 cm long, and was made of INCONEL alloy 601, a material able to work at temperatures as high as 1200 °C, while in the pyrolysis experiments a 309 stainless steel tubular reactor was used.

The system was purged with N_2 in order to remove the oxygen (O_2) originally contained in the installation. The reactor's furnace was warmed up to 1000 °C, and then a CO_2 gas flow of 1375 or 750 mL· min^{-1} was passed through the sample for 2 h. The charcoal reacts with CO_2, producing CO. The gases generated in the whole run were continuously collected in plastic bags, to be immediately analyzed off-line with a gas chromatograph connected to a thermal conductivity and a flame ionization detector (GC-TCD, FID). After 2 h, the input gas stream was again swapped to N_2 and the furnace was cooled down.

2.3. *Analytical Techniques*

The initial feedstock, the solid pyrolysis products, and the commercial reducers were characterized using the following analytical techniques. The proximate analysis was determined by thermogravimetry according to D3173-85 and D3174-82 ASTM standards by means of a LECO TGA-500, and the elemental composition was determined with a LECO TruSpec CHN and a TruSpec S analyzer. The higher heating value (HHV) was determined using a LECO AC-500 automatic calorimetric bomb.

Real density was determined in an AccuPyc 1330T Micromeritics equipment. All the samples were degassed at 120 °C for 18 h before the analyses. The surface areas and the porosity properties were measured via CO_2 adsorption at 0 °C using a surface area analyzer, Quantachrome Nova 4200 apparatus. CO_2 adsorption is a widely used method to analyze materials with narrow micropores, as is the case of carbonized materials. The isotherms were analyzed using the Dubinin Radushkevich equation (DR) for calculating the micropores volume ($cm^3 \cdot g^{-1}$) and the equivalent micropore surface area ($m^2 \cdot g^{-1}$). The pore size distribution was calculated using the Non Local Density Functional Theory (NL DFT) available in the software of the equipment.

3. Results and Discussion

3.1. Yields, and Proximate and Ultimate Analysis of Pyrolysis Solids

The characteristics of the solids obtained in the pyrolysis experiments can be influenced by the operating conditions of the first pyrolysis reactor. Olive tree cuttings and eucalyptus were pyrolyzed at different temperatures and heating rates, therefore, several biochars were obtained. Their yields and characterization are presented in Tables 2 and 3.

Table 2. Proximate and ultimate analyses of the pyrolysis solids obtained at different temperatures (20 $°C \cdot min^{-1}$).

(Weight %)		Olives		Eucalyptus	
		600 °C	750 °C	600 °C	750 °C
Solid yields		22.9	21.3	22.4	22.9
Proximate analysis (ap) [a]	Moisture	2.0	1.7	2.2	1.4
	Volatile matter	11.8	10.0	12.0	8.6
	Ash	9.5	9.2	5.9	6.2
	Fixed Carbon [c]	76.7	79.1	79.9	83.8
Ultimate analysis (daf) [b]	C	86.0	94.5	88.0	90.1
	H	2.1	0.9	1.4	0.8
	N	0.8	1.2	0.6	0.6
	Others (mainly O) [c]	11.1	3.4	10.0	8.5
HHV (ap) [a] ($MJ\ kg^{-1}$)		30.8	28.8	31.6	31.0

[a] ap: as produced basis; [b] daf: dry and ash free basis; [c] By difference.

Table 3. Characteristics of the pyrolysis solids obtained at different heating rates (750 °C).

(Weight %)		Olives			Eucalyptus	
		20 $°C \cdot min^{-1}$	15 $°C \cdot min^{-1}$	3 $°C \cdot min^{-1}$	20 $°C \cdot min^{-1}$	3 $°C \cdot min^{-1}$
Solid yields		21.3	24.6	26.1	22.9	26.0
Proximate analysis (ap) [a]	Moisture	1.7	1.6	1.7	1.4	1.4
	Volatile matter	10.0	10.1	10.6	8.6	8.0
	Ash	9.2	10.2	9.2	6.2	5.3
	Fixed Carbon [b]	79.1	78.1	78.5	83.8	85.3
Ultimate analysis (daf) [c]	C	94.5	92.9	94.6	90.1	96.2
	H	0.9	1.1	0.8	0.8	0.9
	N	1.2	1.4	1.3	0.6	0.9
	Others (mainly O) [b]	3.4	4.6	3.3	8.5	2.0
HHV (ap) [a] ($MJ\ kg^{-1}$)		28.8	27.5	29.5	31.0	31.6

[a] ap: as produced basis; [b] By difference; [c] daf: dry and ash free basis.

Taking a look at the solid yields, in Table 2 it can be seen that temperature has not much influence. However, Table 3 shows an influence of the heating rate: the lower the heating rate, the higher the solid yield. These results are in agreement with the results obtained by other authors [24,25]. The pyrolysis solid yields are in the range of 21–23 wt % at 20 °C· min^{-1} and 26 wt % when the lowest heating rate is used. Therefore the reduction of heating rate has the beneficial effect of increasing the pyrolysis solids yields. Liquid and gas yields obtained in these pyrolysis experiments can be found in [26].

Pyrolysis solids are mainly composed of carbon, with elemental carbon contents around 90 wt %. The moisture content is rather low (≈1.7 wt %), the volatiles content is quite low (≈11 wt %) and the ash content is around 9 and 6 wt % for olives and eucalyptus samples respectively.

It is well known that the higher the fixed carbon and the lower the volatile matter content, the better is the quality of metallurgical reducing agents. Table 2 shows that the fixed and elemental carbon contents of the biocokes of both samples are higher while the volatile matter is lower when the higher pyrolysis temperature (750 °C) is used. Similar tendencies regarding the effect of temperature in the fixed carbon and volatile matter contents of pyrolysis solids have been reported in the literature [27].

Table 3 shows that there is no influence of heating rate on the composition of pyrolysis solids. Although it is a well-known fact that decreasing heating rates intensifies the carbonization process giving rise to carbon-richer products, this tendency is not observed in the results of Table 3. The reason for this may be that 20 °C· min^{-1} is a heating rate slow enough to produce a well carbonized product at 750 °C, so that slower heating rates do not further promote carbonization at such temperature.

Regarding the HHV of the pyrolysis solids it can be seen that it is rather high (≈30 MJ· kg^{-1}) which is in the range of the HHV of commonly used solid fossil fuels, such as bituminous coals [28].

In view of the characteristics of biomass derived pyrolysis solids, many potential applications for them can be proposed. They can be used as rather good quality solid fuel (high HHV, low pollutants (S, N) content and low ash contents compared to fossil fuels), or as sorbent material provided that it is firstly activated, or as soil amendment agent. However, the application proposed in this work is as reducing agent for metallurgical purposes. For this reason, the three previously mentioned commercial reducing agents (metallurgical coke, petroleum coke, and anthracite), used for Zn reduction in rotary kilns by Befesa Zinc Aser S.A. company, were characterized. The detailed specifications that Befesa Zinc Aser S.A. requires from their providers of reducing agents are presented in Table 4. The specifications on dry basis, which are also included in the table, have been calculated considering that the moisture content is 20 wt %, which is as the table shows, the maximum allowable moisture specified by Befesa; these data on dry basis would be the permitted values if the sample did not contain moisture and enable coke samples to be compared regardless of their moisture.

Table 4. Quality specifications for commercial reducing agents required by Befesa Zinc Aser S.A. company 2013.

Material	Parameter	Befesa Zinc Aser Technical Specifications (wt %)	Specifications on Dry Basis (wt %) *
Metallurgical coke	Granulometry	>10 mm: ⩽20% <2 mm: ⩽40% on daily sample	–
	Dry ash	⩽20% on monthly sample	⩽20%
	Moisture	⩽20% on daily sample	–
	Volatile matter	⩽7% on monthly sample	⩽8.75%
	Sulfur	⩽3% on monthly sample	⩽3.75%

Table 4. *Cont.*

Material	Parameter	Befesa Zinc Aser Technical Specifications (wt %)	Specifications on Dry Basis (wt %) *
Petroleum coke	Granulometry	<2 mm: >30% >10 mm: ⩽20% on diary sample	–
	Dry ash Moisture Volatile matter Sulfur	⩽20% on monthly sample ⩽20% on daily sample ⩽15% on monthly sample ⩽3% on monthly sample	⩽20% – ⩽18.75% ⩽3.75%
Anthracite	Granulometry	<2 mm: >30% >10 mm: ⩽20% on diary sample	–
	Dry ash Moisture Volatile matter Sulfur	⩽20% on monthly sample ⩽20% on daily sample ⩽7% on monthly sample ⩽3% on monthly sample	⩽20% – ⩽8.75% ⩽3.75%

* Calculated considering a moisture content of 20 wt % (maximum permitted by Befesa).

Table 5 presents the proximate and ultimate analyses of the commercial reducing agents and the biomass-based reducing agents obtained at 3 °C· min^{-1} and 750 °C, all on dry basis. For better analysis of the results, the Befesa specifications are included in brackets in Table 5.

Table 5. Proximate and ultimate analyses of the commercial reducers and the bioreducers (750 °C and 3 °C· min^{-1}).

Dry Basis (wt %)		Commercial Reducers			Bioreducers	
		Metallurgical Coke	Petroleum Coke	Anthracite	Olives	Eucalyp.
Moisture		11.4 (<20)	6.4 (<20)	18.0 (<20)	1.7	1.4
Proximate analysis	Volatile matter	3.9 (*<8.75*)	10.1 (*<18.75*)	7.2 (*<8.75*)	10.8	8.1
	Ash	12.5 (*<20*)	2.0 (<20)	11.2 (*<20*)	9.4	5.4
	Fixed carbon *	83.6	87.9	81.6	79.9	86.5
Ultimate analysis	C	84.4	83.5	86.6	85.6	90.9
	H	0.5	2.8	0.6	0.8	1.0
	N	1.0	1.3	1.0	1.2	0.8
	S	0.9 (*<3.75*)	5.6 (*<3.75*)	0.6 (*<3.75*)	<0.05	<0.05
	Others *	0.8	4.9	0.0	3.0	1.9
HHV (MJ· kg^{-1})		29.3	35.9	31.1	29.5	31.6

* By difference; ( ) In brackets Befesa specifications on dry basis.

It is a fact worth mentioning that the commercial reducers have very high moisture contents. Such contents certainly do not correspond to the natural inherent moisture of these materials, and it is most probably accidental moisture incorporated into the samples during their transport and/or storage. As a matter of fact, water drops could be seen in the containers in which the commercial reducers were provided. For this reason, the proximate and ultimate analyses of all the samples are presented in Table 5 on dry basis, in order to more fairly compare the intrinsic properties of the commercial and the biomass derived reducers, regardless of the moisture content.

It has also to be mentioned that although the commercial reducers were provided by Befesa Zinc Aser S.A. itself, one of them (petroleum coke) does not totally fulfill the quality requirements specified by the company, since it contains more sulfur (5.6 wt %) than that specified in Table 4 (<3 wt %).

Comparing the results of the commercial reducers and the bioreducers, the following advantages of the bioreducers can be mentioned: they have much lower moisture contents than any of the commercial reducers, lower ash contents than the metallurgical coke and the anthracite, and significantly lower sulfur contents than any of the commercial reducers. There is only one specification not fulfilled in the bioreducer derived from the olives sample: the volatile content is slightly higher than that specified for metallurgical coke and anthracite, though it does meet, by far, the petroleum coke volatile matter specification.

Therefore, it can be concluded that as far as composition is concerned, the olives and eucalyptus derived bioreducers can replace the commercial reducers in metallurgical processes in rotary kilns with the great advantage of having lower ash and sulfur contents.

3.2. Real Density, Specific Surface Area, and Porosity Measurements

Real density, specific surface area, and porosity of the pyrolysis reducers and of the commercial reducers were measured.

The results obtained in the textural characterization of the pyrolysis solids obtained at 20 °C· min^{-1} and different pyrolysis temperatures are presented in Figures 1 and 2 and Table 6.

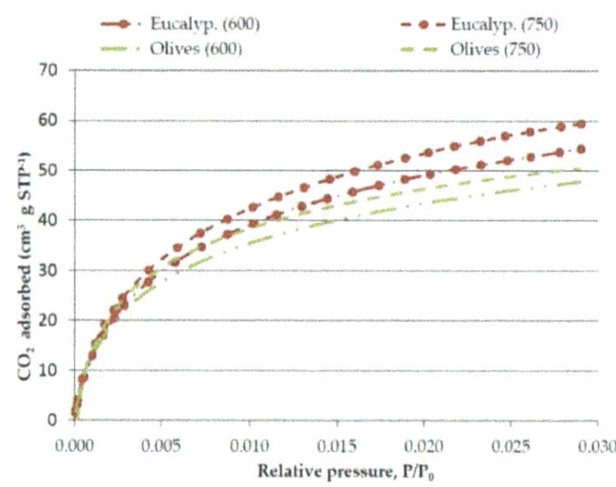

Figure 1. Effect of temperature on CO_2 adsorption isotherms of the bioreducers obtained at 20 °C· min^{-1} from olives and eucalyptus samples.

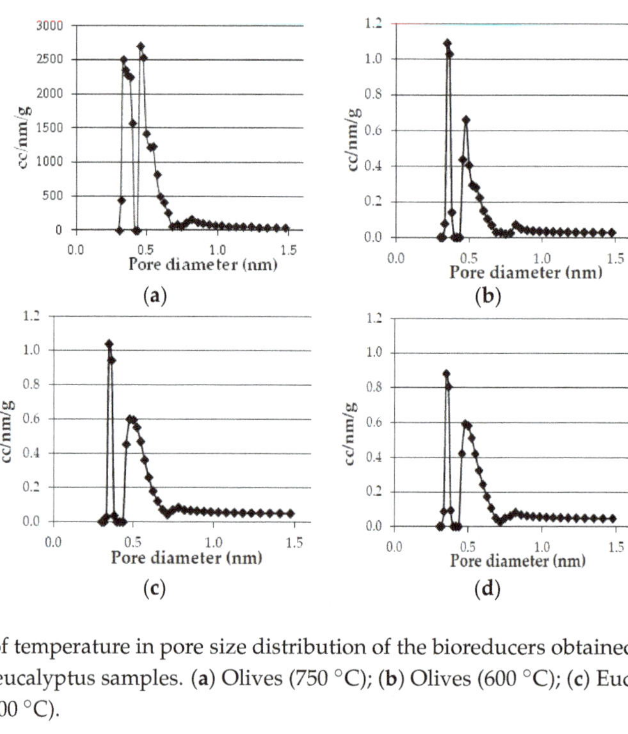

Figure 2. Effect of temperature in pore size distribution of the bioreducers obtained at 20 °C· min^{-1} from olives and eucalyptus samples. (**a**) Olives (750 °C); (**b**) Olives (600 °C); (**c**) Eucalyptus (750 °C); (**a**) Eucalyptus (600 °C).

Table 6. Effect of the temperature in the textural characterization of the bioreducers (20 °C· min^{-1}).

	Temperature (°C)	Real Density (g· cm^{-3})	Micropore Volume (cm^3· g^{-1})	Micropore Equivalent Surface Area (m^2· g^{-1})
Olives	600	1.669	0.15	360
	750	1.926	0.16	375
Eucalyptus	600	1.622	0.19	444
	750	1.849	0.20	476

The CO_2 adsorption isotherms presented in Figure 1 show the CO_2 volume adsorbed (cm^3· g^{-1}) against the relative pressure (P/P_0). The CO_2 adsorption isotherms are clearly of Type I that are typical of microporous materials. Although there is some controversy, it is frequently considered that the micropores volume determined by CO_2 adsorption corresponds to narrow micropores (<0.7 nm).

The pore size distributions of the bioreducers samples calculated by the NL-DFT method are presented in Figure 2, which show that the bioreducers do not have micropores greater than 0.7 nm, therefore, the micropores volume determined corresponds to the total pores volume.

Concerning the effect of the pyrolysis temperature, Table 6 shows that the real density clearly increases as the temperature is raised from 600 °C to 750 °C, while the micropore volume and surface area also increases but very slightly.

The increase of micropore surface area with temperature is contrary to what has been observed by Burhenne, *et al.* [29] who reported a decrease in char surface area when the pyrolysis temperature was raised from 500 °C to 800 °C. This discrepancy may be due to differences in the surface area determination methods: N_2 (used by Burhenne) *versus* CO_2 adsorption (used in this study). Burhenne's reasoning is that at higher temperature an occlusion of most micropores occurs; however it is plausible that what happens is that at higher pyrolysis temperature, narrower micropores are produced, and these, as will be explained next, are difficult to be measured with liquid N_2.

Agirre, *et al.* [30] also found that the surface area increases when the pyrolysis temperature is raised. The reason for the increase of surface area with temperature may be that at higher temperatures greater amount of volatiles are driven out of the solid, giving rise to more micropores.

Concerning the effect of heating rate Figures 3 and 4 and Table 7 show the results obtained on the textural characterization of the solids obtained with olives and eucalyptus samples in pyrolysis at 750 °C with different heating rates.

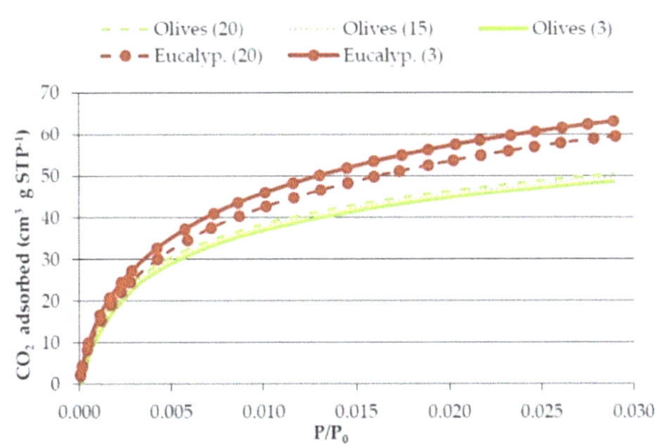

Figure 3. Effect of heating rate on CO_2 adsorption isotherms of the bioreducers obtained at 750 °C from the olives and eucalyptus samples.

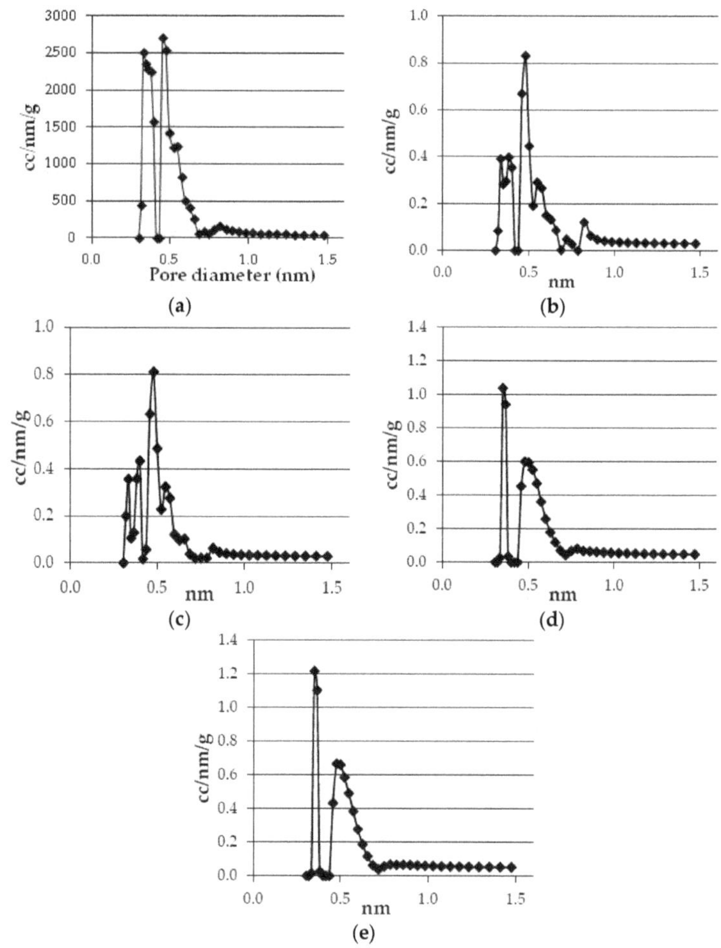

Figure 4. Effect of heating rate in pore size distribution of the bioreducers obtained at 750 °C from the olives and eucalyptus samples. (**a**) Olives 20 $^{\circ}C \cdot min^{-1}$; (**b**) Olives 15 $^{\circ}C \cdot min^{-1}$; (**c**) Olives 3 $^{\circ}C \cdot min^{-1}$; (**d**) Eucalyptus 20 $^{\circ}C \cdot min^{-1}$; (**e**) Eucalyptus 3 $^{\circ}C \cdot min^{-1}$.

Table 7. Effect of heating rate in the textural characterization of the bioreducers (750 °C).

	Heating Rate ($^{\circ}C \cdot min^{-1}$)	Real Density ($g \cdot cm^{-3}$)	Micropore Volume ($cm^3 \cdot g^{-1}$)	Micropore Equivalent Surface Area ($m^2 \cdot g^{-1}$)
Olives	20	1.926	0.16	375
	15	1.867	0.15	362
	3	1.871	0.15	363
Eucalyptus	20	1.849	0.20	476
	3	1.830	0.21	501

There is no clear influence of heating rate on the textural properties. Table 7 shows that in the case of olives the real density, micropore volume, and surface area somewhat decrease with the decrease in the heating rate, but in the case of the eucalyptus sample the density also decreases but the micropore volume and surface area somewhat increase. Figures 3 and 4 also show a different effect of heating rate in olive solids and in eucalyptus solids. It has been reported that the longer the char residence time, the greater its surface area [31,32]. This was not the case in this study, which may be due to the fact that 20 $^{\circ}C \cdot min^{-1}$ is a low enough heating rate under which no further increase in surface area is produced.

If the surface areas of the bioreducers are compared to those presented in the literature, it can be seen that there are great differences. Very low charcoal surface areas have been reported by several authors: Huo, *et al.* [33] reported N_2 BET area < 9 $m^2 \cdot g^{-1}$ for chars obtained at 900 °C from saw dust and straw, Burhenne, *et al.* [29] reported N_2 BET area ≈ 1 $m^2 \cdot g^{-1}$ for spruce wood chars obtained at 800 °C, Senneca [34] also reported N_2 BET areas < 1 $m^2 \cdot g^{-1}$ for char samples obtained from pine wood at 850 °C.

On the contrary, other authors report surface areas in the range of those obtained in this study. Rösler, *et al.* [35] obtained surface areas (measured by CO_2 adsorption) between 325 and 425 $m^2 \cdot g^{-1}$ for biochars obtained at 750 °C, Burhenne, *et al.* [29] obtained BET areas of 200–400 $m^2 \cdot g^{-1}$ for chars produced at 500 °C, and Carrier, *et al.* [36] reported a N_2 BET area of 259 $m^2 \cdot g^{-1}$ for char obtained from sugarcane bagasse at 460 °C and 349–452 $m^2 \cdot g^{-1}$ for that same char but after heating it to 800 °C and 900 °C in a N_2 atmosphere, as well as 441–570 $m^2 \cdot g^{-1}$ after activating the char with steam at 700–900 °C.

The reason for these differences in the surface areas is, most probably, the determination method used. As has been mentioned before, it is difficult to obtain reliable results with N_2 adsorption methods with carbonaceous materials, since it is carried at −196 °C and at such low temperature, N_2 has diffusion problems in the narrow micropores and therefore, it may take a long time to reach thermodynamic adsorption equilibrium, it may take (if reached) even weeks. Another reason for erroneous BET area measurements with liquid N_2 that has been reported in the literature [37], is that there can be pore shrinkage, so that the measured surface areas are apparent, and not real. In this study, in the first place, it was tried to determine the surface areas by liquid N_2 adsorption, but for most of the bioreducers samples the equilibrium was not reached, even after a week, and in those samples that reached equilibrium, the BET area obtained was very low (7–15 $m^2 \cdot g^{-1}$).

The bioreducers obtained in this study have surface areas comparable to those reported by Carrier, *et al.* [36]. In particular, the eucalyptus derived bioreducers present surface areas (476–501 $m^2 \cdot g^{-1}$) equivalent to those reported by Carrier, *et al.* [36] for sugarcane derived chars activated with steam at 700–900 °C for 1 hour (441–570 $m^2 \cdot g^{-1}$). Therefore, the eucalyptus bioreducers could be used as a sorbent material without needing to be activated. However, it has to be mentioned that the surface area of the eucalyptus bioreducers, the same as that of Carrier, *et al.* [36] activated chars, are well below commercial activated carbon values (BET area ≈ 1000 $m^2 \cdot g^{-1}$).

A comparison between the textural properties of the three commercial reducing agents and of the pyrolysis solids obtained from olives and eucalyptus samples at 750 °C and 3 $°C \cdot min^{-1}$ are presented in Table 8.

Table 8. Textural characterization of commercial reducers and bioreducers (750 °C and 3 $°C \cdot min^{-1}$).

	Real Density ($g \cdot cm^{-3}$)	Micropore Volume ($cm^3 \cdot g^{-1}$)	Micropore Equivalent Surface Area ($m^2 \cdot g^{-1}$)
Olives	1.871	0.15	363
Eucalyptus	1.830	0.21	501
Metallurgical coke	1.916	0.01	24
Petroleum coke	1.389	0.07	156
Anthracite	1.793	0.05	122

Concerning real density, Table 8 shows that the bioreducers are comparable to metallurgical coke and anthracite, while petroleum coke has a significantly lower real density. This may be attributed to the lower ash content and the higher micropore volume of petroleum coke.

With respect to surface area, it can be seen that the bioreducers surface areas are much greater than those of the commercial reducers. Therefore, a higher reactivity of the bioreducers compared to the commercial reducers might be expected (this will be discussed in the following section).

Figure 5 shows the CO_2 adsorption isotherms of the three commercial reducing agents and of the pyrolysis solids obtained from olives and eucalyptus samples at 750 °C and 3 °C· min^{-1}. It has to be mentioned that although it is not clearly visible (due to the scale) the commercial reducers isotherms are also Type I, which is typical of microporous materials. However, as might have been expected, according to the surface area data, the commercial reducers' curves are much lower than those of the bioreducers.

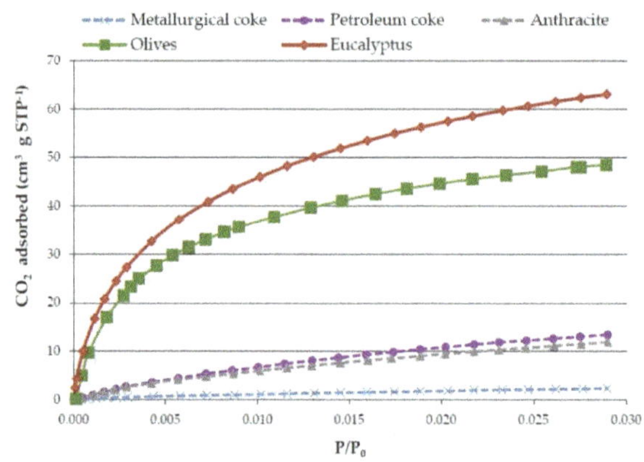

Figure 5. CO_2 adsorption isotherms of commercial reducers and olives and eucalyptus bioreducers (obtained at 750 °C and 3 °C· min^{-1}).

The pore size distributions of the bioreducers and the commercial reducing agents calculated by the NL-DFT method are presented in Figure 6. It can be seen that there are important differences between the curves that correspond to the commercial reducers and the pore size distribution of the bioreducers. First of all, the amount of pore of each size is much smaller in the commercial reducers than in the bioreducers. It has to be born in mind that the y-axis scale is about 10 times lower in the commercial reducers' graphs than in the bioreducers' graphs.

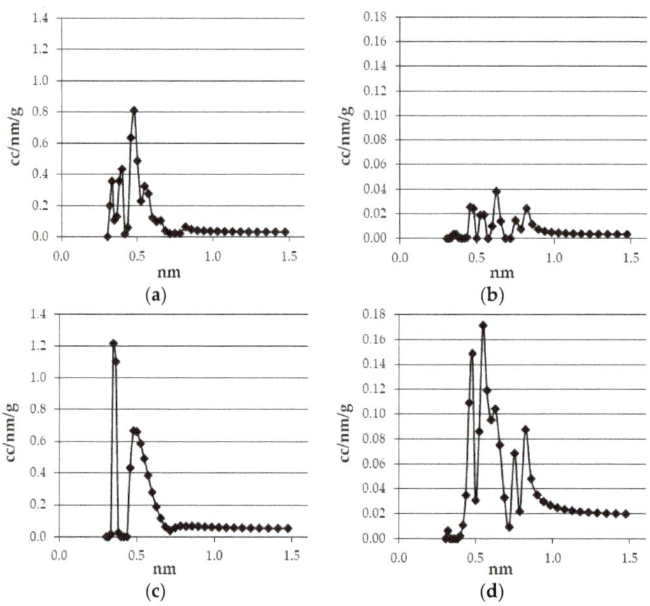

Figure 6. *Cont.*

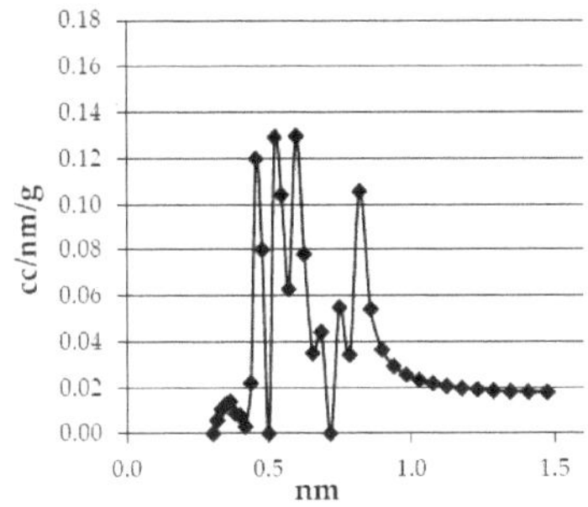

(e)

Figure 6. Pore size distribution of commercial reducers and olives and eucalyptus bioreducers (obtained at 750 °C and 3 °C· min^{-1}). (**a**) Olives; (**b**) Metallurgical coke; (**c**) Eucalyptus; (**d**) Petroleum coke; (**e**) Anthracite.

To summarize, the main conclusion concerning the comparison of the bioreducers with the commercial reducing agents is that, although in terms of composition (proximate and ultimate analyses) both types of reducing agents are quite similar, or even better, the bioreducers, in terms of surface area and porosity, are very different, and this brings about, as will be seen in the following section, different reactivity behaviors.

3.3. *Reactivity*

The reactivity experiments carried out were classified in the following three groups:

- Reactivity tests with the three commercial reducers (metallurgical coke, petroleum coke, and anthracite) using 50 g of sample and 1375 mL· min^{-1} of CO_2 flow.
- Reactivity tests with metallurgical coke using different amounts of sample and two CO_2 flow ratios (1375 mL· min^{-1} and 750 mL· min^{-1}).
- Reactivity tests with the two bioreducers (olives and eucalyptus) using 30 g of sample and 750 mL· min^{-1} of CO_2 flow.

The reasons for performing these three sets of experiments as well as the results obtained are explained in the following section.

3.3.1. Reactivity Tests with the Commercial Reducers

The objective of this first set of reactivity tests was to determine the reactivity behavior of the reducing agents used and provided by Befesa Zinc Aser S.A. (metallurgical coke, petroleum coke, and anthracite). As has been mentioned in Section 2.2.2 the reactivity tests were performed based on the ASTM D 5341-99 standard but with a smaller amount of sample. In this first set of tests, 50 g of sample and 1375 mL· min^{-1} CO_2 flow were used, which are proportional to the specifications of the ASTM standard.

The results obtained of both, CRI and R factor, are presented in Table 9. It can be seen that the metallurgical coke and anthracite have similar reactivities according to both, their CRI and R factor, while both parameters are significantly lower for the petroleum coke. Agirre, *et al.* [30] have reported CRI and R values for a typical commercial petroleum coke (13.7% and 16.5%, respectively) which are comparable but somewhat lower than those obtained in this study (16.5% and 20.0%, respectively). Other authors have also observed that petroleum cokes have lower reactivities than metallurgical cokes [38]. It has been reported that this may be because petroleum coke has somewhat more crystalline structures which are less reactive [39].

Table 9. Reactivity results of commercial reducing agents (50 g and 1375 mL· min^{-1} CO_2).

	Metallurgical Coke	Petroleum Coke	Anthracite
CRI (%)	30.9 ± 0.1	16.5 ± 0.2	31.6 ± 0.7
R factor (%)	47.4 ± 1.8	20.0 ± 0.1	47.4 ± 0.6

After this first set of reactivity tests, it was attempted to determine the reactivity of the pyrolysis solids. However, due to the lower apparent density of these solids (0.45 g cm^{-3} for olives and 0.42 g· cm^{-3} for eucalyptus, compared to 1.11 for metallurgical coke, 0.72 for petroleum coke and 1.31 for anthracite), it was not possible to introduce 50 g of sample in the tubular reactor. Therefore, it was decided to carry out some experiments to investigate if the amount of sample used could have an influence on the results obtained on the reactivity tests. The results are presented in the following section.

3.3.2. Reactivity Tests of the Metallurgical Coke with Different Amounts of Sample and Different CO_2 Flows

The CRI and *R* factor obtained with the metallurgical coke using different amounts of sample and different CO^2 flows are presented in Table 10.

Table 10. Reactivity results of metallurgical coke.

	1375 mL· min^{-1}			750 mL· min^{-1}
	50 g	30 g	15 g	30 g
CRI (%)	30.9 ± 0.1	43.5 ± 1.3	41.0 ± 2.0	34.5 ± 0.5
R factor (%)	47.4 ± 1.8	27.0 ± 0.4	9.8 ± 0.1	45.9 ± 1.2

The reactivity results obtained with 1375 mL· min^{-1} CO_2 flow and with different amounts of sample show that the CRI index significantly increases when the amount of sample is reduced from 50 g to 30 g, but no further increase is observed from 30 g to 15 g. The explanation of this fact may be the following: the greater the amount of sample, the larger is the height occupied by it inside the tubular reactor; the CO_2 stream is fed into the reactor through the bottom, therefore it starts reacting with the coke, and as it goes up through the char bed it is impoverished in CO_2, leading to a lower reaction rate at the upper part of the reactor. This effect is not observed from the 15 g to the 30 g tests probably because with smaller column heights there is CO_2 enough all through the column both with the 15 and the 30 g samples and therefore, the negative effect of increasing the amount of sample is not observed in this range.

Regarding the *R* factor, it clearly decreases when the amount of sample is reduced in the whole mass sample range. This was obviously the expected trend, since as less sample is used, less total CO is produced, and consequently, for the same CO_2 flow, the CO/CO_2 ratio is lower and therefore, the R factor as well.

Comparing the results obtained with 30 g of sample at the two different CO_2 flows, it was observed that the CRI decreases while the *R* factor increases as the CO_2 flow decreases. This was to be expected since, if there is less CO_2 available for the same amount of sample, on the one hand, less coke reacts and therefore, the weight loss is smaller, and consequently the CRI decreases. On the other hand, since the CO_2 concentration is much lower, even though less CO has been produced, the CO/CO_2 ratio increases, and consequently, the R factor increases.

If the results obtained with 30 g and 750 mL· min^{-1} are compared with those obtained with 50 g and 1375 mL· min^{-1}, it can be seen that they are quite similar, which is attributed to the fact that the amount of sample/CO_2 flow ratio used in these two tests is quite similar ($\approx$26).

Therefore, it has been demonstrated that in the reactivity tests the amount of sample used, plays a very important role, but that if the amount of sample/CO_2 flow ratio is kept constant, the influence of the amount of sample can be neglected.

Therefore, the reactivity tests of the bioreducers, which are presented in the following section, were carried out with 30 g of sample and 750 mL·min^{-1} CO_2 flow, so that they could be compared with the reactivity tests of the commercial reducers carried out with 50 g and 1375 mL·min^{-1} CO_2.

3.3.3. Reactivity Tests with the Bioreducers

The results of the reactivity tests of the pyrolysis solids (olives and eucalyptus) carried out using 30 g and 750 mL·min^{-1} CO_2 are included in Table 11. It has to be mentioned that the error of these measurements could not be determined in all cases due to the shortage of sample which made it impossible to repeat some of the tests. In those cases in which the errors were determined, the CRI error was in the range 1–3 and the *R* factor error in the range 2–4.

Table 11. Reactivity results of pyrolysis bioreducers (750 °C) (30 g and 750 mL·min^{-1} CO_2).

	Olives			Eucalyptus
	20 °C·min^{-1}	15 °C·min^{-1}	3 °C·min^{-1}	3 °C·min^{-1}
CRI (%)	87.8	86.8	86.1	93.7
R factor (%)	81.3	82.4	74.4	76.1

It can be seen that the reactivity of the bioreducers is very high. In the case of the olives sample, the CRI values are around 87% therefore only 13% of the initial mass has not reacted in the reactivity test. Taking into account that the ash contents of the olive derived bioreducers is close to 10% (Table 5) it can be concluded that the sample has almost completely reacted with CO_2. The same happens in the case of the eucalyptus derived bioreducer: the CRI value is as high as 94% when it has an ash content of 5.4%. Therefore, the values in Table 11 can be considered to be the maximum achievable reactivities. Burhenne, *et al.* [29] also obtained almost 100% reactivity of spruce wood derived char in tests carried out both in TGA at 800 °C for 3–5 h and in a fixed bed reactor at 800 °C for 30 min. For this reason, a discussion of the influence of the heating rate or type of sample on the reactivity of the bioreducers obtained in this study does not deserve attention.

Comparing Tables 9 and 11 it can be seen that the reactivity of bioreducers is far higher than that of the commercial reducers. This may be attributed to the fact that the surface areas of the commercial reducers are much lower (24–156 m^2·g^{-1}) than those of the bioreducers derived from olives and eucalyptus samples (363–501 m^2·g^{-1}). Huo, *et al.* [33] also obtained much higher reactivities of bioreducers than of petroleum coke or anthracite, and they also reported that there was a relationship between reactivity and surface area.

On the contrary, bioreducers obtained by pyrolysis of fruit tree cuttings at 900 °C by Agirre, *et al.* [30], have CRI values in the range of 20%–50% and *R* factor 35%–70%, both values much lower than those obtained in this study, and comparable to those of commercial reducing agents. There are several reasons to explain these differences: (1) on the one hand, the equipment used for reactivity measurements by Agirre, *et al.*, is very different from the one used in this study; it is a horizontal tubular furnace and the sample is placed in a crucible inside the reactor, therefore the CO_2 stream is only in contact with the upper surface of the sample, while in this study a vertical reactor is used, and therefore the CO_2 stream goes through the whole bioreducer bed; (2) the reaction time once reached at the reaction temperature (1000 °C) used by Agirre, *et al.*, was only 15 min instead of 2 h which is the time specified in the ASTM D 5341-99 and the one used in the reactivity tests of this study; (3) the particle size used by Agirre, *et al.*, (4–5 cm) was larger than that used in this study (1 cm); (4) the type of original biomass from which bioreducers were derived was different (fruit tree cuttings *vs.* olive tree cuttings and eucalyptus); this fact, although less influential than the previous ones, may also be of importance.

In view of the characteristics of the olive and eucalyptus derived bioreducers and in order to analyze their suitability for metallurgical processes, it is worth mentioning that there are three roles that coke plays in metallurgical processes: (1) it has to provide heat or energy; (2) it has to provide a reducing atmosphere by means of the reaction: $C + CO_2 \leftrightarrow 2\,CO$; (3) it has to act as a support medium for the burden if the process is carried out in a blast furnace.

The first two requirements are well fulfilled by the bioreducers obtained in this study; they have rather high heating values (≈30 MJ· kg^{-1}) and they are fully reactive. However, bioreducers cannot fulfill the third requirement, and therefore, due to their missing strength and high reactivity, cannot be used in blast furnaces as support medium, but they can be used tuyére-injected in the lower part of the blast furnaces to substitute typically used fossil fuels (pulverized carbon, oils, natural gas). On the other hand, bioreducers can also be used in the non-ferrous metal industry since no mechanical strength is required, because rotary kilns, where the burden is mixed and rotates with the coke, are frequently used.

Nevertheless, the final decision to determine the suitability of bioreducers for the non-ferrous metallurgical industry depends on the specific characteristics of the process itself. Metal reduction experiments should be carried out with the bioreducers in order to determine if the high reactivities could be a handicap for the process, but such kind of experiments were beyond the scope of this paper. However, it has been reported in the literature, experiments on the reduction of electric arc furnace dusts carried out with charcoal which show that with bioreducers higher zinc volatilization and iron oxide reduction can be achieved in comparison to those obtained with the fossil cokes normally used in industry [15].

4. Conclusions

As far as proximate and ultimate analyses are concerned, olives and eucalyptus derived reducing agents (bioreducers) are of better quality than typical commercial reducing agents used in non-ferrous processes (metallurgical coke, petroleum coke, anthracite), since the former have lower ash and sulfur contents.

The surface area and porosity of the bioreducers obtained in this study (363–501 $m^2 \cdot g^{-1}$ and 0.15–0.21 $cm^3 \cdot g^{-1}$, respectively) are much higher than those of typical commercial reducing agents used in metallurgical processes in rotary kilns (24–156 $m^2 \cdot g^{-1}$ and 0.01–0.07 $cm^3 \cdot g^{-1}$, respectively).

The bioreducers reactivity measurements with CO_2 are very much conditioned by the experimental procedure used. If the ASTM D 5314-99 standard procedure cannot be strictly followed due to sample shortage, a mass sample/CO_2 flow ratio equivalent to that of the ASTM standard should be used.

The bioreducers obtained in this study have no mechanical strength and present extremely high reactivity with CO_2 (close to 100%). Therefore, they cannot be used as top burden in blast furnaces, but they can be used as fuel and reducing agent either tuyére-injected at the lower part of the blast furnace or in non-ferrous metallurgical processes where no mechanical strength is needed, such as in rotary kilns.

Acknowledgments: The authors thank the University of the Basque Country (US 11/21), the Basque Country Government (GIC 10/31), MEC the Spanish Ministry of Science and Innovation (ENE 2011-23950) and BEFESA Steel R&D Company for financial assistance for this work.

Author Contributions: Isabel De Marco and Alexander López-Urionabarrenechea conceived and designed the experiments; Aitziber Adrados, Jon Solar and Naia Gastelu performed the experiments; Aitziber Adrados, Isabel De Marco and Blanca M. Caballero analyzed the data; Alexander López-Urionabarrenechea and Jon Solar contributed reagents/materials/analysis tools; Aitziber Adrados and Isabel De Marco wrote the paper.

Conflicts of Interest: The authors declare no conflict of interest.

References

1. Suopajärvi, H.; Pongrácz, E.; Fabritius, T. The potential of using biomass-based reducing agents in the blast furnace: A review of thermochemical conversion technologies and assessments related to sustainability. *Renew. Sustain. Energy Rev.* **2013**, *25*, 511–528. [CrossRef]
2. Montiano, M.G.; Díaz-Faes, E.; Barriocanal, C.; Alvarez, R. Influence of biomass on metallurgical coke quality. *Fuel* **2014**, *116*, 175–182. [CrossRef]
3. Díez, M.A.; Borrego, A.G. Evaluation of CO_2-reactivity patterns in cokes from coal and woody biomass blends. *Fuel* **2013**, *113*, 59–68. [CrossRef]
4. EPA Overview of Greenhouse Gases. Available online: http://www.epa.gov/climatechange/ghgemissions/gases/co2.html (accessed on 21 December 2015).
5. World Resources Institute Understanding the IPCC Reports. Available online: http://www.wri.org/ipcc-infographics (accessed on 21 December 2015).
6. UNFCCC Background on the UNFCCC: The International Response to Climate Change. Available online: http://unfccc.int/essential_background/items/6031.php (accessed on 21 December 2015).
7. Avila, C.; Pang, C.H.; Wu, T.; Lester, E. Morphology and reactivity characteristics of char biomass particles. *Bioresour. Technol.* **2011**, *102*, 5237–5243. [CrossRef] [PubMed]
8. Demirbaş, A. Carbonization ranking of selected biomass for charcoal, liquid and gaseous products. *Energy Convers. Manag.* **2001**, *42*, 1229–1238. [CrossRef]
9. Elyounssi, K.; Collard, F.-X.; Mateke, J.N.; Blin, J. Improvement of charcoal yield by two-step pyrolysis on eucalyptus wood: A thermogravimetric study. *Fuel* **2012**, *96*, 161–167. [CrossRef]
10. Park, S.-W.; Jang, C.-H.; Baek, K.-R.; Yang, J.-K. Torrefaction and low-temperature carbonization of woody biomass: Evaluation of fuel characteristics of the products. *Energy* **2012**, *45*, 676–685. [CrossRef]
11. Xiao, L.-P.; Shi, Z.-J.; Xu, F.; Sun, R.-C. Hydrothermal carbonization of lignocellulosic biomass. *Bioresour. Technol.* **2012**, *118*, 619–623. [CrossRef] [PubMed]
12. Xu, R.; Ferrante, L.; Hall, K.; Briens, C.; Berruti, F. Thermal self-sustainability of biochar production by pyrolysis. *J. Anal. Appl. Pyrolysis* **2011**, *91*, 55–66. [CrossRef]
13. Díez, M.A.; Alvarez, R.; Barriocanal, C. Coal for metallurgical coke production: Predictions of coke quality and future requirements for cokemaking. *Int. J. Coal Geol.* **2002**, *50*, 389–412. [CrossRef]
14. Griessacher, T.; Antrekowitsch, J. Alternative reducing agents in the recycling of heavy metal-containing residues. *Proc. EMC* **2011**, *4*, 1349–1362.
15. Griessacher, T.; Antrekowitsch, J.; Steinlechner, S. Charcoal from agricultural residues as alternative reducing agent in metal recycling. *Biomass Bioenergy* **2012**, *39*, 139–146. [CrossRef]
16. Demirbas, A. Effects of temperature and particle size on bio-char yield from pyrolysis of agricultural residues. *J. Anal. Appl. Pyrolysis* **2004**, *72*, 243–248. [CrossRef]
17. Angın, D. Effect of pyrolysis temperature and heating rate on biochar obtained from pyrolysis of safflower seed press cake. *Bioresour. Technol.* **2013**, *128*, 593–597. [CrossRef] [PubMed]
18. *Standard Test Method for Measuring Coke Reactivity Index (CRI) and Coke Strength After Reaction (CSR)*; ASTM D 5341-99; ASTM International: West Conshohocken, PA, USA, 2004; Volume 99.
19. Lundgren, M.; Ökvist, L.S.; Björkman, B. Coke reactivity under blast furnace conditions and in the CSR/CRI Test. *Steel Res. Int.* **2009**, *80*, 396–401. [CrossRef]
20. ESYRCE. Encuesta sobre superficies y rendimientos de cultivos. 2015. Available online: http://agricultura.gencat.cat/web/.content/de_departament/de02_estadistiques_observatoris/21_aigua/fitxers_estatics/estudioregadios2008.pdf (accessed on 21 December 2015).
21. García-Maraver, A.; Zamorano, M.; Ramos-Ridao, A.; Díaz, L.F. Analysis of olive grove residual biomass potential for electric and thermal energy generation in Andalusia (Spain). *Renew. Sustain. Energy Rev.* **2012**, *16*, 745–751. [CrossRef]
22. Räisänen, T.; Athanassiadis, D. Basic chemical composition of the biomass components of pine, spruce and birch. Available online: http://www.biofuelregion.se/UserFiles/file/Forest Refine/1_2_IS_2013-01-31_Basic_chemical_composition.pdf (accessed on 21 December 2015).
23. Adrados, A.; Lopez-Urionabarrenechea, A.; Solar, J.; Requies, J.; de Marco, I.; Cambra, J.F. Upgrading of pyrolysis vapours from biomass carbonization. *J. Anal. Appl. Pyrolysis* **2013**, *103*, 293–299. [CrossRef]

24. Ibarrola, R.; Hammond, J.; Shackley, S. Pyrolysis biochar systems for recovering biodegradable materials: A life cycle carbon assessment. *Waste Manag.* **2012**, *32*, 859–868. [CrossRef] [PubMed]
25. Lehmann, J.; Joseph, S. *Biochar for Environmental Management Science and Technology*; Sterling, VA: London, UK, 2009.
26. Adrados, A. *Optimization of the Pyrolysis Process for the Production of a Biomass Derived Reducing Agent and Hydrogen-Rich Gases*; University of the Basque Country (UPV/EHU): Bilbao, Spain, 2014.
27. Imam, T.; Capareda, S. Characterization of bio-oil, syn-gas and bio-char from switchgrass pyrolysis at various temperatures. *J. Anal. Appl. Pyrolysis* **2012**, *93*, 170–177. [CrossRef]
28. U.S. Deparment of Energy Lower and Higher Heating Values of Hydrogen and Other Fuels. Available online: http://www.centreflow.ca/2011/03/15/natural-gas-is-more-energetic/ (accessed on 21 December 2015).
29. Burhenne, L.; Damiani, M.; Aicher, T. Effect of feedstock water content and pyrolysis temperature on the structure and reactivity of spruce wood char produced in fixed bed pyrolysis. *Fuel* **2013**, *107*, 836–847. [CrossRef]
30. Agirre, I.; Griessacher, T.; Rösler, G.; Antrekowitsch, J. Production of charcoal as an alternative reducing agent from agricultural residues using a semi-continuous semi-pilot scale pyrolysis screw reactor. *Fuel Process. Technol.* **2013**, *106*, 114–121. [CrossRef]
31. Dutta, B.; Raghavan, G.S.V.; Ngadi, M. Surface characterization and classification of slow and fast pyrolyzed biochar using novel methods of pycnometry and hyperspectral imaging. *J. Wood Chem. Technol.* **2012**, *32*, 105–120. [CrossRef]
32. Hasan Khan Tushar, M.S.; Mahinpey, N.; Khan, A.; Ibrahim, H.; Kumar, P.; Idem, R. Production, characterization and reactivity studies of chars produced by the isothermal pyrolysis of flax straw. *Biomass Bioenergy* **2012**, *37*, 97–105. [CrossRef]
33. Huo, W.; Zhou, Z.; Chen, X.; Dai, Z.; Yu, G. Study on CO_2 gasification reactivity and physical characteristics of biomass, petroleum coke and coal chars. *Bioresour. Technol.* **2014**, *159*, 143–149. [CrossRef] [PubMed]
34. Senneca, O. Kinetics of pyrolysis, combustion and gasification of three biomass fuels. *Fuel Process. Technol.* **2007**, *88*, 87–97. [CrossRef]
35. Rösler, G.; Griessacher, T.; Antrekowitsch, J. The influence of charcoal properties on its applicability in metallurgical operations. In Proceedings of the 4th International Symposium on Energy from Biomass and Waste, Venice, Italy, 12–15 November 2012.
36. Carrier, M.; Hardie, A.G.; Uras, Ü.; Görgens, J.; Knoetze, J.H. Production of char from vacuum pyrolysis of South-African sugar cane bagasse and its characterization as activated carbon and biochar. *J. Anal. Appl. Pyrolysis* **2012**, *96*, 24–32. [CrossRef]
37. Jones, J.C. Letter: The BET surface areas of carbonised biomass. *Fuel Process. Technol.* **2012**, *106*, 784. [CrossRef]
38. Menéndez, J.A.; Pis, J.J.; Díez, M.A.; Barriocanar, C.; Casal, M.D.; Parra, J.B.; Álvarez, R. Adición de coque de petróleo a mezclas coquizables. Modificación de la calidad del coque metalúrgico. *Rev. Metal.* **1995**, *31*, 235–241. [CrossRef]
39. Ruiz, O.; Romero-Palazón, E.; Díez, M.A.; Marsh, H. Co-carbonization of green petroleum cokes with Cortonwood coal: Influence on structure, reactivity and microstrength of resultant cokes. *Fuel* **1990**, *69*, 456–459. [CrossRef]

2

Influence of Radiation Sterilization on Properties of Biodegradable Lactide/Glycolide/Trimethylene Carbonate and Lactide/Glycolide/ε-caprolactone Porous Scaffolds with Shape Memory Behavior

Piotr Rychter [1], Natalia Śmigiel-Gac [1,2], Elżbieta Pamuła [3], Anna Smola-Dmochowska [2], Henryk Janeczek [2], Wojciech Prochwicz [1] and Piotr Dobrzyński [1,2,*]

Academic Editor: Marek M. Kowalczuk

1 Faculty of Mathematics and Natural Sciences, Jan Dlugosz University in Czestochowa, Armii Krajowej 13/15 Av., Czestochowa 42-218, Poland; p.rychter@ajd.czest.pl (P.R.); nsgac@cmpw-pan.edu.pl (N.Ś.-G.); w.prochwicz@ajd.czest.pl (W.P.)
2 Centre of Polymer and Carbon Materials, Polish Academy of Sciences, Curie-Sklodowska 34 St., Zabrze 41-800, Poland; asmola@cmpw-pan.edu.pl (A.S.-D.); hjaneczek@cmpw-pan.edu.pl (H.J.)
3 Department of Biomaterials, Faculty of Materials Science and Ceramics, AGH University of Science and Technology, Al Mickiewicza 30, Kraków 30-059, Poland; epamula@agh.edu.pl
* Correspondence: p.dobrzynski@ajd.czest.pl

Abstract: The aim of the study was the evaluation of gamma irradiation and electron beams for sterilization of porous scaffolds with shape memory behavior obtained from biodegradable terpolymers: poly(L-lactide-*co*-glycolide-*co*-trimethylene carbonate) and poly(L-lactide-*co*-glycolide-*co*-ε-caprolactone). The impact of mentioned sterilization techniques on the structure of the scaffolds before and after the sterilization process using irradiation doses ranged from 10 to 25 kGy has been investigated. Treatment of the samples with gamma irradiation at 15 kGy dose resulted in considerable drop in glass transition temperature (T_g) and number average molecular weight (M_n). For comparison, after irradiation of the samples using an electron beam with the same dose, no significant changes in structure or properties of examined scaffolds have been noticed. Higher doses of irradiation via electron beam caused essential changes of the scaffolds' pores resulting in partial melting of their surface. Nevertheless, obtained results have revealed that sterilization with electron beam, when compared to gamma irradiation, is a better method because it does not affect significantly the physicochemical properties of the scaffolds. Both used methods of sterilization did not influence the shape memory behavior of the examined materials.

Keywords: biodegradable polymer; gamma sterilization; electron beam sterilization; radiation degradation; scaffold; shape memory

1. Introduction

One of the criteria which must to be met for biomaterials is the possibility of their sterilization leading to destruction of pathogenic microorganisms. For a majority of biomedical devices, sterilization is mandatory and a validation before their clinical use is necessary. However, in the case of materials prepared from biodegradable polyesters or polyester-carbonates possessing susceptibility to hydrolysis and low thermal resistance, fulfilling these criteria is challenging.

Even when low-temperature methods for sterilization processes are being used, influence of this treatment later results in changes of final physical properties of this material and, consequently, on its application [1,2]. With this respect, it is crucial to select appropriate methods of sterilization to avoid any undesirable damages of material.

Since, the three-dimensional (3D) scaffolds are designed for cell growth within their entire volume, they possess unique properties, such as relatively poor mechanical properties, very high porosity, and specific surface area, which provide the optimal condition for growing cells [3–5]. Proper choice of a sterilization method is especially difficult when the examined material must keep its thermally-induced shape memory properties.

Biomaterials of this type are capable of returning from a temporary to a previously-programmed shape as a result of a mechanical response to thermal stimulation at a temperature close to body temperature [6–8]. This means that scaffolds with temporary shape, which was accomplished via two steps (first—compression at higher temperature; second—rapid cooling at low temperature) must be sterilized in such a way as to avoid too fast a relaxation of this material to its permanent (initial) shape. With this respect, the sterilization should proceed as fast as possible, at the lowest possible temperature, much lower than the temperature inducing the return of the sample to its permanent shape.

These limitations result in a narrow choice of sterilization methods. To obtain the required decontamination sterilization by ethylene oxide (EtO) should be undertaken at temperatures between 45 and 60 °C, which is not suitable for the examined scaffolds [9].

According to literature data, the vast majority of polymeric materials with shape memory behavior have been sterilized using a gas plasma sterilization process at room temperature [5,10,11]. However, in the case of this method, when used for sterilization of porous scaffolds, the increase in pore interconnection has been observed. This phenomenon may be explained by the fact that foams have been exposed to variable pressure during the sterilization cycle (from atmospheric to 54–67 Pa) that should have promoted a collapse of thinner pore walls [11]. Additionally, the method of sterilization should assure sterility in the entire volume of the material which is not always possible to achieve during plasma or chemical sterilization of highly porous scaffolds.

Following the literature, the most popular sterilization method of porous scaffolds exhibiting shape memory behavior is immersion of the samples in 70% ethanol (v/v in water) followed by UV irradiation (λ = 254 nm) for 10 min [12,13]. However, this method does not provide total decontamination and is classified rather as a disinfection, than sterilization, because of its inability to destroy hydrophilic viruses or bacterial spores [14].

Based on the already existing reports related to sterilization of highly porous scaffolds with shape memory properties dedicated for the cell growth purposes, the use of radiation methods seems to be the most effective. Radiation sterilization, among other advantages like short time of treatment, proceeded at room temperature, or lack of toxic residues after sterilization, also possess high decontamination efficiency even for materials which are difficult for internal penetration of the sterilization agent, like porous scaffolds.

Current investigations devoted to optimization of sterilization methods of polymeric scaffolds or nanofibers demonstrate ambiguous and contradictory results. It may be explained by the fact, that the scaffolds' composition and microstructure of the used polymer chain are key factors influencing the degradation reactions occurring upon irradiation [15]. Sterilization of scaffolds made from poly(ε-caprolactone) using gamma irradiation caused an essential decrease in average molecular weights and glass transition temperature, leading to deterioration in their mechanical properties. However, it is worth noting that the mentioned treatment did not influence cell attachment and growth [3].

It has also been shown [15] that a copolymer containing dioxepanone (DXO) units and a high amount of L-lactide was more prone to degradation by chain-scission than a material containing higher amount of caproyl units. Usefulness of radiation methods to sterilize porous materials obtained from polylactide and poly(trimethylene carbonate) blends has been also proved, despite the fact of lowering their glass transition, molecular weight, and the deterioration of thermo-mechanical properties [16].

Comparative studies of the type of sterilization method on the properties of acrylate-based copolymers exhibiting shape memory behavior have revealed better usefulness of gamma irradiation as compared to sterilization via low temperature plasma techniques. The acrylic networks sterilized by plasma elicited a strong cytotoxic response [16]. The influence of applied sterilization methods on the biodegradable materials is dependent both on the structure and composition of polymeric materials but also on the scaffold's morphology.

The aim of this study is examination of the usefulness of radiation methods for the sterilization of scaffolds made from terpolymers L-lacide/glycolide/ε-caprolactone and L-lactide/glycolide/TMC (trimethylene carbonate) exhibiting shape memory behavior. Results related to effectiveness of proposed scaffolds in minimally invasive surgery for bone defects treatment have already been published by us [17,18].

2. Results and Discussion

2.1. *Scaffolds Formulation and Procedure of Investigations*

In this study porous scaffolds with shape memory behavior designed for large bone defects treatment via minimally invasive surgery approach were tested. The scaffolds were formed using two different types of microblock terpolymers obtained via ring opening polymerization (ROP): poly(L-lactide-co-glycolide-co-ε-caprolactone) LA/GL/CL and poly(L-lactide-co-glycolide-trimethylene carbonate) LA/GL/TMC with similar molecular weight and polydispersity.

The first type of terpolymers has been obtained in two stage reaction (Scheme 1). At the first stage, the hydroxyl-terminated oligocarbonate shave been synthesized in order to be used afterward, as a macroinitiator for the synthesis of terpolymer LA/GL/TMC. Obtained terpolymer was built of lactidyl (LL) and carbonate microblocks (TT) bonded mainly by short lactidyl/glicolidyl (GGLL, LLGG) and glycolidyl (GG) sequences) [8], (see also NMR spectra, Supplementary Materials, Figures S1 and S2).

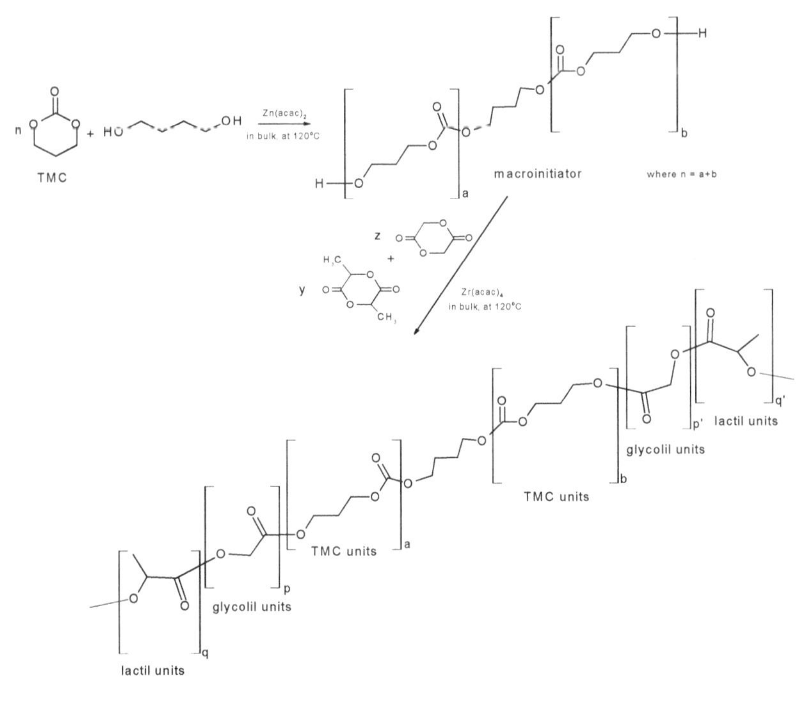

Scheme 1. The two-step synthesis procedure of L-lactide/glycolide/trimethylene carbonate terpolymers (LGT21 and LGT40).

The third copolymer (LGC) has been obtained during one stage reaction via ROP of L-lactide, glycolide and ε-caprolactone (Scheme 2) proceeded in bulk and initiated with zirconium(IV) acetylacetonate ($Zr(acac)_4$). The chain of the synthesized terpolymer consisted mainly of longer lactidyl LLLL microblocks with embedded shorter glicolidyl/caproil and glycolil/caproil sequences (CLGGCL, CLGGGCL, CLGCL) [19], (see also NMR spectra, Supplement Materials, Figure S3).

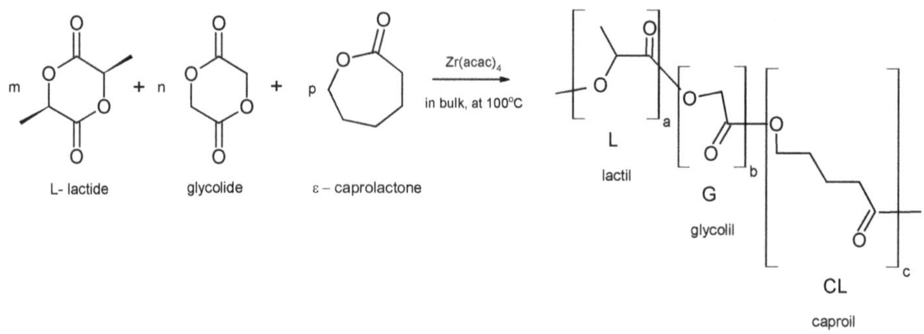

Scheme 2. L-lactide/glycolide/caprolactoneterpolymer (LGC) synthesis.

In addition to the scaffolds containing caproil units LA/GL/CL (Table 1, LGC), two types of scaffolds consisting of different amount of carbonate units in polymer chains have been treated with radiation sterilization (Table 1, LGT50 and LGT20). A detailed description of the polymer's microstructure, as well as methods of its determination allowing it to obtain optimal properties for shape memory effect have been already reported [7,8,19].

Scaffolds were produced via solvent casting/particulate leaching technique using the same porogen (sodium chloride particles) [17,20]. Obtained scaffolds had a similar average porosity and diameter of pores. Principal properties of the used terpolymers, as well as scaffolds made from them have been presented in Table 1.

To obtain a temporary shape of the scaffolds, cylindrically shaped materials with diameter *ca.* 10 mm have been circumferentially compressed at temperatures close to T_g of the particular terpolymer. Next, after fast cooling, the diameter of compressed scaffolds was *ca.* 7 mm (compression about 30%). Properties of the scaffolds with temporary shape have been presented in Table 2. As soon as possible, compressed scaffolds were sterilized with gamma radiation and electron beam, applying the following doses: 10, 15, and 25 kGy.

2.2. *Influence of Gamma Irradiation on the Properties of Matrices and Scaffolds' Morphology*

Changes in properties of the polymeric matrix and scaffolds after absorption of γ irradiation doses are presented in Table 2 (where the numbers of scaffolds correspond to those shown in Table 1, with additional indication C—in a compressed temporary shape).

Gamma irradiation caused relatively strong degradation of copolymers' chains resulting in a significant decrease in molecular weight and increase in molar mass dispersity. This effect was particularly noticeable for terpolymers containing carbonate units. In the case of terpolymer LGT40 relative molecular mass loss (dM_n) exceeded 40% of initial average M_n. Meanwhile, molecular weight of the sample LGT21, containing lower amounts of carbonate units, decreased up to 35% (Table 2). As a consequence, decreasing molecular weight together with a growing dose of irradiation resulted in a gradual drop in glass transition temperature. It is worth noting that DSC (differential scanning calorimetry) analysis has proved no changes in melting temperature of crystalline regions coming from the ordered phase of lactidyl microblocks in all tested terpolymers (Figure 1).

Table 1. Properties of investigated scaffolds and used polymeric matrices.

No.	Properties of Polymeric Matrix				Properties of Scaffold			
	Composition (mol. %)	M_n (kg/mol)	Đ	T_g (°C)	Average Porosity P (%)	Average Pore Diameter (μm)	Scaffold Diameter (mm)	Compressive Strength of Dried Scaffolds (MPa)
LGT21	L-lactide—68 Glycolide—11 TMC units—21	36.8	2.3	47	86 ± 5	395 ± 59	10.1 ± 0.2	0.8 ± 0.1
LGT40	L-lactide—46 Glycolide—14 TMC units40	23.9	1.9	28	88 ± 9	410 ± 67	10.0 ± 0.3	0.6 ± 0.1
LGC	L-lactide—79 Glycolide—10 Caprolactone—11	35.1	2.2	55	83 ± 3	380 ± 45	10.2 ± 0.3	0.9 ± 0.2

Notes: M_n—average number molar mass determined with GPC; Đ—molar mass dispersity; T_g—glass transition temperature determined with DSC (II run for amorphous sample obtained by quenching from melt). Presented data regarding scaffold properties were calculated as an average of five samples of each type of the scaffolds ± standard deviation (SD).

Table 2. Effect of γ irradiation dose on properties of terpolymers and scaffolds at their temporary shape.

No.	Properties of Polymeric Matrix					Properties of Scaffold at Temporary Shape		
	Dose (kGy)	M_n (kg/mol)	dM_n (%)	Đ	T_g (°C)	Average Porosity P (%)	Scaffold Diameter (mm)	Compressive Strength of Dried Scaffolds (MPa)
LGT21 C	0	36.8	-	2.3	47	55 ± 5	6.8 ± 0.3	1.1 ± 0.1
	10	32.7	11	2.4	46.6	58 ± 7	7.3 ± 0.4	1.1 ± 0.2
	15	29.8	19	2.5	46.4	57 ± 6	7.4 ± 0.3	1.0 ± 0.2
	25	23.6	36	2.7	45.6	58 ± 7	7.4 ± 0.4	0.7 ± 0.1
LGT40 C	0	23.9	-	1.9	28	58 ± 5	7.0 ± 0.4	0.6 ± 0.1
	10	19.4	19	2.5	26.7	61 ± 7	7.8 ± 0.5	0.4 ± 0.1
	15	17.1	29	2.5	26.5	62 ± 4	7.8 ± 0.3	0.4 ± 0.2
	25	13.2	45	2.8	25	61 ± 5	7.9 ± 0.4	0.2 ± 0.1
LGC C	0	35.1	-	2.2	55	53 ± 3	6.7 ± 0.3	0.9 ± 0.2
	10	31.7	10	2.5	53.6	57 ± 4	7.4 ± 0.3	0.8 ± 0.2
	15	29,8	15	2.9	53.5	58 ± 5	7.6 ± 0.5	0.8 ± 0.1
	25	26,2	25	2.9	52.6	58 ± 4	7.4 ± 0.4	0.5 ± 0.2

Notes: M_n—average number molar mass determined with GPC; dM_n—relative molecular mass loss; Đ—molar mass dispersity; T_g—glass transition temperature determined with DSC (II heating run for amorphous sample obtained by quenching from melt). Presented data regarding scaffold properties were calculated as an average of five samples of each type of the scaffolds ± SD.

Growing destruction of terpolymers' chains combined with an increasing level of absorbed irradiation reflects an essential decrease in mechanical properties of sterilized scaffolds (Table 2). The highest decrease in M_n denoted for scaffold LGT40 C, when treated with a 25 kGy dose, resulted in a decrease in compression strength of the material and reached only 1/3 of its initial value. In other cases, after treatment with the same dose of irradiation, the lowering mechanical properties were also significant. Scanning electron microscopy analysis of scaffolds' surfaces after γ irradiation has revealed no changes in morphology of pore surface. Pictures demonstrating the highest degradation rate are presented in Figure 2 (sample LGC C) and Figure 3 (sample LGT40 C). Sterilization via gamma irradiation did not considerably affect scaffold morphology (Table 2). Dimensions of compressed samples remained practically unchanged.

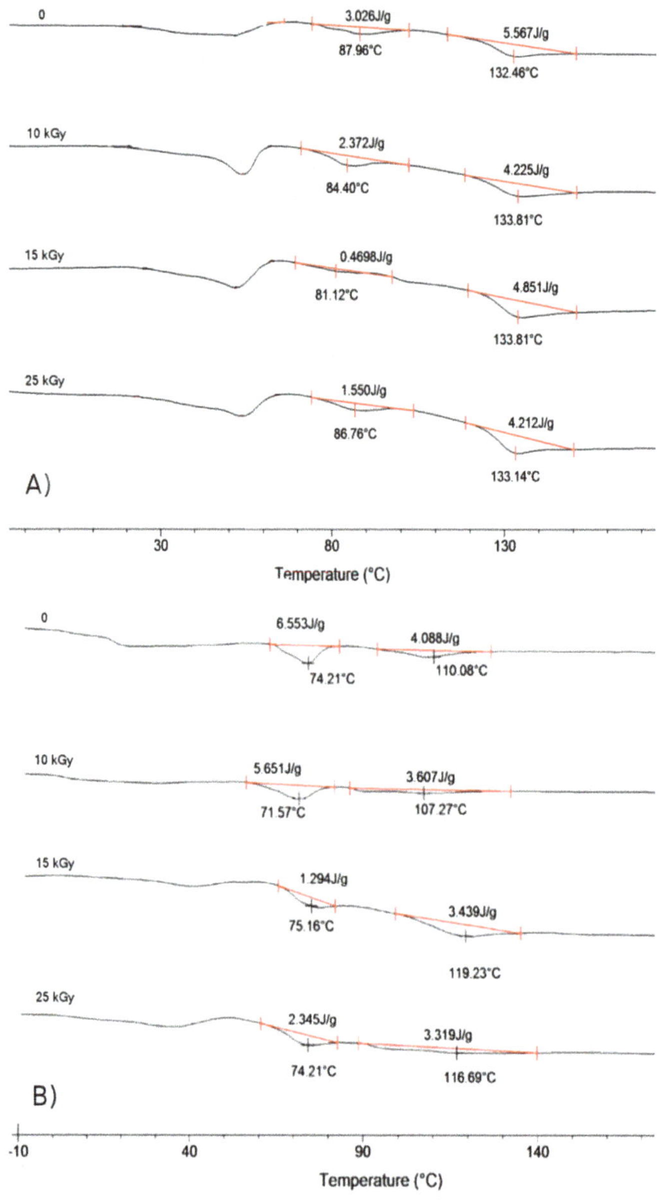

Figure 1. *Cont.*

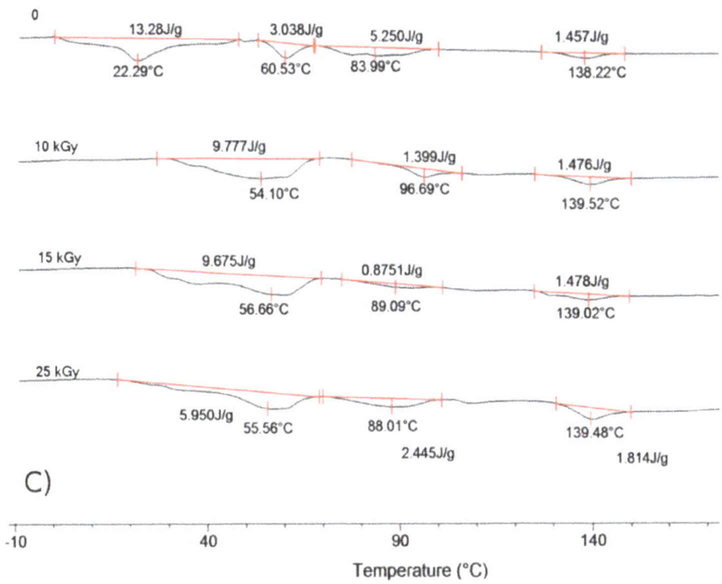

Figure 1. DSC traces for copolymer (**A**) LGT21 C; (**B**) LGT40 C; and (**C**) LTG C. I-heating run at 20°C/min after γ irradiation with dose: 0 kGy, 10 kGy, 15 kGy, and 25 kGy.

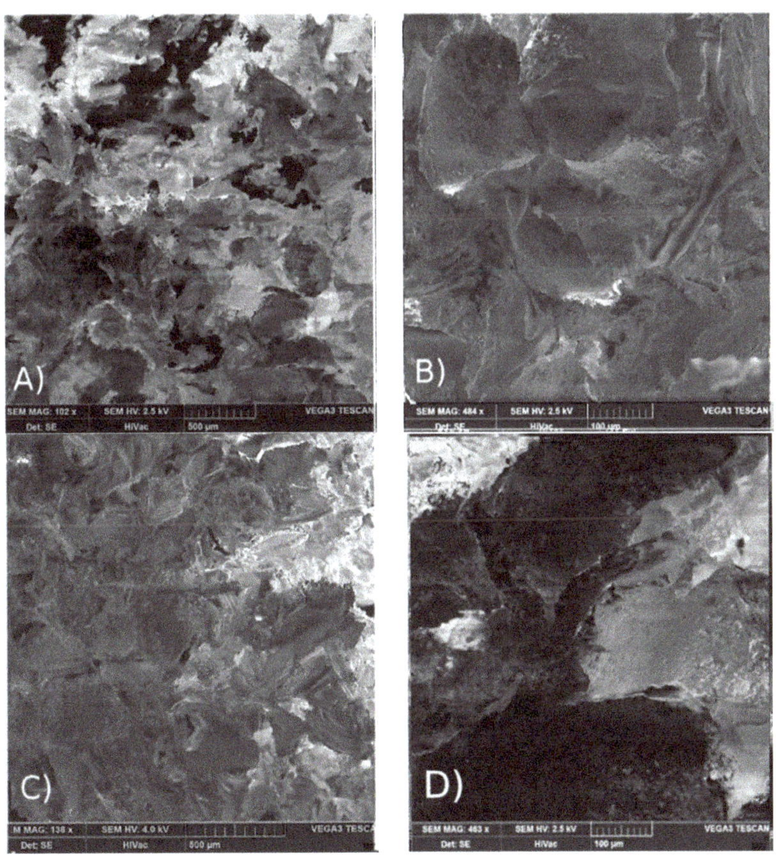

Figure 2. SEM pictures of porous surface of scaffold LGC C: (**A**,**B**) before sterilization; (**C**,**D**) after γ irradiation with a 25 kGy dose.

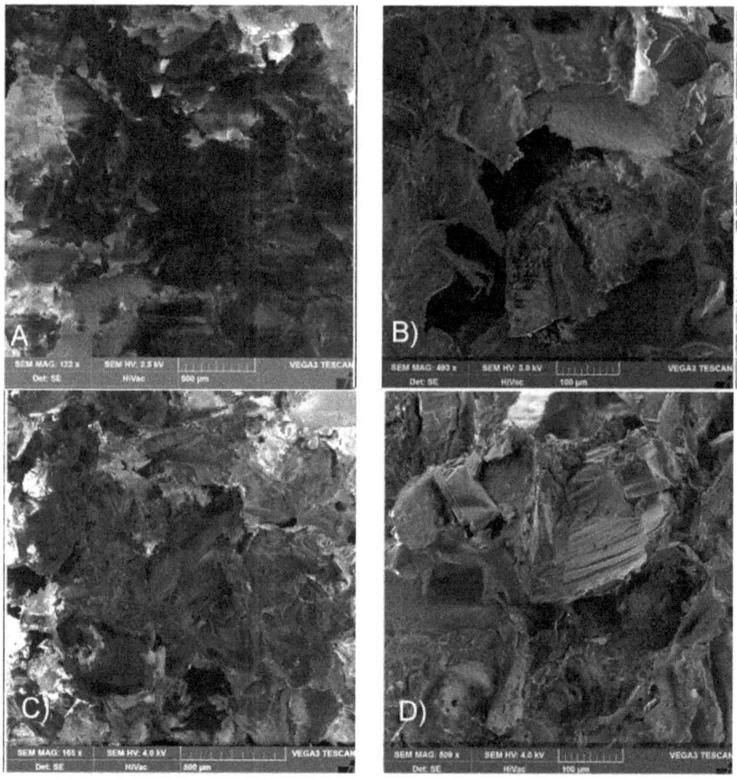

Figure 3. SEM pictures of porous surface of scaffold LGT40 C; (**A**,**B**) before sterilization and (**C**,**D**) after γ sterilization with irradiation dose of 25 kGy.

2.3. Influence of Electron Beam Sterilization on the Properties of Matrices and Scaffolds Morphology

Changes in properties of polymeric matrix and scaffolds after treatment with various doses of electron beam are presented in Table 3.

Table 3. Effect of absorbed dose of electron beam on properties of terpolymers and scaffolds at their temporary shape.

No.	Properties of Polymeric Matrix					Properties of Scaffold at Temporary Shape		
	Dose (kGy)	M_n (kg/mol)	dM_n (%)	Đ	T_g (°C)	Average Porosity P (%)	Scaffold Diameter (mm)	Compressive Strength of Dried Scaffolds (MPa)
LGT21 C	0	36.8	-	2.3	47	55 ± 5	6.8 ± 0.3	1.1 ± 0.1
	10	34.1	7	2.4	47	57 ± 6	7.2 ± 0.5	1.1 ± 0.2
	15	32.8	11	2.4	46.3	56 ± 4	7.1 ± 0.3	1.0 ± 0.2
	25	25.8	30	3.2	45.8	57 ± 4	7.2 ± 0.4	0.9 ± 0.2
LGT40 C	0	23.9	-	1.9	28	58 ± 5	7.0 ± 0.4	0.6 ± 0.1
	10	22.0	8	2.0	25	57 ± 6	7.1 ± 0.5	0.6 ± 0.2
	15	20.2	15	2.0	25	60 ± 7	8.3 ± 0.5	0.7 ± 0.3
	25	19.2	20	2.3	24.5	63 ± 4	8.5 ± 0.5	0.5 ± 0.2
LGC C	0	35.1	-	2.2	55	53 ± 3	6.7 ± 0.3	0.9 ± 0.2
	10	33.7	4	2.2	53	54 ± 4	7.1 ± 0.2	0.9 ± 0.1
	15	33.1	6	2.3	53.5	59 ± 4	7.4 ± 0.3	0.9 ± 0.2
	25	30.8	12	2.7	52.6	60 ± 6	7.6 ± 0.4	0.7 ± 0.3

Notes: M_n—average number molar mass determined with GPC; dM_n—relative molecular mass loss; Đ—molar mass dispersity; T_g—glass transition temperature determined with DSC (II heating run for amorphous sample obtained by quenching from melt). Presented data regarding scaffold properties were calculated as the average of five samples of each type of the scaffolds ± SD.

Decrease in molecular weight of all the samples sterilized with electron beam were lower as compared to sterilization by gamma irradiation (Tables 2 and 3; see also Supplementary Materials Figures S4–S6). The highest drop in M_n of the polymers has been noticed (similarly like in the case of γ irradiation) for terpolymers containing carbonate microblocks. Contrary to results obtained using γ irradiation, noticed in this case is a decrease in molecular weight which was similar for both scaffolds containing TMC units. In the case of sample LGT40, when treated with a 25 kGy dose, the decrease in M_n was about 20% as compared to initial molecular mass. This drop in M_n was half in comparison with the samples sterilized with γ irradiation. The decrease in molecular weight for sample LGT21 reached 30% of initial value. Relatively low changes of molecular weight (12% drop of M_n) for scaffolds LGC C composed of LA/GL/CL units and, as a consequence, a noticeable decrease in their glass transition temperature has been observed (Figure 4).

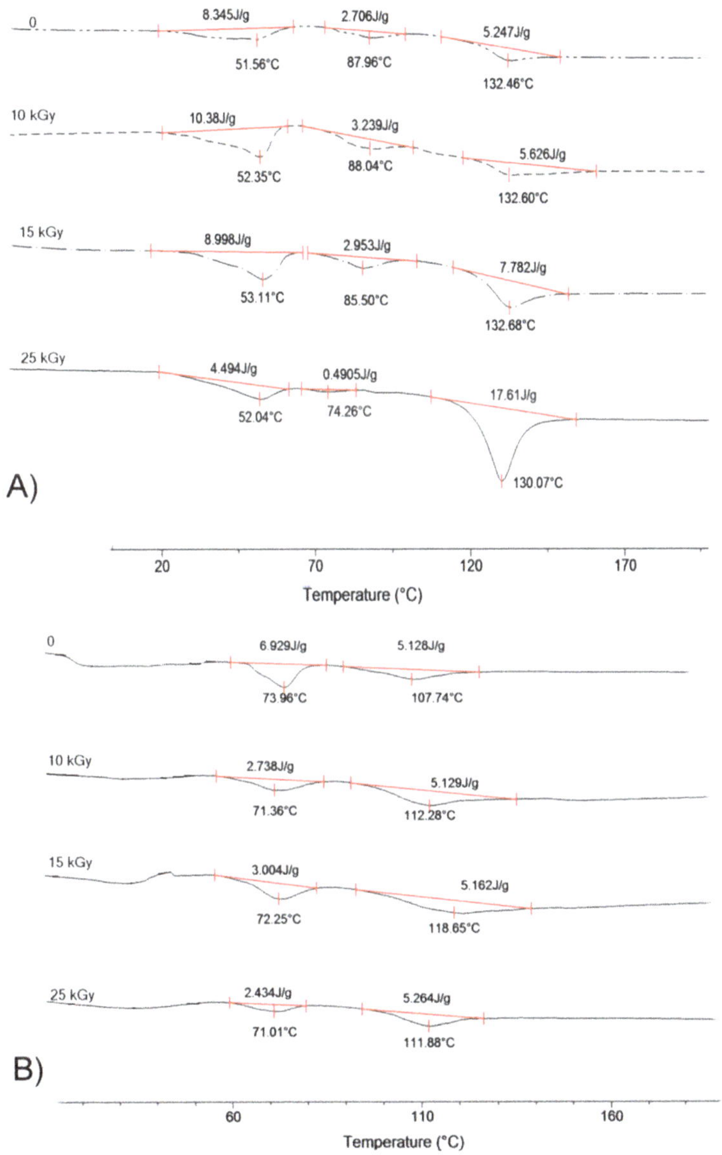

Figure 4. *Cont.*

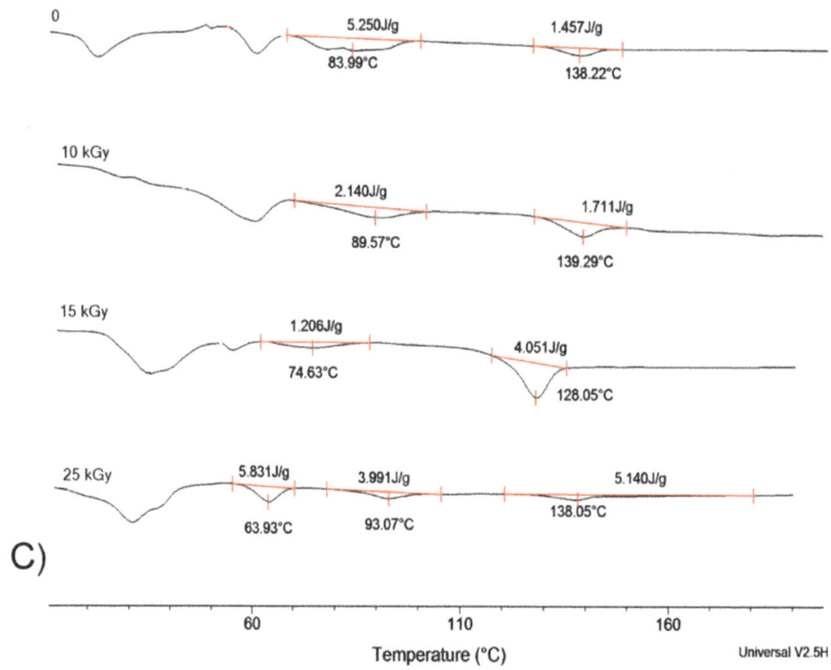

Figure 4. DSC trace for copolymer (**A**) LGT21 C; (**B**) LGT40 C; and (**C**) LGC C. I-heating run at 20 °C/min after electron beam irradiation with dose: 0 kGy, 10 kGy, 15 kGy, and 25 kGy.

Sterilization of polymeric matrices LGT21 C and LGT40 with electron beam at doses 15 kGy and even higher caused a slight increase in melting enthalpy (Figure 4). The most probable explanation of this phenomenon is the increase of matrix arrangement caused by growing mobility of terpolymer chains induced by the temperature rise of the material caused by the action of electrons. This phenomenon might have also affected the surface morphology of examined scaffolds. The influence of electron beam on the changes of surface properties of polylactide has been already reported [21]. In the presented study, this effect has much stronger impact on the samples, because prepared 3D scaffolds possess a unique, porous structure with very high interconnectivity, built from thin-walled biodegradable terpolymer.

Among SEM pictures presented in Figure 5, we can observe, that in the case of scaffold LGT40 C only samples irradiated with the lowest dose, present the morphology identical with initial ones. The dose of 15 kGy, and especially 25 kGy, caused significant changes in the surface of pores, resulting in melting of the polymeric matrix. This may be caused by the fact that this terpolymer exhibits the lowest glass transition temperature among all other samples (higher amount of TMC units as compared to sample LGT21 C).

The largest changes have been observed for scaffold LGC C (Figure 5H,I) made from terpolymer with the highest T_g (determined by DSC method). In spite of a high T_g value of this terpolymer, the observed "flowing" effect may be caused by the presence in the polymer chain caproil sequences resulting in an increase of chain mobility even in lower temperatures during heating, which comes from the growing level of absorbed electron energy (T_g is a temperature range rather, than a specific temperature).

2.4. Effect of Sterilization on Shape Memory Properties

Effect of sterilization process on the shape memory properties of the scaffolds (return time and extent of compressed samples) has been investigated by comparative study, in which results related to sterilized scaffolds were compared to non-irradiated ones (Table 4). The recovering temperatures were selected so as to initialize the return of shape, and at the same time were close to body temperature.

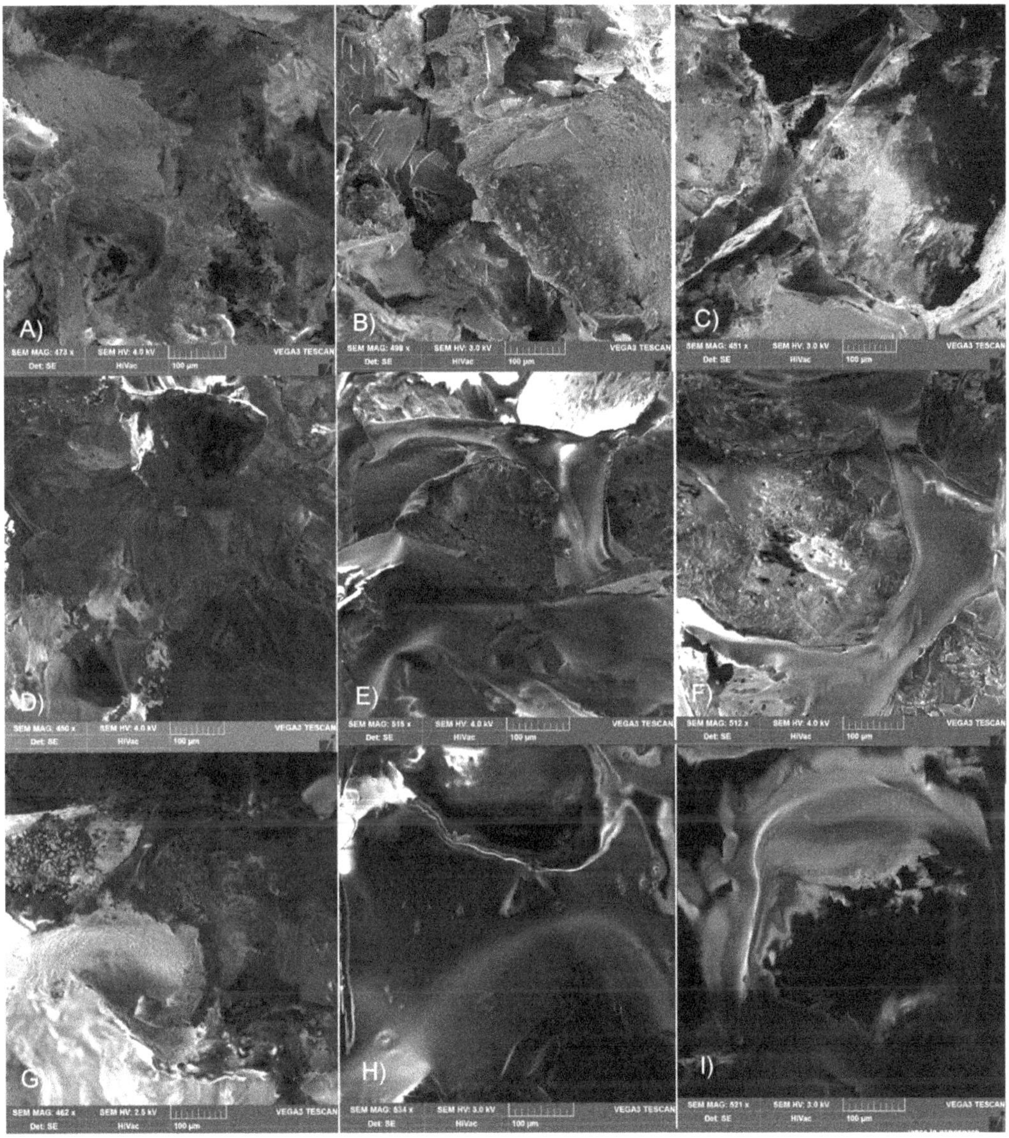

Figure 5. SEM pictures of scaffolds pore surface after electron beam sterilization made of terpolymer LGT21 C with dose (**A**) 10 kGy; (**B**) 15 kGy; (**C**) 25 kGy; terpolymer LGT40 C with dose (**D**) 10 kGy; (**E**) 15 kGy; (**F**) 25 kGy and terpolymer LGC C with dose (**G**) 10 kGy; (**H**) 15 kGy; and (**I**) 25 kGy.

Although the sterilization effect at the 25 kGy irradiation dose has been noticed for all tested samples, in the majority of cases this impact was rather minor. No visible changes on the pore surface after returning to the temporary shape was observed (Supplementary Materials, Figure S7) and the shape recovery ratio was high. Scaffolds made with terpolymer LGT40 exhibiting the lowest glass transition temperature, in the case of gamma irradiation, demonstrated almost twice as slow are turn to permanent shape, and the extent of the return was much lower when compared to a reference sample. The main reason of this behavior was probably a very high decrease in molecular weight of the terpolymer, which is related to shortening chain length and destruction of its initial microstructure which is responsible for the shape memory effect (chain entanglement, phase separation, or presence of hydrogen bonds).

Scaffolds made from terpolymer LGT40 and sterilized with electron beam practically did not exhibit shape memory behavior. This was caused by their total relaxation as a consequence of the heating effect generated during the sterilization process. As a result of this process, directly after irradiation, a slight return of scaffold's diameter from initial 7.7 to 8.5 mm has been observed.

Table 4. Influence of sterilization method on shape memory behavior of the scaffolds.

The Type of Sterilization	No.	Dose (kGy)	T_r (°C)	e_m (mm)	D_s (mm)	e_r (mm)	t_R (s)	R_r (%)
Unsterilized	LGT21C	0	37	10.1	6.8	9.7	600	96
	LGT40 C	0	30	10.0	7.0	9.6	320	96
	LGC C	0	41	10.2	6.7	9.4	340	94
Gamma Irradiation	LGT21 C	25	37	10.1	7.4	9.1	690	89
	LGT40 C	25	30	10.0	7.9	8.6	520	86
	LGC C	25	41	10.2	7.4	9.2	380	90
Electron Beam	LGT21 C	25	37	10.1	7.2	9.6	620	95
	LTG40 C	25	30	10.0	8.5	8.5	>1000	nd
	LGC C	25	41	10.2	7.6	9.1	410	89

Notes: No.—type of the scaffold according to Table 2; T_r—recovering temperature, e_m—diameter of sample before compression; D_s—diameter sample after sterilization; e_r—diameter of sample after recovery; t_R—shape recovery time; R_r—maximal shape recovery ratio (Equation (1)). Presented data were calculated as an average of three samples of each type of the scaffolds. nd—not determined.

3. Experimental Section

3.1. Terpolymers Synthesis and Characterization

3.1.1. Materials and Chemicals

L-Lactide, glycolide (Glaco Ltd., Beijing, China) were purified by recrystallization from ethyl acetate solution and dried in a vacuum oven at room temperature; ε-caprolactone (Fluka, Steinheim, Germany) was dried with calcium hydride and distilled under argon atmosphere before use. Trimethylene carbonate (Ingelheim Boehringer, Ingelheim am Rhein, Germany), dry chloroform and methanol (Aldrich, Steinheim, Germany) were used as received. Initiators and catalysts: zirconium (IV) acetylacetonate—($Zr(acac)_4$), zinc(II) acetylacetonate monohydrate, 1,4-butanediol ($Zn(acac)_2xH_2O$), (Aldrich, Steinheim, Germany) were used as received.

3.1.2. Synthesis of L-lactide/glycolide/trimethylene Carbonate Terpolymers

The first stage to produce terpolymers containing carbonate units was the synthesis of hydroxyl-terminated oligo(trimethylene carbonate), which were obtained by ROP of TMC. The oligomerization was conducted in bulk, at 120 °C with presence of 1,4-butandiol as initiator and $Zn(acac)_2x\ H_2O$ as a catalyst with catalyst/monomer ratio (C/M) of 1:3000 (Scheme 1). After 2 h monomer conversion more than 99% was obtained. The detailed course of the oligomerization was described previously [8,18]. Obtained oligotrimethylene carbonate, with average number molecular mass M_n = 5100 g/mol, molecular mass dispersity Đ = 2.1(both results determined by GPC) was used as a macroinitiatorin copolymerization of L-lactide with glycolide (with lactide/glycolide/initiator molar ratio as; 70/10/0.4 for LTG21 or 55/10/0.9 for LTG40). This reaction was also conducted in bulk at 120 °C with the use $Zr(acac)_4$ as a catalyst with C/M molar ratio as 1:2500. After 12 h, a terpolymer LTG21 was obtained with a yield of about 98%, M_n = 36 800 g/mol and LTG40 with yield 91%, M_n = 23,900 g/mol. This product were dissolved in chloroform, precipitated with cold methanol, and then vacuum-dried to a constant weight. Examples of obtained chromatograms (Figures S4 and S5) and NMR spectra (Figures S1 and S2) are published in Supplementary Materials.

3.1.3. Synthesis of L-lactide/glycolide/ε-caprolactone Terpolymer

Terpolymerization of glycolide with L-lactide and ε-caprolactone was performed in bulk at 100 °C, in one pot ROP reaction with $Zr(acac)_4$ as an initiator and initiator/monomers molar ratio I/M as 1: 1000 (Scheme 2) according to previously described methods [19]. After 48 h, terpolymer LTC was obtained with a yield of about 96%.This product was dissolved in chloroform, precipitated with cold methanol, and then vacuum-dried to a constant weight. An example of the obtained chromatogram (Figure S5) and NMR spectrum (Figure S3) are published in Supplementary Materials.

3.1.4. Terpolymers and Scaffolds Characterization

Average number molecular weights (M_n) and molar mass dispersity (Đ) of the synthesized terpolymer samples were measured using gel permeation chromatography (GPC) (Viscotek apparatus Rimax, Malvern Instruments, Worcestershire, England, chloroform, temperature 35 °C, flow 1 mL/min, using two Viscotek 3580 columns, refractive detector, calibration with polystyrene standards—Easi Vial PS-M calibration kit (Agilent, Santa Clara, CA, USA), with M_p from 1000 to 500,000 g/mol).

The composition and microstructure of the terpolymers' chains were determined with NMR measurements. The NMR spectra of the terpolymers were recorded at 600 MHz with the Avance II Bruker Ultrashield Plus Spectrometer (Bruker Corporation, Anaheim, CA, USA) and a 5 mm sample tube. Dried deuterated chloroform was used as solvent and tetramethylsilane was used as the internal standard. The ^{1}H NMR spectra were obtained with 32 scans, a 2.65 s acquisition time, and an 11 μs pulse width. The ^{13}C NMR spectra were recorded at 150 MHz with the same spectrometer and conditions as the proton spectra. The acquisition time was 0.9 s, the pulse width was 9.4 μs, the delay between pulses was 2 s, and the spectral width was 36,000 Hz. During the conducted analysis of the terpolymer chain structure, the signals assignment previously published by us was used [8,19,22].

Thermal properties, such as glass transition temperature and heat of melting, were examined by DSC with a DuPont 1090B apparatus calibrated with gallium and indium (heating and cooling rate of 20 °C/min in the range from −100 to 220 °C, DuPont, Newark, DE, USA) according to the ASTM E 1356-08 standard [23].

Compressive strength and elastic modulus of the scaffolds were determined under compression using the Instron 4200 tensile testing machine (Illinois Tool Works, Glenview, IL, USA) with a ramp speed of 0.02 cm/s, at room temperature.

3.2. Formulation of Scaffolds and Methods of Their Characterization

The scaffold preparation procedure was described in detail elsewhere [24]. In brief, the scaffolds were produced by the classical solvent casting/particulate leaching technique. Sieved sodium chloride particles (POCh, Gliwice, Poland) of about 300–480 μm, were mixed with 10% (*w*/*v*) terpolymer solution in methylene chloride (Aldrich, Steinheim, Germany) to end up with a porosity of 85% after salt leaching. The mixture was transferred into a special Teflon form (diameter of mold cavity 10 mm and depth 8 mm) and quickly cooled down in the freezer to −80 °C. Then, the samples were lyophilized. The solid salt/polymer composites in the barrel forms were transferred into 1000 mL glass container filled with ultra-high quality water (UHQ-water, produced in UHQ PS system, Purelab, ELGA LabWater, Buckinghamshire. England). To enhance salt leaching, the water flow was assured by using of a magnetic stirrer operating at a speed of 50 rpm. The washing was stopped until the conductivity of the water was about 2 μS/cm, which usually took 4–5 days. The samples were dried in air and vacuum.

For scaffold structure characterization, thin slices were cut from a representative sections of the scaffolds, and images were captured on a scanning electron microscope (SEM, model Tescan VEGA 3SBU, Tescan Orsay Holding, Brno, Czech Republic). Accelerating voltage range was 1–3 kV to minimize the impact of beam energy on the observed samples (the samples were not coated with conductive layer that may have an impact on the morphology of the surface), and then observed under magnifications of about ×100 and ×500. All SEM images were performed at 23 °C under high vacuum, using secondary electron mode.

The scaffolds' porosity was calculated from the density of the solid copolymer, the mass, and the dimensions of each scaffold, according to the ASTM standard D-3574-08 [25].

3.3. Deformation Scaffolds to Temporary Shape

Temporary shape formation of the scaffolds was done on a device designed for the experiment purpose by adaptation of self-centering four-jaw chuck. The central place where the scaffold was fixed was heated by infrared lamp for maintaining the required temperature. The device allowed for precise compression of small polymeric samples simultaneously in three directions due to regulation of jaws by screwing or unscrewing. After deformation process, fixed in the device sample was immediately cooled with compressed air, receiving in this way temporary shape. Such received samples, to avoid self-return to permanent shape caused by their relaxation, have been kept in a freezer at −60 °C prior to use.

3.4. Measurements of Scaffolds' Shape Memory Behavior

The free-strain recovery test was conducted under isothermal conditions, in water. The samples of scaffolds were heated to the required temperature rapidly by immersing in the water bath set at 20 °C, or 37 °C, or 40 °C and remained at that temperature measuring the recovered changes in length as a function of time.

The ability of the sample to recover was quantified by the shape recovery ratio parameter (R_r), defined as:

$$R_r = (e_r/e_m) \times 100\% \tag{1}$$

where, e_m is the final diameter of tested sample with temporary shape before sterilization; and e_r is the measured residual diameter after sterilization and one complete cycle of the shape memory experiment.

3.5. Conditions of Sterilization

The scaffolds, packed in argon atmosphere into rigid and tightly closed polystyrene packaging were transported in thermostatic containers at temperature below 8 °C to the Institute of Nuclear Chemistry and Technology in Warsaw for sterilization. The irradiation of the scaffolds was performed in a Gamma Chamber 5000 apparatus (BRIT, Navi Mumbai, India), using isotopes of ^{60}Co as a gamma radiation emitter at room temperature. A mean dose rate was 5.59 kGy/h. Sterilization using electron beams was conducted by means of the electron accelerator Elektronika 10/10 (Institute of Nuclear Chemistry and Technology, Warsaw, Poland) generating a beam with energy of 10 MeV and average power of 10 kW at room temperature too.

4. Conclusions

During designing and formation of scaffolds for cells growth purposes, one of the most important factors to be taken into consideration is choice of optimal, effective method of sterilization which does not, or eventually causes at a minimal level, undesirable side effects. In the case of materials exhibiting shape memory behavior induced thermally, it can be particularly difficult, considering that the temporary shape of the scaffolds must be kept after sterilization. With this respect, radiation sterilization seems to be the most proper method for these purposes.

Obtained results have proved that by keeping appropriate storage and transport conditions of scaffolds with temporary shape it is possible to sterilize them and maintain their designed shape. Generally, if the dose is not more than 10 kGy, sterilization with use of both methods, gamma radiation and electron beam, will not considerably affect the shape memory behavior. In the case of sterilized scaffolds prepared from terpolymers with relatively high glass transition temperature (above 40 °C), parameters describing the rate and extent of the return from temporary to permanent shape, even at high dose of irradiation, were comparable to non-treated samples.

Problems with maintaining shape memory behavior appeared only in the case when scaffolds prepared from terpolymer with lower T_g were irradiated by means of electron beam. For this type of scaffold, their ability to return to permanent shape was lost as a result of partial self-decompression and relaxation of the samples during sterilization. The reason of this phenomenon is heat generated as a result of conversion of the kinetic energy of the electrons. This process at doses of 15 kGy and higher caused essential changes in morphology of the pores' surface.

Nevertheless, it should be kept in mind that the methods of sterilization proposed in the presented study (especially gamma irradiation) also possess few drawbacks, including strong degradation of polymer and drop in T_g, resulting in a decrease in the mechanical properties of the scaffolds. By contrast, one of the advantages of γ-radiation use is the fact that it does not cause any noticeable changes of the pores' surface, even at the highest dose of irradiation. For comparison, sterilization of the scaffolds by means of electrons beam caused much less negative influence on mechanical properties of treated samples. However, from the other side, using this type of sterilization, at the dose of 15 kGy, the effect of slight melting of pores' surface of terpolymers containing caproil segments and high amount of carbonate units was observed.

Based on the results obtained, it can be concluded that the most optimal method for sterilization of porous, 3D scaffolds was electron beam irradiation at the dose not higher than 10 kGy. The higher dose of irradiation induces the higher, undesirable changes in the structure of scaffolds and as a consequence, decline in their shape behavior properties. It is also worth remembering that the sterilization is intended to destroy pathogens; however, their residues containing antigens, foreign proteins which are not removed during this treatment. With this respect, it is crucial to keep rules of good manufacturing practice during synthesis of terpolymers and preparation of the scaffolds afterward in order to limit microbiological contamination as much as possible.

Further investigations related to the possibilities of application presented in this study, such as scaffolds designed for large bone defect treatment, is being planned. Using the same procedure of scaffold preparation and methods of their sterilization (not exceeding a dose of 10 kGy), we would like to examine whether the sterility at the sterility assurance level (SAL) of 10^{-6} is possible. It is worth mentioning that, as presented in this study, scaffolds sterilized with electron beam at 10 kGy dose were successfully applied for evaluation of their cytotoxicity and shape memory properties in *in vitro* conditions [17].

Acknowledgments: The work was performed within the framework of the research project No. NCN-2011/01/B/ST5/06296 funded by the National Science Centre Poland.

Author Contributions: Anna Smola synthetized and characterized the terpolymers used in investigations. Piotr Rychter, Natalia Śmigiel-Gac prepared the scaffolds and performed the mechanical testing, microstructural characterization of the scaffolds and shape memory behavior parameters under the supervision of Piotr Dobrzynski and Elzbieta Pamula. The SEM observation was carried out by Wojciech Prochwicz and DSC measurements and analysis of the thermal properties were made by Henryk Janeczek. All authors discussed the results and implications and approved the final manuscript.

Conflicts of Interest: The authors declare no conflict of interest.

References

1. Athanasiou, K.; Niederauer, G.G.; Agrawal, C.M. Sterilization, toxicity, biocompatibility and clinical applications of polylactic acid/polyglycolic acid copolymers. *Biomaterials* **1996**, *17*, 93–102. [CrossRef]
2. Shearer, H.; Ellis, M.J.; Perera, S.P.; Chaudhuri, J.B. Effects of common sterilization methods on the structure and properties of poly(D,L-lactic-co-glycolic acid) scaffolds. *Tissue Eng.* **2006**, *12*, 2717–2727. [CrossRef] [PubMed]

3. Cottam, E.; Hukins, D.W.L.; Lee, K.; Hewitt, C.; Jenkins, M.J. Effect of sterilisation by gamma irradiation on the ability of polycaprolactone (PCL) to act as a scaffold material. *Med. Eng. Phys.* **2009**, *31*, 221–226. [CrossRef] [PubMed]
4. Rainer, A.; Centola, M.; Spadaccio, C.; Gherardi, G.; Genovese, J.; Licoccia, S.; Trombetta, M. Comparative study of different techniques for the sterilization of poly-L-lactide electrospun microfibers: Effectiveness *vs.* material degradation. *Int. J. Artif. Organs* **2010**, *33*, 76–85. [PubMed]
5. De Nardo, L.; Alberti, R.; Cigada, A.; Yahia, L.; Tanzi, M.C.; Farè, S. Shape memory polymer foams for cerebral aneurysm reparation: Effects of plasma sterilization on physical properties and cytocompatibility. *Acta Biomater.* **2009**, *5*, 1508–1518. [CrossRef] [PubMed]
6. Behl, M.; Lendlein, A. Shape-memory polymers. *Mater. Today* **2007**, *10*, 20–28. [CrossRef]
7. Zini, E.; Scandola, M.; Dobrzynski, P.; Kasperczyk, J.; Bero, M. Shape memory behavior of novel (L-Lactide-Glycolide-Trimethylene carbonate) terpolymers. *Biomacromolecules* **2007**, *8*, 3661–3667. [CrossRef] [PubMed]
8. Smola, A.; Dobrzynski, P.; Cristea, A.; Kasperczyk, J.; Sobota, M.; Gebarowska, K.; Janeczek, H. Bioresorbable terpolymers based on L-lactide, glycolide and trimethylene carbonate with shape memory behaviour. *Polym. Chem.* **2014**, *5*, 2442–2452. [CrossRef]
9. Mendes, G.C.C.; Brandão, T.R.S.; Silva, C.L.M. Ethylene oxide sterilization of medical devices: A review. *Am. J. Infect. Control* **2007**, *35*, 574–581. [CrossRef] [PubMed]
10. Lendlein, A.; Behl, M.; Hiebl, B.; Wischke, C. Shape-memory polymers as a technology platform for biomedical applications. *Expert Rev. Med. Devices* **2010**, *7*, 357–379. [CrossRef] [PubMed]
11. De Nardo, L.; Moscatelli, M.; Silvi, F.; Tanzi, M.C.; Yahia, L.; Farè, S. Chemico-physical modifications induced by plasma and ozone sterilizations on shape memory polyurethane foams. *J. Mater. Sci. Mater. Med.* **2010**, *21*, 2067–2078. [CrossRef] [PubMed]
12. Tseng, L.F.; Mather, P.T.; Henderson, J.H. Shape-memory-actuated change in scaffold fiber alignment directs stem cell morphology. *Acta Biomater.* **2013**, *9*, 8790–8801. [CrossRef] [PubMed]
13. De Nardo, L.; Bertoldi, S.; Cigada, A.; Tanzi, M.C.; Haugen, H.J.; Farè, S. Preparation and characterization of shape memory polymer scaffolds via solvent casting/particulate leaching. *J. Appl. Biomater. Funct. Mater.* **2012**, *10*, 119–126. [CrossRef] [PubMed]
14. Holy, C.E.; Cheng, C.; Davies, J.E.; Shoichet, M.S. Optimizing the sterilization of PLGA scaffolds for use in tissue engineering. *Biomaterials* **2001**, *22*, 25–31. [CrossRef]
15. Odelius, K.; Plikk, P.; Albertsson, A.C. The influence of composition of porous copolyester scaffolds on reactions induced by irradiation sterilization. *Biomaterials* **2008**, *29*, 129–140. [CrossRef] [PubMed]
16. Yakacki, C.M.; Lyons, M.B.; Rech, B.; Gall, K.; Shandas, R. Cytotoxicity and thermomechanical behavior of biomedical shape-memory polymer networks post-sterilization. *Biomed. Mater.* **2008**, *3*. [CrossRef] [PubMed]
17. Rychter, P.; Pamula, E.; Orchel, A.; Posadowska, U.; Krok-Borkowicz, M.; Kaps, A.; Smigiel-Gac, N.; Smola, A.; Kasperczyk, J.; Prochwicz, W.; *et al.* Scaffolds with shape memory behavior for the treatment of large bone defects. *J. Biomed. Mater. Res. Part A* **2015**, *103*, 3503–3515. [CrossRef] [PubMed]
18. Kot, M.; Wawryło, L.; Rychter, P.; Prochwicz, W.; Smola, A.; Dobrzyński, P. Hydrolitic degradation of biodegradable scaffolds based on L-lactide/glycolide/TMC and L-lactide/glycolide/e-caprolactone ter polymers. *Eng. Biomater.* **2015**, *18*, 10–19.
19. Dobrzynski, P. Synthesis of biodegradable copolymers with low-toxicity zirconium compounds. III. Synthesis and chain-microstructure analysis of terpolymer obtained from L-lactide, glycolide, and ε-caprolactone initiated by zirconium(IV) acetylacetonate. *J. Polym. Sci. Part A Polym. Chem.* **2002**, *40*, 3129–3143. [CrossRef]
20. Pamuła, E.; Dobrzyński, P.; Bero, M.; Paluszkiewicz, C. Hydrolytic degradation of porous scaffolds for tissue engineering from terpolymer of L-lactide, ε-caprolactone and glycolide. *J. Mol. Struct.* **2005**, *744–747*, 557–562. [CrossRef]
21. Cairns, M.L.; Dickson, G.R.; Orr, J.F.; Farrar, D.; Hardacre, C.; Sa, J.; Lemoine, P.; Mughal, M.Z.; Buchanan, F.J. The potential of electron beam radiation for simultaneous surface modification and bioresorption control of PLLA. *J. Biomed. Mater. Res. Part A* **2012**, *100*, 2223–2229. [CrossRef] [PubMed]
22. Gębarowska, K.; Kasperczyk, J.; Dobrzyński, P.; Scandola, M.; Zini, E.; Li, S. NMR analysis of the chain microstructure of biodegradable terpolymers with shape memory properties. *Eur. Polym. J.* **2011**, *47*, 1315–1327. [CrossRef]

23. *Standard Test Method for Assignment of the Glass Transition Temperatures by Differential Scanning Calorimetry*; ASTM E 1356; ASTM International: West Conshohocken, PA, USA, 2014.
24. Pamula, E.; Menaszek, E. *In vitro* and *in vivo* degradation of poly(L-lactide-co-glycolide) films and scaffolds. *J. Mater. Sci. Mater. Med.* **2008**, *19*, 2063–2070. [CrossRef] [PubMed]
25. *Standard Test Methods for Flexible Cellular Materials—Slab, Bonded, and Molded Urethane Foams*; ASTM D3574-08; ASTM International: West Conshohocken, PA, USA, 2008.

3

Acoustic Behavior of Subfloor Lightweight Mortars Containing Micronized Poly (Ethylene Vinyl Acetate) (EVA)

Luiza R. Brancher [1], Maria Fernanda de O. Nunes [2,*], Ana Maria C. Grisa [1], Daniel T. Pagnussat [2] and Mára Zeni [1]

Academic Editor: Prabir Sarker

[1] Laboratory of Materials Chemistry Research, University of Caxias do Sul, Rua Francisco Getúlio Vargas, 1130, Caxias do Sul RS 95070-560, Brazil; lrbrancher@ucs.br (L.R.B.); amcgrisa@ucs.br (A.M.C.G.); mzandrad@ucs.br (M.Z.)
[2] Laboratory of Constructive Technology, University of Caxias do Sul, Avenida Frederico Segala, 3099, Caxias do Sul RS 95010-550, Brazil; dtpagnussat@ucs.br
* Correspondence: mfonunes@ucs.br

Abstract: This paper aims to contribute to acoustical comfort in buildings by presenting a study about the polymer waste micronized poly (ethylene vinyl acetate) (EVA) to be used in mortars for impact sound insulation in subfloor systems. The evaluation method included physical, mechanical and morphological properties of the mortar developed with three distinct thicknesses designs (3, 5, and 7 cm) with replacement percentage of the natural aggregate by 10%, 25%, and 50% EVA. Microscopy analysis showed the surface deposition of cement on EVA, with preservation of polymer porosity. The compressive creep test estimated long-term deformation, where the 10% EVA sample with a 7 cm thick mortar showed the lowest percentage deformation of its height. The impact noise test was performed with 50% EVA samples, reaching an impact sound insulation of 23 dB when the uncovered slab was compared with the 7 cm thick subfloor mortar. Polymer waste addition decreased the mortar compressive strength, and EVA displayed characteristics of an influential material to intensify other features of the composite.

Keywords: recycled aggregates; lightweight mortar; poly (ethylene vinyl acetate); impact noise

1. Introduction

Following the rapid growing of vertical constructions, users search, among other options, those which will bring them better acoustic comfort, since one of the main causes of discomfort in residential buildings is the lack of acoustic insulation of impact noise.

Impact noise is the noise produced from a mechanical impact and transmitted by the building structure. Usually, the impact noise is better perceived at the story below the floor where people are walking, dragging furniture, falling objects, *etc.* The usual materials employed in reducing this kind of noise and consequently increase acoustic insulation in floor systems employ principles related to the dampening of mechanical impact. However, this efficiency in dampening should not impart huge modifications to the rigidity of the floor system so as not to cause instability when a user is walking.

In this sense several studies have been developed in the search for incorporating different polymer wastes into new materials to qualify the acoustic performance in buildings. The influence of factors, such as the particle size distribution of the material incorporated into different kinds of composites, varies according to the kind of material, but as a general principle the lower the particle size, the higher is the efficiency of the material towards insulation of impact sound [1–3].

António *et al.* [4] studied cement-cork screeds with three varying cement proportions, three different thicknesses and their application for the reduction of impact sound in floors. In this study, as the cement content increased, the impact reduction declined, particularly when the cement-cork screed was tested as floor covering. The best results were obtained for the 3 cm thick sample. However, the samples with highest cement content give similar impact sound reduction for the three thicknesses tested.

D'Alessandro *et al.* [2] tested lightweight concrete containing polymers derived from recycled sheets of electric wires with small, 1-m^2 size samples. They explain that even if the developed lightweight concrete cannot be considered a traditional resilient material, as the ones to which the method is addressed, the high percentage of polymer makes the product elastic enough. The performed investigations show that it is possible to use polymeric residues in a new product showing promising acoustic and thermal performance if compared with common lightweight screeds. They conclude that the high stiffness of the material is the most important variable in impact sound reduction.

In addition to particle size, the hot or cold press process also influences, the cold processes leading to better acoustic performance since the integrity of polymer materials is maintained [5].

Studies report that the use of polymer wastes in conventional cementitious composites reduces the mechanical strength of the material as the polymer percentage is increased [6]. However, studies found a different relationship with high performance concrete. Self-Compacting Concrete (SCC) is a new class, sophisticated high performance concrete based on recycled end-of-life vehicle tires endowed with environmental, economic and performance benefits. Najim and Hall [7] indicate that the Self-Compacting Rubberized Concrete (SCRC) composite could be used for applications requiring deformable concrete with high flowability and low/medium strength (<35 MPa). The rubber aggregate was used to partially substitute fine and coarse aggregates in proportions of 5%, 10% and 15%. They concluded that SCRC has mechanical properties sufficient for structural applications, but with superior dynamic properties relative to conventional mixes.

Borzog-Haddad *et al.* [8] claim that several studies have demonstrated that structural elements made of post-consumer recycled polymers can be used in construction. Nevertheless, some precautions should be observed due to the importance of accelerated creep test methods for polymeric products. Creep is a deformation of a viscoelastic material under a constant load, there being a strong relationship between time, creep strain and applied stress. The magnitude of creep depends on the type of polymer and the manufacturing processes, which cause a large difference in the creep behavior among different polymeric products. Therefore, the creep of each polymeric product should be evaluated so that appropriate reduction factors can be applied in design.

However, this material can have non-structural applications, which do not require the high mechanical strength of constructive elements, for which the increase in polymer amount could mean higher efficiency in thermal and acoustic insulation [9–11].

The footwear industry of South Brazil produce every month huge amounts poly (vinyl ethylene acetate) (EVA) wastes, a thermosetting polymer which is not biodegradable and which cannot be reprocessed, which renders it a low-cost raw-material. Tutikian *et al.* [9] tested the impact sound reduction of composites made with two types of recycling processes with and without treatment to the EVA powder generated during the process. The authors found similar results for the two cases, which indicates that a strict control in the production of the composites is not required, which could represent some reduction in production costs.

In this context, this study aims at evaluating the influence of micronized EVA waste on the physical, mechanical and morphological properties of subfloor mortars in order to improve their properties of impact noise insulation in residential buildings. The impact sound insulation test was carried out with subfloor samples with 50% EVA waste in order to check the efficiency of the highest possible incorporation of polymer.

2. Procedure

A methodology was established to characterize the subfloor samples behavior considering the use with a concrete slab of typical Brazilian residential buildings.

EVA micronized waste was supplied by a footwear industry from Rio Grande do Sul State/Brazil. CP IV cement was supplied by Votorantim and as natural aggregate medium sand was employed, classified as small aggregate. Sand was previously dried in an oven at the temperature of 100 °C being kept in a sealed vessel until use.

2.1. Methods of Preparation

The preparation of the subfloor mortar was carried out in agreement with the NBR 13276 [12] Brazilian Standard. In order to determine the amounts to be employed, a proportion of 1:4 (binder: aggregate) by volume and corresponding amount of 1:6.5 by mass was established. For the subfloor mortar, the proportion by mass was prepared by replacing 0%, 10%, 25% and 50% of the natural aggregate by the EVA micronized waste having average particle size 401.45 μm and specific weight 1.11 g/cm^3, according to the NBR NM 23 Brazilian Standard and unit mass 0.16 g/cm^3 in accordance with the NBR NM 45 [13] Brazilian Standard.

Table 1 lists the proportions employed for each mortar sample and the water/binder ratio which was not obeyed since the polymer has a huge capacity for water absorption.

Table 1. Proportions employed in the preparation of samples.

Samples	Proportions (Mass)	Water/Binder Ratio
Reference Mortar	1:6.5	0.92
Mortar 10% EVA	1:0.068:5.82	0.55
Mortar 25% EVA	1:0.17:4.85	0.97
Mortar 50% EVA	1:0.34:3.23	0.65

After homogenization, cylindrical 5 cm diameter × 10 cm height and prismatic test specimens 4 cm × 4 cm × 16 cm were molded in order to perform mechanical and physical testing. Besides, 20 cm × 20 cm surface area plates were prepared, of three different thicknesses: 3 cm, 5 cm and 7 cm. Test specimens remained for 28 days under cure in a climatized room at a temperature of 23 ± 2 °C and a relative humidity of 80% ± 5%.

2.2. Characterization

2.2.1. Water Absorption, Voids Index and Actual Specific Weight

Absorption tests after water immersion at 23 °C (±2 °C) temperature, voids index after saturation in water and actual specific weight were performed in triplicate in accordance with the NBR 9778 [14] Brazilian Standard, after 28 days under cure.

Test specimens were kept in a Quimis® oven at a temperature of 60 °C (±5) °C for 72 h. After drying test specimens were kept submerged in water in a climatized room at a temperature of 23 ± 2 °C, they were weighed in a Marte® scale until mass stability. The test procedure includes drying the sample at a temperature of 105 ± 5 °C for a period of 72 h. The mass of the dry sample in grams is registered; thereafter, it is immersed in water at 23 ± 2 °C and kept for 72 h in this condition. After saturation, the sample is placed in a container full of water which must be gradually brought to boiling and held for 5 h. Allow the sample to cool naturally to 23 ± 2 °C and the mass of the sample immersed in water at hydrostatic balance is registered.

2.2.2. Field Emission Scanning Electronic Microscopy

For the morphological analysis of the samples the same proportion of the mortar samples of 1:4 by volume was kept. Morphological analysis of the samples was performed in a Tescan® model Mira

3 field emission scanning electronic microscope aiming at evaluating the polymer interface at the cementitious matrix microstructure.

2.2.3. Compressive Strength

The test for compressive strength was carried out on cylindrical test specimens in triplicate, in accordance with NBR 7215 [15] Brazilian Standard in a EMIC—PC200I hydraulic press in order to evaluate the mechanical strength of the mortars modified by the EVA micronized waste.

2.2.4. Compressive Creep

For the compressive creep test Mitutoyo dial indicators of 0.01 accuracy were used to measure the subfloor creep, in accordance with the procedures of the ISO 20392 [16] Standard. Mortar plates were compressed by a 200 kg/m² steel plate until stabilization of the dial indicators. Creep was estimated by regression analysis for a 13-year period, which is the minimum useful life of a subfloor in accordance with Brazilian Standard NBR 15575 [17]. The creep of reference samples and that of 10% and 50% EVA micronized waste in replacement of natural aggregate was evaluated in order to check if the thickness reduction throughout the useful life of the floor system keeps the same dampening capacity, with porosity preservation.

2.2.5. Impact Noise

The impact noise test was performed in accordance with the ISO 140-7 Standard [18] in a two-floor prototype of dimensions 2.55 m × 4.90 m × 2.58 m. The masonry of the upper floor was build from ceramic blocks and that of the lower floor from concrete blocks, both of 19 cm thickness. Rooms are separated by a 13 cm thickness structural concrete slab and have brickwork walls covered with plaster and paint.

The noise for the tests was generated with a normalized Bruel and Kjaer type 3207 tapping machine and the measurements were carried out with a Bruel and Kjaer type 2270 sound level analyzer, which has a ½″ microphone.

On the upper floor is placed the standard tapping machine which is in charge of emitting the mechanical impact on the subfloor, which is the origin of noise in the lower floor, which is captured by a microphone (Figure 1). Five measurements were obtained for each four positions of the instrument, at the bands of one-third octave between 100 Hz and 3.15 kHz and the result is obtained by the weighed standardized impact noise level $L'_{nT,w}$. The test was carried out with subfloor samples with 50% EVA waste in order to check the efficiency of the highest possible incorporation of polymer.

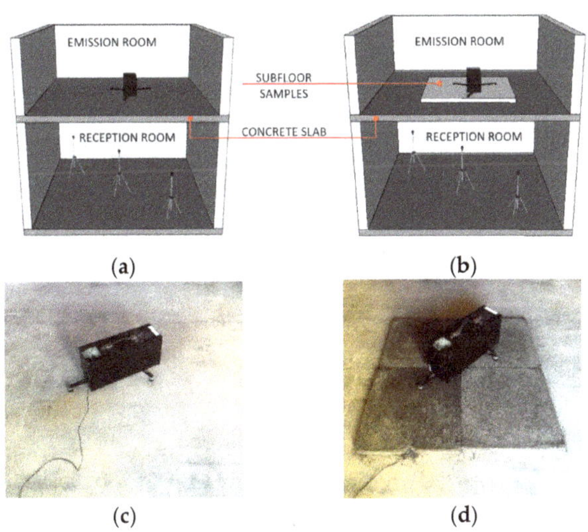

Figure 1. Rooms of the tests and position of the equipment: (**a**) concrete slab; (**b**) subfloor samples; (**c**) experimental setup without samples; and (**d**) with samples.

3. Results and Discussion

3.1. Absorption of Water, Voids Index and Actual Specific Weight

By adding the polymer waste, the voids in the sample augment as a function of EVA porosity. Consequently its specific weight is reduced which contributes to reduce overburden of structure of a multi-flooring building. The increase in EVA waste content promotes increase in voids index which assures dispersion of energy at the subfloor surface layer when a percussion action is exerted on the floor, which contributes for reducing the mechanical impact to be transmitted by the slab.

Table 2 lists results for water absorption, voids index and actual specific weight, with the respective standard deviation.

Table 2. Results for water absorption, voids index and actual specific weight.

Samples	Water Absorption (%)	Voids Index (%)	Actual Specific Weight (kg/m^3)
Reference	9.19 ± 0.18	17.95 ± 0.19	2380 ± 19
10% EVA	11.61 ± 0.26	20.81 ± 0.30	2260 ± 17
25% EVA	11.48 ± 0.76	19.49 ± 0.99	2110 ± 21
50% EVA	15.60 ± 1.62	22.45 ± 1.70	1850 ± 20

The higher water absorption of the mortar produced with EVA may be explained by the pore size and pore network formed due to the packing of grains of each sample. A discrepancy was observed for water absorption results and voids index for the 25% EVA mortar relative to the other waste percentages. This can be explained in the light of better packing of grains in this range of replacement of sand by the polymer.

3.2. Field Emission Scanning Electron Microscopy

The mortar morphology (Figure 2a) evidenced that there is no chemical interaction among the sands grains and the cement resulting from the aggregate property of being inert.

Mortars (Figure 2b–d) show the same morphological feature no matter the EVA content. EVA has a porous structure and cement deposits superficially on EVA, favoring the increase in mortar's voids index.

(a) (b) (c) (d)

Figure 2. Samples Morphology: (**a**) reference mortar; (**b**) mortar 10%; (**c**) mortar 25%; and (**d**) mortar 50% EVA wastes.

Whenever there is no intrusion of cement particles within the EVA aggregate pores, the low rigidity of the system is kept as such, preserving dissipation of mechanical energy by the material internal dampening, with the consequent reduction in impact noise.

Energy dissipation of the mechanical impact is explained by the increased contact time derived from the dampening of kinetic energy by the elastic nature of the material. Under these circumstances the excitation spectrum keeps the same force amplitude with different distributions throughout time [19].

3.3. Compressive Strength

Table 3 lists data for compressive strength of the reference mortar and of mortars having 10%, 25% and 50% EVA wastes.

Table 3. Data for compressive strength of the mortars prepared in this study.

Samples	Compressive Strength (MPa)
Reference Mortar	5.44 ± 0.12
Mortar 10%	4.07 ± 1.45
Mortar 25%	2.07 ± 0.16
Mortar 50%	1.44 ± 0.25

Addition of EVA wastes represented reduction in compressive strength of the material. Variations from 5.44 MPa for the reference mortar down to 1.44 MPa for the mortar incorporating 50% EVA (Table 2) were observed. This behavior can be explained in the light of the high porosity of the EVA waste, which heightens the mortar voids index, reducing its mechanical strength.

Herrero *et al.* [20] studied the influence on mortars of different amounts and particle size ranges in end-of-cycle rubber. In mechanical testing it was observed that reductions in strength are more apparent for mortars of higher polymer amounts (60%) and lower particle size (0–0.6 mm). The mortar containing the highest amount of micronized polymer (50%) used in the present research with particle size of nearly 0.4 mm had the lowest mechanical strength, these results being similar to those reported by Herrero *et al.* [10].

3.4. Compressive Creep

Figure 3 illustrates the results for compressive creep, which evidences the deformation of reference, 10% and 50% EVA plates for three different thicknesses for a 13-year period.

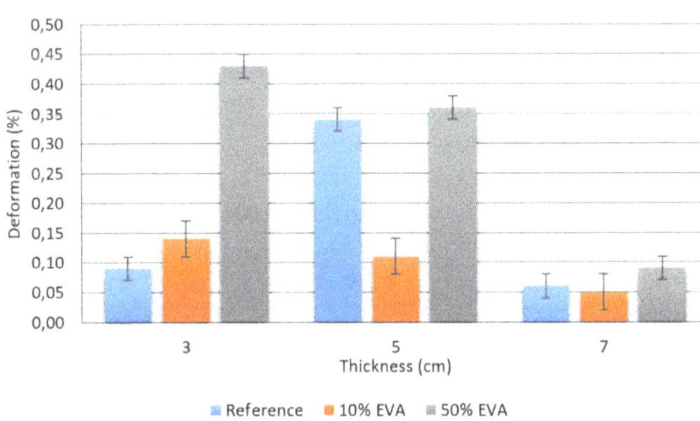

Figure 3. Creep results for subfloor samples.

The compressive creep deformation results were all less than 0.5%, and these values can be considered favorable to the use of the material as a non-structural element, since structural elements present satisfactory results between 20%–30% [21].

One can also compare the results with the compressive creep of resilient materials commonly used to reduce impact noise. Gnip *et al.* [22] analyzed the deformation under constant compression of mineral wool samples with seven different thicknesses and densities. Deformation was lower in samples with greater thickness, with 0.3% for 80 and 150 mm. For the thinner samples of 30 and 40 mm, the deformation was between 1% and 1.3%.

The ability of aggregates to restrain movement of a cement paste depends upon the extensibility of the paste, the degree of cracking of the paste, the compressibility of the aggregate and changes in the aggregate moisture content. Generally, concretes that have aggregates that are hard, dense and have low absorption and high modulus of elasticity are desirable when concrete with low creep is needed [23].

The lower height creep percent mortar was the 7 cm thickness, 10% EVA sample. In view of the evaluation of mortars in subfloor systems, lower creep assures the integrity of the plate throughout time and reduced diminishing of voids index, which collaborates with mechanical impact dampening during its use [19].

3.5. Impact Noise

Analysis of the acoustic test results were performed by comparing the noise level curve (L'_{nT}) of the concrete slab as well as of each subfloor samples tested. The performance rating was obtained in $L'_{nT,w}$.

The impact noise tests were performed with 50% EVA waste only, in order to assess the viability of the material with highest possible residue content. Furthermore, the procedure using larger samples and thorough testing with other proportions would generate more waste.

Table 4 lists results for impact noise level for samples of 50% EVA waste.

Table 4. Results for impact noise level for samples of 50% EVA waste.

Sample	Sample Thickness (cm)	Impact Noise Level $L'_{nT,w}$ (dB)
50% EVA 3 cm	3	51
50% EVA 5 cm	5	58
50% EVA 7 cm	7	56
Concrete slab	13	79

Taking into consideration that the highest $L'_{nT,w}$ level accepted by Brazilian standards is 80 dB, all the samples tested complied with such requirement. It was observed that the 3 cm thickness sample exhibited impact noise propagation lower than those of the other, thicker samples. It is well-known that as the slab thickness increases, the propagation of the impact noise of a concrete plate is reduced [3]. Therefore, a better acoustic performance is expected for the thicker plates. On receiving the mechanical impact, the subfloor degradation occurs as a result of its limited thickness, as seen in the 3 cm thickness sample which was not able to keep its integrity after the impact. Hence, the particles of the lower portion of the plate detached and dampened the mechanical impact, with consequent reduction in subfloor impact noise.

The results of the impact noise levels can be grouped into three different zones, determined by frequency ranges up 400 Hz, between 400 Hz and 1.6 kHz and above 1.6 kHz (Figure 4). The results at low frequencies, up 400 Hz corresponded to bass sounds measured in the reception room. In this situation, the 3 cm thickness sample presented higher sound reduction comparing to concrete slab. For 160 Hz frequency range the samples with 5 and 7 cm thickness showed sound levels as high as the ones measured with the concrete slab. This could be due to the self-resonance-frequency effect, and it occurs at low frequencies and is directly proportional to the superficial density of the solid material [24]. Thus, the higher the superficial density of a specific material the higher the level of acoustic pressure.

Within middle frequency interval, from 400 Hz and 1.6 kHz, the levels of acoustic pressure reduced when the frequency was higher, *i.e.*, the higher the frequency band, the more acute is the sound. Moreover, the higher the frequency, the lower the measured sound pressure level measured in the reception room with EVA waste subfloor samples.

In the zone of higher frequencies, from 1.6 kHz, there was an increase of the sound insulation, especially for reduced sound level with 7 cm thickness subfloor sample. In this sample, a mortar detachment of particles was observed. This effect was caused by loose particles of reduced size which can, under certain circumstances even increase the efficiency of a floor system. However, these particles can lead to moves that jeopardize the integrity of the whole floor system [25].

It should be emphasized that this effect can be suppressed with a final coating material, which despite being considered a complementary solution, in this case can significantly contribute to the attenuation of impact sound transmission. This could be attributed to the fact that the time of impact percussion in final covering floors is smaller than the impact time of a percussion carried out in the uncovered floor [19].

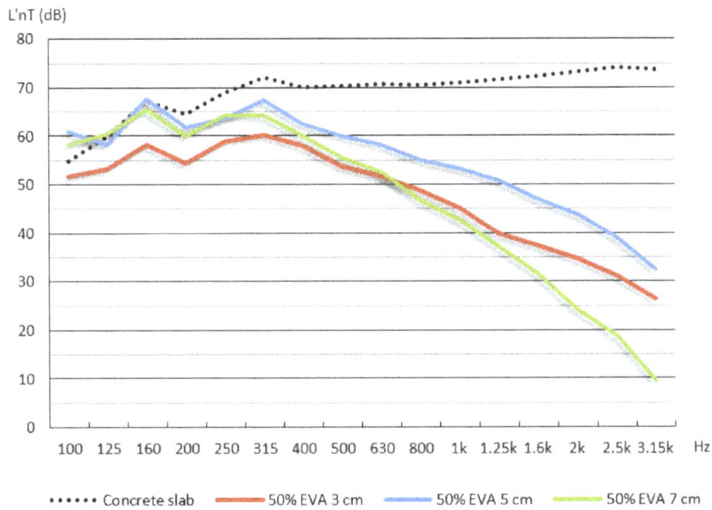

Figure 4. Impact noise level one-third octave band frequency for samples of 50% EVA waste.

The effect of substitution of natural aggregates for lightweight aggregate to reduce the impact noise has also been observed in other studies. Branco and Godinho [3] found significant reductions in noise levels of lightweight mortars with cork, polystyrene and clay, mainly in the frequency bands above 400 Hz, when compared with conventional mortar subfloor. In another study, Olmeda *et al.* [26] incorporated petroleum coke to produce lightweight mortars and also identified sound levels reductions with increased frequency bands. The most significant differences were of 1.6 kHz.

4. Conclusions

Samples having incorporated the EVA polymer waste had favorable features for impact noise insulation. Addition of the waste favored the rise in voids index which contributes to reduce the subfloor actual specific weight rendering lighter the system, with benefits not only to the building general structure in terms of reduction of structural overcharge, but also to impact noise attenuation transmitted by the slab at each floor. In view of the non-utilization of the cementitious composite in structural buildings, reduction of compressive strength resulting from waste incorporation is not a determining factor for the suggested use according to the present study. The long term lower subfloor creep, as estimated by compressive creep tests assures the system integrity, allowing the subfloor to keep the property of impact noise dampening, providing better acoustic comfort to users of residential buildings.

Acknowledgments: This work was funded by University of Caxias do Sul, through the Research group of Lightweight Concrete. Additional support was provided by National Council of Scientific and Technological Development (CNPq), Research Support Foundation of Rio Grande do Sul State (FAPERGS) and Basf Brazil Ltda.

Author Contributions: Maria Fernanda de O. Nunes, Ana Maria C. Grisa, Daniel T. Pagnussat and Mára Zeni designed the research and analyzed the results. Luiza R. Brancher, Maria Fernanda de O. Nunes and Ana Maria C. Grisa performed the experiments. Luiza R. Brancher, Maria Fernanda de O. Nunes and Ana Maria C. Grisa finalized the text writing.

Conflicts of Interest: The authors declare no conflict of interest.

References

1. Maderuelo-Sanz, R.; Martín-Castizo, M.; Vílchez-Gómez, R. The performance of resilient layers made from recycled rubber fluff for impact noise reduction. *Appl. Acoust.* **2011**, *72*, 823–828. [CrossRef]
2. D'Alessandro, F.; Asdrubali, F.; Baldinelli, G. Multi-parametric characterization of a sustainable lightweight concrete containing polymers derived from electric wires. *Constr. Build. Mater.* **2014**, *68*, 277–284. [CrossRef]
3. Branco, F.G.; Godinho, L. On the use of lightweight mortars for the minimization of impact sound transmission. *Constr. Build. Mater.* **2013**, *45*, 184–191. [CrossRef]
4. António, J.; Moreira, A.; Tadeu, A. Impact sound transmission provided by concrete layers incorporating cork granules. *Noise Control Eng. J.* **2013**, *61*, 458–468. [CrossRef]
5. Ribeiro, W.B.; Rizzo, M.V.; Bortoluz, N.; Zeni, M.; Nunes, M.F.O.; Grisa, A.M.C. Characterization of polyurethan skin agglomerates for acoustic insulation from impact noise. *Mater. Res.* **2014**, *17*, 210–215. [CrossRef]
6. Pendhari, S.S.; Kant, T.; Desai, Y.M. Application of polymer composites in civil construction: A general review. *Compos. Struct.* **2008**, *84*, 114–124. [CrossRef]
7. Najim, K.B.; Hall, M.R. Mechanical and dynamic properties of self-compacting crumb rubber modified concrete. *Constr. Build. Mater.* **2012**, *27*, 521–530. [CrossRef]
8. Bozorg-Haddad, A.; Iskander, M.; Chen, Y. Compressive strength and creep of recycled HDPE used to manufacture polymeric piling. *Constr. Build. Mater.* **2012**, *26*, 505–515. [CrossRef]
9. Tutikian, B.F.; Nunes, M.F.O.; Leal, L.C.; Marquetto, L. Hormigón ligero con agregado reciclado de EVA para atenuación del ruido de impacto. *Mater. Constr.* **2013**, *63*, 309–316. [CrossRef]
10. Grisa, A.M.C.; Rauber, R.; Nunes, M.F.O. Evaluation of physical properties and acoustic performance the agglomerates of wood pellets polyurethane residue. *Mod. Environ. Sci. Eng.* **2015**, *1*, 99–103. [CrossRef]
11. Gil, L. New cork-based materials and applications. *Materials* **2015**, *8*, 625–637. [CrossRef]
12. *NBR 13276: Argamassa Para Assentamento e Revestimento de Paredes e Tetos—Preparo da Mistura e Determinação do índice de Consistência*; ABNT: Rio de Janeiro, Brazil, 2005.
13. *NBR NM 45: Agregados—Determinação da Massa Unitária e do Volume de Vazios*; ABNT: Rio de Janeiro, Brazil, 2006.
14. *NBR 9778: Argamassa e Concreto Endurecidos—Determinação da Absorção de água, Índice de Vazios e Massa Específica*; ABNT: Rio de Janeiro, Brazil, 2005.
15. *NBR 7215: Cimento Portland—Determinação da Resistência à Compressão*; ABNT: Rio de Janeiro, Brazil, 1997.
16. *ISO 20392. Thermal-Insulating Materials—Determination of Compressive Creep*; ISO: Geneva, Switzerland, 2007.
17. *NBR 15575-1. Edifícios Habitacionais—Desempenho. Parte 1: Requisitos Gerais*; ABNT: Rio de Janeiro, Brazil, 2013.
18. *ISO 140-7. Acoustics—Measurement of Sound Insulation Inbuilding Elements—Part 7 Field Measurements of Impact Sound Insulation of Floors*; ISO: Geneva, Switzerland, 1998.
19. Hopkins, C. *Sound Insulation*; Elsevier Ltd.: Burlington, UK, 2012.
20. Herrero, S.; Mayor, P.; Hernández-Olivares, F. Influence of proportion and particle size gradation of rubber from end-of-life tires on mechanical, thermal and acoustic properties of plaster–rubber mortars. *Mater. Des.* **2013**, *47*, 633–642. [CrossRef]
21. Domingo-Cabo, A.; Lázaro, C.; López-Gayarre, F.; Serrano-López, M.A.; Serna, P.; Castaño-Tabares, J.O. Creep and shrinkage of recycled aggregate concrete. *Constr. Build. Mater.* **2009**, *23*, 2545–2553. [CrossRef]
22. Gnip, I.; Vaitkus, S.; Vejelis, S. Predicting the deformability of mineral wool slabs under constant compressive stress. *Constr. Build. Mater.* **2009**, *23*, 675–680. [CrossRef]
23. Meyerson, R. Compressive Creep of Prestressed Concrete Mixtures with and Without Mineral Admixtures. Master's Thesis, Virginia Polytechnic and State University, Blacksburg, VA, USA, 2001.
24. Harris, D. *Noise Control Manual for Residential Buildings*; Builders' Guides; McGraw-Hill Education: Hightstown, NJ, USA, 1997.
25. Schiavi, A.; Prato, A.; Pavoni Belli, A. The "dust spring effect" on the impact sound reduction measurement accuracy of floor coverings in laboratory. *Appl. Acoust.* **2015**, *97*, 115–120. [CrossRef]
26. Olmeda, J.; Frías, M.; Olaya, M.; Frutos, B.; Sánchez de Rojas, M.I. Recycling petroleum coke in blended cement mortar to produce lightweight material for Impact Noise Reduction. *Cem. Concr. Compos.* **2012**, *34*, 1194–1201. [CrossRef]

4

Sustainable High Quality Recycling of Aggregates from Waste-to-Energy, Treated in a Wet Bottom Ash Processing Installation, for Use in Concrete Products

Philip Van den Heede [1], Niels Ringoot [1], Arno Beirnaert [1], Andres Van Brecht [2], Erwin Van den Brande [2], Geert De Schutter [1] and Nele De Belie [1,*]

Academic Editor: Jorge de Brito

[1] Magnel Laboratory for Concrete Research, Ghent University, Technologiepark Zwijnaarde 904, Ghent B-9052, Belgium; philip.vandenheede@ugent.be (P.V.H.); nielsringoot10@hotmail.com (N.R.); arno_beirnaert@hotmail.com (A.B.); geert.deschutter@ugent.be (G.D.S.)

[2] Indaver nv, Dijle 17a, Mechelen B-2800, Belgium; andres.van.brecht@indaver.be (A.V.B.); erwin.van.den.brande@indaver.be (E.V.B.)

* Correspondence: nele.debelie@ugent.be

Abstract: Nowadays, more efforts towards sustainability are required from the concrete industry. Replacing traditional aggregates by recycled bottom ash (BA) from municipal solid waste incineration can contribute to this goal. Until now, only partial replacement has been considered to keep the concrete workability, strength and durability under control. In this research, the feasibility of a full aggregate replacement was investigated for producing prefabricated Lego bricks. It was found that the required compressive strength class for this purpose (C20/25) could be achieved. Nevertheless, a thorough understanding of the BA properties is needed to overcome other issues. As BA is highly absorptive, the concrete's water demand is high. This workability issue can be dealt with by subjecting the fine BA fraction to a crushing operation to eliminate the porous elements and by pre-wetting the fine and coarse BA fractions in a controlled manner. In addition, a reactive NaOH washing is needed to avoid formation of longitudinal voids and the resulting expansion due to the metallic aluminum present in the BA. Regarding the long-term behavior, heavy metal leaching and freeze-thaw exposure are not problematic, though there is susceptibility to acetic and lactic acid attack and maybe increased sensitivity to alkali-silica reaction.

Keywords: municipal solid waste incineration; bottom ash; concrete; aggregate replacement; prefabricated Lego brick

1. Introduction

Yearly, 3,331,000 tons of municipal solid waste are generated in Flanders. Around 880,000 tons are incinerated to gain energy and reduce the waste volume with 90% (Waste-to-Energy) [1,2]. The incineration results in bottom ash (BA), boiler ash and air pollution control residues. BA is mostly disposed of as landfill. The high waste volume, cost of landfill and lack of disposal sites force the industry to recycle the BA as secondary raw material [3]. This would also be more in agreement with European Directive 2008/98/EC [4], as it saves valuable natural resources and gives waste a financial value [5]. To obtain an end-of-waste status *cf.* ED 2008/98/EC [4], several criteria need to be fulfilled. The material should serve a purpose. The technical requirements and legislation of this purpose should be met. There needs to be a market for it and there should be no negative environmental effects [4]. In this context, companies involved in municipal solid waste incineration (MSWI) intend to recover the BA as aggregates for concrete. The successful production of so-called concrete Lego bricks is seen as the minimal aim. These are prefabricated unreinforced blocks which are used to build modular partition walls at industrial and rural sites. For several years now, the waste management company Indaver nv in Belgium has been trying to tune the properties of BA at its incineration plant to meet this goal. In the past, this already resulted in the publication of papers on the proper characterization and treatment of the Belgian BA (e.g., Vandecasteele *et al.* [6], Van Gerven *et al.* [7]).

According to literature, BA from Belgium [8,9], the Netherlands [10], France [11], Germany [5], Italy [3,12], Spain [2] and South-Korea [13] show quite some similarities in composition. They usually consist of bottle glass (around 15%, composed of Na_2O, CaO, $CaCO_3$ and SiO_2 [2]), as well as metals (ferrous and non-ferrous, such as Al, Cu and Zn), ceramics (brick, plaster and mortar) and organic residues (bone fragments, charcoal, plant fibers and polymers). The latter elements have only a small contribution to the total mass (each with an amount of about 2%) [5,10]. Evidently, the variety in constituting materials implies a heterogeneous chemical composition of the BA: 30%–70% silica, 10%–15% aluminum oxide, 20% sodium oxide and 10%–15% calcium oxide. A smaller yet still important fraction of (heavy) metals is also present [5,10]. BA is lighter than gravel (2210 kg/m^3 *vs.* 2470 kg/m^3). The water absorption is 4–10 times the value of gravel (4%–10%). Both characteristics are due to a high porosity [3,11]. BA is highly angular and has a weaker abrasion strength than limestone [14]. BA concrete is usually characterized by a lower compressive strength, a lower Young's modulus, a lower workability and a higher porosity than concrete with limestone as aggregate. Nevertheless, even without modifications and with full replacement of the aggregates by BA, a concrete strength class of C20/25 can be assured [15].

In the fresh state, extensive aggregate replacement by BA may have a negative effect on the concrete workability. This is mainly due to the higher water absorption of the BA. However, when using rather high water-to-cement (W/C) ratios, workability may not be a real issue. According to Pera *et al.*, a W/C ratio of 0.60–0.65 is ideal to achieve both an acceptable workability and a sufficient mechanical performance [11]. Moreover, the use of a superplasticizer (SP) can easily improve the workability if needed [16].

In hardened state, most damage phenomena are induced by the expansive reaction of metallic Al in an alkaline environment, which increases the porosity and reduces the strength. This effect exists for all hydroxide-forming elements present in the BA [5,11]. Alkali-silica reaction (ASR) is considered not detrimental because of the higher porosity of the BA concrete. A literature survey on other expected damage phenomena (e.g., freeze-thaw, acid attack, *etc.*) provides little or no information on the performance of BA concrete.

Literature on the leaching behavior of concrete containing BA from MSWI is rather scarce. Since the material is immobilized within the cementitious matrix of the concrete, leaching values are normally lower than when the individual grains are in direct contact with the leaching solution. Regarding the leaching of (heavy) metals, Sorlini and Ginés found values that meet the national criteria (Italy for Sorlini and Spain for Ginés) [17,18].

In the past, tests were mainly performed in view of a partial replacement of traditional aggregates by BA [10,11,16,17] and they included variations in cement type [17], W/C ratio [8,12], workability and compaction [8,19]. The concrete quality can also be improved by optimizing the BA properties using opto-mechanical glass separation [11], reactive washing with NaOH (=Lye treatment) [11], vitrification [12] and sintering [17].

2. Research Relevance and Methodology

In this research, the focus was on obtaining a qualitative concrete with replacement of traditional crushed limestone aggregates by BA for the production of prefabricated concrete products. The main novel character of the research lies in the fact that full instead of partial aggregate replacement is aimed for. The main benefit of such a practice is rather evident. If we are able to incorporate more BA from MSWI in concrete, less of the material will get landfilled, its valorization potential will increase significantly and the concrete industry will have to rely less on natural resources to ensure its production capacity. In other words, there will be considerable advantages from both an economic and environmental point of view. However, these benefits only exist if the production of this BA concrete is practically feasible in the casting stage and if the hardened concrete is sufficiently strong and durable for its intended field of application. Therefore, the following research approach was adopted. After careful characterization of the physical and chemical properties of the BA, several test mixtures with varying W/C ratios were made to find an optimal balance between workability and strength of the concrete. The effects of pre-treating the BA in various ways were carefully evaluated. A dedicated study of the expected long-term behavior for one potentially suitable mixture was done and a final optimized concrete mix design was proposed. Prefabricated Lego bricks were made using both mix designs and were compared with a Lego brick made of traditional concrete.

3. Materials and Methods

3.1. Concrete Constituents

The inert fraction of concrete traditionally consists of sand and aggregates. For the aggregates, it is often useful to use both a fine and a coarse fraction to achieve a better packing. In this research, the aim was a full replacement of the traditional fine and coarse aggregates by BA processed at the MSWI plant of Indaver nv. In terms of phase composition, this consisted of glass remains, a stony fraction (ceramics, porcelain, bricks, *etc.*), amorphous slags and some ferrous (e.g., steel) and non-ferrous (e.g., Al, Cu, Pb, Zn) metals that cannot be removed in the treatment installation. The glass remains (bottle glass, grains of glass with devitrification products) normally represent the largest fraction. The proportioning with the other fractions is somewhat seasonally bound.

Different types of fine and coarse BA aggregates were considered, *i.e.*, sieved and aged BA 0/6 and 2/6, crushed BA 0/6, sieved and washed BA 0/6 and 6/20, and crushed BA 6/20. The crushed and sieved and washed fractions originated from the same 6/50 fraction which was obtained after an initial processing stage at Indaver nv. The other fractions were the result of a natural aging process of a 2/6 batch. More details on how the 6/50 fraction and the aged 2/6 batch were obtained can be found in Vandecasteele *et al.* [6]. The crushing, sieving, washing and aging techniques that were applied afterwards to obtain the 0/6, 2/6 and 6/20 fractions were similar to the ones mentioned in that previous research. Crushing was done by means of a commercial mobile cone crusher that allows for setting the required maximum aggregate size. A mobile installation was also used for sieving the material to the desired sieve aperture. The material to be sieved was usually wet due to its outdoor unsheltered storage. No additional water was added during sieving. Washing operations were done in most cases with process water that resulted from drying the filter cake at the Indaver incineration plant. Logically, this process water contains fines that adhere somewhat to the BA. The aging process consisted of piling the BA on 5–10 m high heaps in open air on a paved floor at the incineration plant with the collection of percolating rainwater, exposing them to wind and rain.

The BA-based aggregates should replace the traditional crushed limestone 2/6 and 6/20 which are commonly used for concrete manufacturing in Belgium. The remaining portion of the inert fraction, *i.e.*, the sand, was a natural river sand 0/4.

Two different cement types were used for concrete manufacturing, either a traditional Portland cement CEM I 52.5 N MF or a low alkali (LA) high sulfate resistant CEM I 52.5 N HES LA HSR. The total cement content for each concrete composition amounted to 350 kg/m^3.

The mixing water consisted of tap water at a temperature of 20 °C. The amount of water used for the concrete varied with the applied W/C ratio: 0.65, 0.60 or 0.55. When lowering the water content substantially, the use of a SP was often imperative. Initially, polycarboxylic ether-based MasterGlenium 51 con. 35% (dry matter mass percentage: 35%, density at 20 °C: 1100 kg/m^3) was used for this purpose. Later on, a switch was made to polycarboxylic ether-based MasterGlenium ACE 30 con. 30% SPL (dry matter mass percentage: 35%, density at 20 °C: 1070 kg/m^3) to reduce the delay in setting time.

3.2. *Physical and Chemical Characterization of the Aggregates*

3.2.1. Water Absorption over 24 h and Mass Density

The water absorption over 24 h (WA_{24}) of all studied aggregates was measured in accordance with NBN EN 1097-6 using a calibrated pycnometer. The apparent mass density ρ_a (kg/m^3) and the relative mass density after oven drying ρ_{rd} (kg/m^3) were determined in accordance with the same standard.

3.2.2. Water Absorption as a Function of Time

To evaluate the water absorption as a function of time, a hydrostatic weighing technique similar to the one proposed by Garcia-Gonzalez *et al.* and Tegguer was applied [20,21]. Per aggregate type, a sample of at least 1000 g was inserted in a cylindrical strainer. This strainer was hung from a balance and submerged in a basin filled with demineralized water. The weight increase was monitored over a time period of 10 days.

3.2.3. Particle Size Distribution

The particle size distributions of the aggregates and sand were determined *cf.* NBN EN 933-1. The sieving column consisted of standardized sieves with following apertures: 63, 45, 40, 31.5, 22.4, 20, 16, 14, 12.5, 10, 8, 6.3, 4, 2, 1, 0.5, 0.25, 0.125 and 0.063 mm. The cumulative percentage of material passing through each sieve was calculated to plot the particle size distributions.

3.2.4. Aggregate Crushing Value

The strength of an aggregate can be characterized by its crushing value. In this research, this parameter—also known as the static compressive strength—was determined in accordance with Belgian standard NBN B11-205, for both the BA and the limestone aggregates. A cylindrical metal mold with an inner diameter of 150 mm was filled with a dried aggregate sample of a certain particle size range. After adding the sample, the mold was fitted with a no-friction plunger which was then loaded to 400 kN in 4 min. Before unloading again, this maximum load onto the sample was maintained for 2 min. Finally, the crushed fraction of the aggregate <2 mm was separated from the sample by means of a sieving operation. The mass ratio of the sample without the crushed fraction over the initial sample before loading counts as the static compressive strength of the aggregate. Normally, this test is to be conducted on coarse aggregate fractions only (minimum aggregate diameter: 6 mm). Nevertheless, the static compressive strengths of the fine BA and limestone aggregates were determined in the same manner. Per type of sample, the test was repeated three times.

3.2.5. Chemical Composition

The chemical properties of the BA were determined using Inductively Coupled Plasma (ICP), Ion Chromatography (IC) and Flow Injection Mercury System (FIMS) techniques available at the laboratories of Indaver nv.

3.3. *Concrete Mix Design and Production*

Once the total cement content (=350 kg/m^3) and the W/C ratio (0.65, 0.60 or 0.55) were set for the concrete compositions under investigation, the required amounts of sand, fine and coarse aggregates were determined. This was done by approximating the optimal particle size distribution curve of Fuller using the least-of-squares method. This approach normally results in a minimal spacing between the different particles of the inert fraction.

In the initial research stage only the fine aggregates were replaced by BA. In the next stage, the same was done for the coarse aggregates. As such, the individual effects of each BA could be studied unambiguously. In the final stage, both the fine and coarse limestone aggregates were replaced.

Concrete was produced in accordance with NBN EN 12390-2. First, the fine and coarse aggregates, the sand and the cement, were mixed for 1 min. Then, the water was added whereupon mixing continued for 2 min. After this step, the workability of the concrete was evaluated by means of the slump and flow test (NBN EN 12350-2, NBN EN 12350-5) and the required amount of SP was determined. This amount was added while mixing for another 2 min. The SP dosing step was repeated until the proper slump and flow class were obtained. Finally, the test samples were cast and optimally cured in a climate chamber at 20 ± 2 °C and 95% ± 5% relative humidity (RH).

3.4. Strength Performance

The compressive strength was determined in accordance with NBN EN 12390-2. This was done after seven days, 28 days and 56 days of optimal curing. Per testing age three cubic specimens ($n = 3$) with a 150 mm side were subjected to the test. The characteristic value of the compressive strength f_{ck} at 28 days was obtained in correspondence with EN 1990 *cf.* Equation (1).

$$f_{ck} = \overline{x} - k_n \times s\,(N/mm) \tag{1}$$

where x is the mean value of the compressive strength of n samples (N/mm^2); k_n is a factor equal to 3.37 for $n = 3$ and a variation coefficient V_X unknown from prior knowledge; and s is the standard deviation on the individual values of the compressive strength of n samples. With the characteristic compressive strength known, the proper compressive strength class as specified in NBN EN 206-1 can be assigned.

3.5. Susceptibility to Expansion

The susceptibility to expansion was evaluated by monitoring the height of the cubic samples intended for the compressive strength tests as a function of time (after 1, 7, 28 and 56 days) and relative to the their original height just after casting (=150 mm). Note that this is only a rudimentary assessment approach to get a first idea on the expansive behavior. The change in length of only one cube side reflects the total expansion only to a certain extent since the initial early age expansion in other directions is hindered by the cubic mold. Moreover, the duration of dihydrogen gas generation after reaction with the metallic Al and the setting time of the concrete also govern the amount of expansion as well. Further investigation on these influencing parameters is certainly still necessary in the future.

3.6. Porosity

The porosity was determined using a vacuum saturation technique followed by hydrostatic weighing *cf.* NBN B 05-201. Since the amount of water absorbed during the vacuum saturation is mainly limited to the pores in contact with the outer sample surface, the resulting porosity only reflects the open/permeable porosity. Per concrete mixture, the test was performed on nine cylindrical samples (diameter: 100 mm, height: 50 mm) taken from cubes (side: 150 mm). After 28 days of optimal curing, the samples were weighed and dried at a temperature of 40 ± 5 °C until constant mass (Δm after 24 h < 0.1%). Next, the samples were placed in a normalized vacuum tank. A vacuum with a residual pressure of 2.7 kPa was applied for 2.5 h. While keeping this vacuum condition, water was introduced at a rate of 50 mm/h until complete immersion of the samples. Subsequently, the air pressure was restored and the samples were kept under water for 24 h. Afterwards, they were weighed hydrostatically as well as above water. After the weighing, they were dried at 105 ± 5 °C until constant mass (Δm after 24 h < 0.1%) whereupon the whole procedure was repeated. Then, the open/permeable porosity φ was calculated for the two pre-drying temperatures (Equation (2)).

$$\varphi = \frac{m_s - m_d}{m_s - m_l} \times 100\,(\%) \tag{2}$$

where m_d is the oven dry mass after pre-drying at 40 °C or 105 °C (g); m_s is the water saturated mass after vacuum saturation (g); and m_l is the mass under water after vacuum saturation (g).

3.7. Long-Term Behavior

3.7.1. Susceptibility to Leaching

After 28 days of optimal curing, six cubes (side: 100 mm) were conserved in a closed tank filled with demineralized water. Conform CMA/2/II/A, the volume of the reagents should be two to five times the sample volume. Conform NEN 7345, this is four times the sample volume. In this research, the latter volume ratio was chosen. A water sample was taken after 30 days of immersion and examined using ICP, IC and FIMS.

3.7.2. Resistance to ASR

The susceptibility to ASR was evaluated using the modified Oberholster test *cf.* the Walloon technical guideline STM D424 ST6. This method was chosen because, in contrast with the original Oberholster test or South-African NBRI (National Building Research Institute) accelerated test method, it allows for a test directly on the concrete mixtures under investigation instead of on mortar bars containing the aggregates to be studied after an initial processing (crushing and sieving). Moreover, the modified Oberholster test holds the advantage of being a rather fast assessment technique (exposure period: 20 days) in comparison with some of the other methods designed for concrete (exposure periods of up to 12 months *cf.* RILEM TC 106-AAR [22]). Per mixture, six cylinders (diameter: 50 mm, height: 150 mm) taken from cubes (side: 150 mm) were tested. After 28 days of optimal curing the samples were immersed for 24 h in water at 80 °C. Afterwards, the samples were stored in a 40 g NaOH per L water solution at 80 °C. The expansion was continuously monitored on dial gauges in contact with the upper surface of the cylinders. Normally, an expansion exceeding 0.1% implies a potential ASR sensitivity.

3.7.3. Resistance to Acid Attack

Six cylinders (diameter: 113 mm, height: 150 mm) were drilled and cut from six cubes (side: 150 mm). After 28 days, the cylinders were exposed to 15 cycles involving intermittent exposure to acid. One cycle started by keeping the samples 24 h in a dry environment (minimal 60% $\pm$ 5% RH) and continued with 24 h of immersion in an acid solution (30 g acetic acid and 30 g lactic acid per liter), all taking place at a temperature of 20 $\pm$ 2 °C and 60% $\pm$ 5% RH. After each drying period, the mass of the samples was determined. After the last cycle, the mass was determined once again. The damage was evaluated based on the observed mass decrease and visual appearance.

3.7.4. Resistance to Freeze-Thaw Attack

Four cubes (side: 150 mm), optimally cured for 23 days, were subjected to a freeze-thaw test in compliance with NBN B15-231, which in turn refers to NBN B 05-203. The samples were submerged in demineralized water at a temperature of 20 $\pm$ 2 °C until constant mass (Δm 24 h < 0.05%) *cf.* NBN B 15-215. After saturation was observed, the samples were placed in a laboratory freezing chamber and were subjected to 14 freeze-thaw cycles which consisted of: (i) cooling down till 0 °C in 1–2 h; (ii) progressive freezing from 0 °C to −15 °C in 5 h (at a mean rate of 3 $\pm$ 0.5 °C/h and always in the range of 2–4 °C/h); (iii) maintaining a constant freezing temperature of −15 $\pm$ 2 °C for 10–11 h; (iv) defrosting in water; (v) keeping the samples in water until the end of the cycle at 24 h.

Two non-destructive test methods were used to evaluate the freeze-thaw resistance: a visual check-up and ultrasonic pulse velocity measurements using a commercial device. The visual damage assessment was done in accordance with NBN B 27-009/A1.

3.8. Microscopic Characterization on Thin Sections

To explain the porosity, a microscopic investigation on thin sections was necessary. These thin sections were prepared as follows: $45 \times 30 \times 15$ mm^3 prisms were cut from the concrete and 45×30 mm^2 faces were glued onto glass slides with a thickness of 2.9 mm. Then, the samples were cut and polished until the height of the concrete specimens and the glass equaled 10.1 mm. In a next step, the samples were impregnated under vacuum with a fluorescent epoxy. After impregnation, the excess epoxy was ground away and an object glass was glued onto the polished surface. In a final step, the glass slides were cut off and the concrete samples on the object glasses were polished until thin sections with a 25 μm thickness were obtained. A cover glass was glued onto them for protection. All thin sections were examined with a Leica DM LP polarizing microscope. Images were taken with a Leica DFC295 camera in fluorescent light mode.

3.9. Optimization of the Bottom Ash

3.9.1. Pre-Wetting

To limit the high water absorption during concrete mixing and the resulting workability issues, the BA could be pre-wetted. In total, three pre-wetting techniques were examined. All of them were based on their expected water absorption as a function of time (Section 3.2.2). The first technique focused on pre-wetting the BA to a level corresponding with the deflection point in the absorption curve. This level is met as soon as the initial high water uptake rate ends. The time needed for this counts as the required immersion time span of the aggregates prior to the concrete mixing. After immersion, a self-made strainer bucket was used to drain the water. Note that this method was only applied for the fine BA. The second and third methods also relate to the water absorption behavior as a function of time. For these two methods, the expected contact time with water was the decisive criterion and not the change in water absorption rate. As the concrete mixing process and specimen casting takes about 6 min after adding the water, it makes sense to pre-wet the BA with the amount of water that is expected to be absorbed within the 6 min timeframe. This can be done in two ways. Either the aggregates can be pre-wetted in the mixer (this second method is commonly used when manufacturing concrete with lightweight aggregates), or the additional amount of water can be added afterwards as a surplus to the mixing water (the third method). Both the fine and coarse BA were treated like this.

3.9.2. Reactive Washing

The applied reactive washing procedure was based upon literature findings [23] and existing know-how at Indaver nv. To passivate Al inclusions, the BA was washed with 1 M NaOH. The BA was completely immersed in this solution for two weeks. The metallic Al reacts with the NaOH solution and forms non-active Al and H_2. The mixture was stirred from time to time to ensure an optimal interaction between the BA and the solution and to guarantee a proper release of H_2.

To evaluate the effectiveness of this reactive washing, the amount of reactive Al before and after the treatment was determined. Therefore, the BA samples were subjected to a test described in CUR recommendation 116 [24]. Per sample, the BA was combined with the NaOH solution in a closed system flask. The H_2 being released was captured in a second flask initially full of water. Since a portion of the water is displaced by the H_2, the amount of metallic Al can be calculated from Equation (3) after measuring the initial and final water weight.

$$v_{H_2} = \frac{3}{2}\frac{m_{Al}}{M_{Al}} v_0 \ (L) \quad (3)$$

where v_{H2} is the volume of entrapped H_2 (L); M_{Al} is the molar mass of Al (27 g/mol); v_0 is the molar volume (22.4 L/mol); and m_{Al} is the mass of reactive Al (g). Before applying the BA after the reactive washing in concrete, they were first rinsed with water until the pH had decreased sufficiently. For the fine and coarse BA fractions, the pH decreased from 12.8 to 9.0–10.0 and from 12.8 to 8.0, respectively.

3.10. Manufacturing of the Lego Bricks

Three 1600 × 800 × 400 mm^3 Lego bricks were cast: one made of traditional concrete, one made of the initial BA concrete with full replacement of both the fine and coarse aggregate fractions and one made of the optimized BA concrete, also with full aggregate replacement. The mold for the Lego brick was provided by CB Recycling BV. The concrete was manufactured in a concrete mixer with a total volume capacity of only 200 L. As a consequence, the Lego bricks had to be cast in three subsequent stages. Each layer of concrete was carefully compacted using a vibration needle.

4. Results and Discussion

4.1. Physical and Chemical Characterization of the Bottom Ash

4.1.1. Particle Size Distributions

The particle size distributions of crushed limestone 2/6 and BA 0/6 are within the same range (Figure 1). Only the fraction <1 mm is more present in BA 0/6, whereas the limestone has a steeper grading curve, indicating a pure and narrow range of grain diameters. The crushed fraction has more fines than the sieved and washed fraction. This is probably due to the dust generation during crushing.

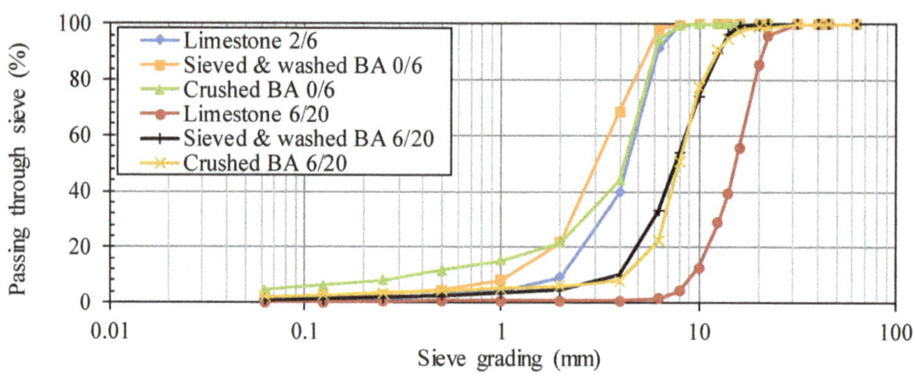

Figure 1. Influence of aggregate type (limestone *vs.* bottom ash (BA)), fraction (2/6 or 0/6 *vs.* 6/20) and BA pre-treatment (sieving and washing *vs.* crushing) on the particle size distribution.

When looking at the particle size distributions of the coarse fractions, a clear shift to the left can be observed between limestone 6/20 and BA 6/20. In accordance with the proper definitions for minimum and maximum aggregate size *cf.* NBN EN 12620, this BA fraction rather meets the requirements of an aggregate 2/14 instead of those of an aggregate 6/20. Comparison between the crushed and sieved and washed BA 6/20 reveals only very limited differences in particle size distribution.

Note that there are several factors with an influence on the difference in particle size distributions, e.g., the grading and strength of the original untreated material that is subjected to the processing as well as the effectiveness of the equipment used for crushing, sieving and washing.

4.1.2. Aggregate Crushing Value

When making the comparison between the recorded aggregate crushing values or static compressive strengths for the sieved and washed and crushed BA and the limestone aggregates, it is immediately clear that the BA clearly shows a lower mechanical performance (Table 1).

Table 1. Influence of aggregate type (limestone *vs.* BA), fraction (2/6 *vs.* 6/10 and 6/20) and BA pre-treatment (sieving and washing *vs.* crushing) on the static compressive strength.

Limestone 2/6	**Limestone 6/20**	
Fraction 2/6	Fraction 6/10	Fraction 10/20
63.8% ± 0.9%	75.1% ± 0.7%	82.7% ± 0.3%
Sieved & washed BA 0/6	**Sieved and washed BA 6/20**	
Fraction 2/6	Fraction 6/10	Fraction 10/20
42.2% ± 0.2%	55.5% ± 1.0%	63.3% ± 0.5%
Crushed BA 0/6	**Crushed BA 6/20**	
Fraction 2/6	Fraction 6/10	Fraction 10/20
50.3% ± 0.6%	58.8% ± 0.3%	65.8% ± 0.8%

This statement holds true for both the fine and coarse aggregates. The difference ranges between 13.6% and 21.6%, depending on the aggregate type and fraction. It is worth mentioning that the crushed BA fractions are all characterized by a significantly higher static compressive strength than the sieved and washed fractions. It is likely that the crushing operations performed in the processing installations of Indaver nv reduced the weaker particles in size, to values lower than the particle sizes considered during the static compressive strength tests. Thus, only the stronger aggregates remain. In the case of the sieved and washed BA there was no elimination of the weaker elements as such.

4.1.3. Mass Density and Water Absorption over 24 h

From the apparent mass densities recorded (Table 2), it is clear that BA is a lighter aggregate than the limestone. This statement holds true for both the fine and the coarse fractions. The relation between the surface dry mass density ρ_{rd} and the apparent mass density ρ_a gives an indication of the material's porosity. As such, the third column of the table clearly shows that the BA is more porous than the reference material. In case of the fine fraction, the crushed BA has a significantly lower porosity than the sieved and washed BA. On the other hand, the porosities of the crushed and sieved and washed coarse fractions are not significantly different from each other.

Table 2. Influence of aggregate type (limestone *vs.* BA), fraction (2/6 or 0/6 *vs.* 6/20) and BA pre-treatment (sieving and washing *vs.* crushing) on the apparent mass density (ρ_a), surface dry mass density (ρ_{rd}) and water absorption after 24 h (WA_{24}).

Aggregate Type	ρ_a (kg/m^3)	ρ_{rd} (kg/m^3)	$(\rho_a - \rho_{rd})/\rho_a \times 100$ (%)	WA_{24} (%)
Limestone 2/6	2750 ± 31	2644 ± 37	3.85 ± 0.28	1.5 ± 0.1
Sieved and washed BA 0/6	2246 ± 13	1793 ± 13	20.17 ± 0.44	11.2 ± 0.3
Crushed BA 0/6	2536 ± 6	2144 ± 11	15.46 ± 0.59	7.2 ± 0.3
Limestone 6/20	2762 ± 18	2710 ± 16	1.87 ± 0.06	0.7 ± 0.0
Sieved and washed BA 6/20	2649 ± 7	2285 ± 11	13.72 ± 0.20	6.0 ± 0.1
Crushed BA 6/20	2697 ± 22	2329 ± 23	13.65 ± 0.25	5.8 ± 0.1

More or less similar conclusions can be drawn from the water absorption after 24 h (Table 2: WA_{24}). This parameter is to a high extent correlated to the ratio of apparent and surface dry mass density. It is immediately clear that the BA has a higher inherent water absorption than the corresponding reference limestone aggregates. The value is about eight times the value of crushed limestone. Fine BA obtained by crushing has a substantially lower water absorption than the sieved and washed fine BA. Its value is comparable to the ones of the coarse BA. As indicated by the static compressive strength values (Section 4.1.2), crushing removes more of the weaker elements in the fine BA. These are probably the more porous elements responsible for the high water absorption. Thus, the crushing operation as pre-treatment seems beneficial for the fine fraction in order to control that property. In general, the fine fraction still absorbs more water than the coarse one. This is a recurring phenomenon for all types of aggregate materials due to their higher surface area. Nevertheless, in case of a highly absorptive BA aggregate, this particular behavior of the fine fraction may result in workability issues when used in concrete (see Section 4.2.1).

4.1.4. Water Absorption as a Function of Time

Figure 2 shows the water absorption as a function of time. The reference limestone aggregate materials barely absorb water within the considered timeframe of around 240 h. The fine and coarse BA fraction are much more absorptive and show a clear deflection point after around 2.5 h of immersion in water. At this point, the fine fraction has already reached 70% of its total absorption capacity, while the coarse BA reaches 50% of its capacity. Note that the fine BA already absorbs over 60% of its total absorption capacity within the first 15 min which is a very relevant timeframe within the concrete manufacturing process.

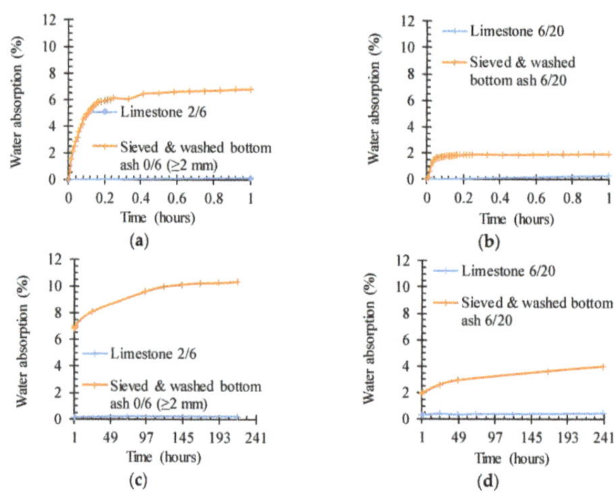

Figure 2. The higher water absorption as a function of time for sieved and washed BA 0/6 (⩾2 mm) and 6/20 in comparison with limestone 2/6 and 6/20. (**a**) fine, between 0 and 1 h; (**b**) coarse, between 0 and 1 h; (**c**) fine, between 1 and 241 h; (**d**) coarse, between 1 and 241 h.

4.1.5. Chemical Composition

The presence of (heavy) metals is of relevance in view of the concrete performance (e.g., the susceptibility to expansion, see Section 4.2.3) and leaching-related environmental problems (see Section 4.4.1). The chemical composition was determined for the sieved and washed BA 0/6, the crushed BA 0/6 and the sieved and washed BA 6/20 (Table 3). Unfortunately, there was no crushed BA 6/20 available at the time of the chemical analysis at the laboratories of Indaver nv. The (heavy) metal content is higher for sieved and washed BA 0/6 than for crushed BA 0/6 and sieved and washed BA 6/20. Also, the amount of Al is lower in case of the latter two BA aggregates, which indicates a lower susceptibility to Al-induced expansion. On the other hand, Fe, Cu and Zn are less present in the sieved and washed 0/6 fraction. This either indicates that these elements were less present in the MSW that was incinerated to obtain this batch of BA or that the separation of these ferrous and non-ferrous elements was more successful.

Table 3. Influence of the fraction (0/6 *vs.* 6/20) and pre-treatment (sieving and washing *vs.* crushing) on the chemical composition of the BA, especially in view of the (heavy) metal content.

Element	Sieved and Washed BA 0/6	Crushed BA 0/6	Sieved and Washed BA 6/20
Al (μg/L)	31,590	5757	3920
Ba (μg/L)	61	148	58
Cu (μg/L)	61	213	117
Zn (μg/L)	13	1332	167
Fe (mg/L)	<0.4	3.0	0.7
Ca (mg/L)	156	137	146
K (mg/L)	45	51	58
Na (mg/L)	85	51	58

4.2. Concrete Mixtures

4.2.1. Mixture Proportions and Fresh State Properties

Lowering the W/C ratio normally results in the application of a higher SP dosage to achieve a sufficient workability (slump class S3–S4). For the reference concrete (Table 4: mixtures REF1–REF4), SP only needed to be added in case of a W/C ratio of 0.55. In case of a replacement of the fine limestone 2/6 by fine BA (Table 4: mixtures BA1–BA12 and BA24), the concrete workability was more problematic. Much higher SP dosages (up to 8 mL/kg cement) were necessary and still it remained difficult to achieve the same workability as for the reference concrete, even if the W/C ratio amounted to 0.60 or 0.65. The use of crushed BA instead of sieved and washed BA turned out less problematic somehow. The insufficient workability can be attributed to the fact that the BA has a high porosity and will thus absorb a lot of water in the beginning. Nevertheless, when also considering mixtures BA13–BA16 with replacement of only the coarse fraction by BA 6/20 (Table 4), which is also quite porous (see Table 2), workability was much less of an issue. Still, when considering full replacement of the limestone by fine and coarse BA (Table 4: mixtures BA17–BA23 and BA25–BA28), the detrimental effect of the fine fraction kept on dominating the workability and the required SP dosage. For the W/C ratio of 0.55, this dosage amounted to no less than 36 mL/kg cement, which is far above the dosage allowed for that type of SP. Moreover, the high dosages would increase the price of these mixtures for rather low-value applications (*i.e.*, prefabricated Lego bricks) quite dramatically.

4.2.2. Strength Performance

For the reference concrete (REF1–REF4), adequate strength results (>33 N/mm²) are already obtained after only seven days of curing (Table 5), including for the highest W/C ratio considered (=0.65). For the lower W/C ratios of 0.60 and 0.55, these initial strength results amount to no less than 38 and 44 N/mm², respectively. This strength performance is still bound to increase because of the still-ongoing hydration of the cement. The most significant strength increase could be observed between 7 and 28 days of curing. As a consequence, the 28-day compressive strength class of this traditional concrete type is at least C30/37, which makes it very suitable for prefabricated Lego brick production, which requires a strength class of only C20/25.

In case only the fine aggregate fraction has been replaced with BA (Table 5: mixtures BA1–BA12 and BA24), the concrete performed worse in terms of compressive strength. Still, the 28-day strength class was at least C20/25 in most cases. Mixture BA10 was the only exception. This can mainly be attributed to the 20 min immersion of the fine BA in water prior to the mixing process. The pre-treatment with water was very beneficial for the concrete workability (Table 4: slump class S4, without SP addition). However, the strength decrease caused by it (Table 5: C16/20) proved rather unacceptable given the intended purpose of the concrete. Lowering the W/C ratio from 0.65 over 0.60 to 0.55 obviously turned out beneficial for the strength performance. For these W/C ratios, a 20 min immersion in water as a pre-treatment still resulted in concrete with strength class C20/25 (Table 5: mixtures BA11–BA12).

The replacement of only the coarse limestone by BA 6/20 (Table 5: mixtures BA13–BA15) clearly had a less detrimental effect on the strength. Just as for the reference concrete mixtures, the compressive strength is already higher than 30 N/mm² after only seven days of optimal curing. Their 28-day strength class was at least C25/30 and can reach C30/37 for W/C ratios of 0.60 and 0.55. As such, a more or less similar strength performance to the reference concrete can be assumed. Only the strength gain between 28 days and 56 days of optimal curing is more pronounced for the BA concrete.

Table 4. Influence of the mixture proportions (ratio sand, fine and coarse aggregates, water-to-cement (W/C) ratio and superplasticizer (SP) dosage), replacement of limestone by BA and BA pre-treatment on the concrete workability.

Mixture	Natural Aggregates			BA		Water	Cement CEM I 52.5 N	SP	W/C Ratio	Workability	
	Sand 0/4	Limestone 2/6	Limestone 6/20	BA 0/6/BA 2/6	BA 6/20					Slump	Flow
	(kg/m³)	(kg/m³)	(kg/m³)	(kg/m³)	(kg/m³)	(kg/m³)	(kg/m³)	(mL/kg Cement)	-	-	-
REF1	439.00	596.00	695.00	-	-	227.50	350.00	-	0.65	S4	-
REF4	439.00	596.00	695.00	-	-	227.50	350.00	-	0.65	S4	F5
REF3	451.00	613.00	714.00	-	-	210.00	350.00	-	0.60	S4	F5
REF2	464.00	627.00	733.00	-	-	192.50	350.00	1 [a]	0.55	S3	F4
BA2	333.00	-	674.00	664.00 [b]	-	227.50	350.00	3 [c]	0.65	S3	-
BA3	282.00	-	700.00	690.00 [b, d]	-	227.50	350.00	3 [c]	0.65	S3	-
BA4	471.00	-	686.00	519.00 [e]	-	227.50	350.00	2 [c]	0.65	S4	-
BA6	436.00	-	717.00	525.00 [e]	-	227.50	350.00	1 [c]	0.65	S2	-
BA7	436.00	-	717.00	525.00 [e]	-	227.50	350.00	1 [a]	0.65	S2	-
BA10	436.00	-	717.00	525.00 [b]	-	227.50 [f]	350.00	-	0.65	S4	-
BA1	342.00	-	350.00	683.00 [b]	-	210.00	350.00	2 [c]	0.60	S1	-
BA5	342.00	-	665.00	683.00 [b]	-	210.00	350.00	8 [c]	0.60	S4	-
BA8	448.00	-	737.00	539.00 [e]	-	210.00	350.00	3 [a]	0.60	S2	-
BA11	448.00	-	737.00	539.00 [b]	-	210.00 [f]	350.00	-	0.60	S3	-
BA24	397.00	-	703.00	557.00 [g]	-	210.00	350.00	-	0.60	S4	F4
BA9	460.00	-	757.00	554.00 [e]	-	192.50	350.00	6 [a]	0.55	S3	-
BA12	460.00	-	757.00	554.00 [b]	-	192.50 [f]	350.00	-	0.55	S2	-
BA13	613.00	444.00	-	-	633.00 [h]	227.50	350.00	-	0.65	S3	F4
BA14	630.00	456.00	-	-	650.00 [h]	210.00	350.00	-	0.60	S3	F4
BA16	617.00	540.00	-	-	594.00 [i]	210.00	350.00	-	0.60	S4	F3
BA15	647.00	469.00	-	-	668.00 [h]	192.50	350.00	6 [a]	0.55	S3	F4
BA17	517.00	-	-	347.00 [j]	797.00 [h]	227.50	350.00	6 [a]	0.65	S3	F4
BA20	517.00	-	-	347.00 [j]	797.00 [h]	227.50	350.00	5 [a]	0.65	S3	F4
BA21	517.00	-	-	347.00 [j]	797.00 [h]	227.50 [k]	350.00	-	0.65	S4	F5
BA22	517.00	-	-	347.00 [j]	797.00 [h]	227.50 [l]	350.00	-	0.65	S2	F4
BA23	517.00	-	-	347.00 [j]	797.00 [h]	227.50	350.00 [m]	3 [a]	0.65	S2	F4
BA26	488.00	-	-	324.00 [j]	793.00 [h*]	227.50 [n]	350.00	-	0.65	S4	F4
BA27	488.00	-	-	324.00 [j]	793.00 [h*]	227.50 [o]	350.00	6 [a]	0.65	S2	F1
BA28	543.00	-	-	343.00 [j, p]	727.00 [h, p]	227.50	350.00	2 [a]	0.65	S2	F4
BA18	531.00	-	-	357.00 [j]	819.00 [h]	210.00	350.00	10 [a]	0.60	S2	F3
BA25	547.00	-	-	400.00 [g]	717.00 [i]	210.00	350.00	5 [a]	0.60	S3	F4
BA19	545.00	-	-	366.00 [j]	841.00 [h]	192.50	350.00	36 [a]	0.55	S3	F2
Lego brick 1	488.00	-	-	366.00 [j]	808.00 [h*]	227.50	350.00	10 [a]	0.65	S2	F1
Lego brick 2	439.00	596.00	695.00	-	-	227.50	350.00	-	0.65	S3	F5
Lego brick 3	667.00	-	-	382.00 [g, p]	566.00 [h*]	227.50 [n]	350.00	-	0.65	S4	F5

Notes: [a] MasterGlenium ACE; [b] BA 0/6 sieved and aged; [c] MasterGlenium 51; [d] glass and metals removed; [e] BA 2/6 sieved and aged; [f] BA submerged for 20 min in water; [g] BA 0/6 crushed; [h] BA 6/20 sieved and washed; [i] BA 6/20 crushed; [j] BA 0/6 sieved and washed; [k] wetted before mixing: 5 m% BA 0/6 and 2 m% BA 6/20; [l] extra water added to mixing water: 5 m% BA 0/6 and 2 m% BA 6/20; [m] CEM I 52.5 N HES LA HSR; [n] wetted before mixing: 2.5 m% BA 0/6 and 1 m% BA 6/20; [o] extra water added to mixing water: 2.5 m% BA 0/6 and 1 m% BA 6/20; [p] after reactive washing with NaOH; * indicates a new delivery.

Table 5. Influence of W/C ratio, replacement of limestone by BA and pre-treatment of the BA on the compressive strength after 7, 28 and 56 days and the strength class.

Mixture	W/C (-)	Mix Details	$f_{c\ 7\ days}$ (N/mm^2)	$f_{c\ 28\ days}$ (N/mm^2)	$f_{c\ 56\ days}$ (N/mm^2)	Strength Class (-)
REF1	0.65	-	33.49 ± 1.38	41.00 ± 0.43	44.85 ± 0.41	C30/37
REF4	0.65	-	33.61 ± 0.42	40.45 ± 0.61	41.72 ± 0.92	C30/37
REF3	0.60	-	38.54 ± 0.81	48.60 ± 1.14	49.37 ± 1.18	C30/37
REF2	0.55	a	44.29 ± 0.55	52.51 ± 1.43	55.15 ± 0.75	C35/45
BA2	0.65	b, c	23.43 ± 0.87	29.18 ± 0.71	31.26 ± 1.01	C20/25
BA3	0.65	b, c, d	29.69 ± 1.16	32.85 ± 0.41	33.26 ± 0.54	C25/30
BA4	0.65	c, e	32.75 ± 1.26	37.23 ± 0.29	38.32 ± 1.07	C25/30
BA6	0.65	c, e	28.62 ± 1.53	34.49 ± 0.46	38.30 ± 1.32	C25/30
BA7	0.65	a, e	28.69 ± 1.21	34.73 ± 0.42	40.46 ± 0.91	C25/30
BA10	0.65	b, f	18.88 ± 0.45	22.98 ± 0.68	26.66 ± 0.53	C16/20
BA1	0.60	b, c	25.52 ± 0.63	31.35 ± 0.81	34.95 ± 0.69	C20/25
BA5	0.60	b, c	30.39 ± 1.54	36.51 ± 1.54	38.10 ± 0.50	C25/30
BA8	0.60	a, e	33.40 ± 0.46	37.81 ± 0.53	42.16 ± 1.42	C25/30
BA11	0.60	b, f	20.82 ± 0.80	25.73 ± 0.14	29.24 ± 0.34	C20/25
BA24	0.60	g	30.34 ± 0.32	36.93 ± 1.38	39.10 ± 2.05	C25/30
BA9	0.55	a, e	36.62 ± 0.97	42.15 ± 1.28	43.20 ± 1.26	C30/37
BA12	0.55	b, f	22.35 ± 0.67	27.71 ± 0.70	31.83 ± 0.67	C20/25
BA13	0.65	h	34.40 ± 0.56	40.90 ± 2.13	46.32 ± 1.24	C25/30
BA14	0.60	h	37.40 ± 0.71	42.82 ± 1.35	50.30 ± 0.84	C30/37
BA16	0.60	i	35.23 ± 1.01	39.85 ± 0.86	47.42 ± 0.80	C25/30
BA15	0.55	a, h	38.34 ± 0.93	40.97 ± 0.53	47.73 ± 2.08	C30/37
BA17	0.65	a, h, j	32.93 ± 1.74	38.72 ± 0.61	40.18 ± 1.30	C25/30
BA20	0.65	a, h, j	26.26 ± 0.71	31.51 ± 1.10	33.15 ± 0.55	C20/25
BA21	0.65	h, j, k	22.03 ± 0.70	29.86 ± 0.92	30.55 ± 0.51	C20/25
BA22	0.65	h, j, l	25.05 ± 0.13	33.11 ± 1.06	34.80 ± 1.71	C20/25
BA23	0.65	a, h, j, m	26.67 ± 0.30	34.81 ± 0.21	34.92 ± 0.74	C25/30
BA26	0.65	h*, j, n	26.23 ± 0.74	32.47 ± 1.17	34.59 ± 0.10	C20/25
BA27	0.65	a, h*, j, o	21.74 ± 0.37	26.03 ± 0.94	28.11 ± 0.23	C20/25
BA28	0.65	a, h, j, p	30.23 ± 1.14	37.02 ± 0.46	38.62 ± 0.40	C25/30
BA18	0.60	a, h, j	28.38 ± 0.66	34.53 ± 0.76	33.67 ± 1.91	C25/30
BA25	0.60	a, g, i	32.15 ± 0.18	37.38 ± 0.95	38.98 ± 0.88	C25/30
BA19	0.55	a, h, j	29.72 ± 2.90	35.80 ± 1.11	36.18 ± 4.37	C25/30

Notes: [a] MasterGlenium ACE; [b] BA 0/6 sieved and aged; [c] MasterGlenium 51; [d] glass and metals removed; [e] BA 2/6 sieved and aged; [f] BA submerged for 20 min in water; [g] BA 0/6 crushed; [h] BA 6/20 sieved and washed; [i] BA 6/20 crushed; [j] BA 0/6 sieved and washed; [k] wetted before mixing: 5 m% BA 0/6 and 2 m% BA 6/20; [l] extra water added to mixing water: 5 m% BA 0/6 and 2 m% BA 6/20; [m] CEM I 52.5 N HES LA HSR; [n] wetted before mixing: 2.5 m% BA 0/6 and 1 m% BA 6/20; [o] extra water added to mixing water: 2.5 m% BA 0/6 and 1 m% BA 6/20; [p] after reactive washing with NaOH; [*] indicates a new delivery.

In case of a full replacement of both the fine and coarse aggregate fraction by BA (Table 5: mixtures BA17–BA23 and BA25–BA28), the earlier observed rather negative strength effects inherent to the fine fraction keep on dominating the overall strength performance of the concrete. Nevertheless, it remains possible to achieve a strength class for the concrete of at least C20/25, even when additional water is added to the BA before mixing to achieve a sufficiently workable concrete without adding SP (mixtures BA21–BA22). Lowering the W/C ratio from 0.65 to 0.60 and 0.55 makes it easier to obtain a strength class of C25/30.

Note that the lower strength performance of the BA concrete can mainly be attributed to the lower aggregate crushing value or static compressive strength of the BA aggregates (Section 4.1.2) and the higher porosity of the material. A more in-depth evaluation of the latter property follows in Section 4.2.3.

4.2.3. Susceptibility to Expansion

All concrete specimens with BA showed signs of expansion (Figure 3a). This behavior is attributed to the presence of metallic Al in the BA. When placed in an alkaline environment (pH > 7), the metallic Al dissolves and emits dihydrogen gas [11]. Obviously, the cement matrix (pH 13) provides these required alkaline conditions. The dihydrogen-induced expansion leads to the formation of longitudinal voids, which were, in this study, mainly concentrated near the troweled surface of the test samples and perpendicular to the casting direction (Figure 3b).

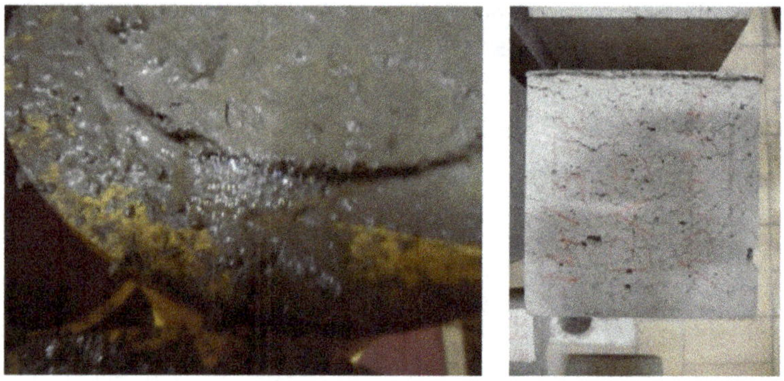

(a) (b)

Figure 3. Reactive Al induced expansion near the troweled surface of a concrete cylinder (**a**); longitudinal expansion voids observed on a concrete test cube (**b**) when containing BA.

There is a clear difference in expansive behavior between the reference concrete and concrete in which the fine and/or the coarse aggregate fractions were replaced with BA. For all three W/C ratios (Figure 4a: 0.55, Figure 4b: 0.60, Figure 4c: 0.65), the expansive behavior of the BA concrete is evident. Whereas the reference concrete never shows a significant change in length for the distance between the bottom and the troweled surface of the test cubes, the concrete with replacement of the fine aggregate by BA is characterized by an expansion of a little less than 4%. Replacement of only the coarse aggregate fraction by BA 6/20 gives less expansion (around 1%), be it only for W/C ratios of 0.65 and 0.60. For a W/C ratio of 0.55, the expansion also equaled around 4%. The recorded expansion in case of a full replacement of the limestone by the appropriate proportioning of fine and coarse BA represents an accumulative effect of the two alternative aggregate fractions. This accumulative effect is the most pronounced for the lowest W/C ratio considered, *i.e.*, 0.55.

The BA concrete with the highest W/C ratio probably shows less expansion because it requires much less SP. The presence of SP usually tends to postpone the time of setting. This means that there is more time for the dihydrogen-induced longitudinal voids to form and, thus, also more time for the resulting expansion to develop. Nevertheless, further testing is still needed to find further confirmation for this.

4.2.4. Selection of the Most Suitable Test Mixture

The selection of the most suitable BA concrete mixture for further testing was mainly governed by the achievable balance between workability (without extensive use of expensive SP) and strength performance. Compressive strength testing revealed that even with a W/C ratio of 0.65, the minimum required strength class for prefabricated Lego bricks (C20/25) could still be met (Section 4.2.2). Knowing this, it seems no problem to stick to this high W/C ratio as it also helps to compensate for the workability issues inherent to the presence of mainly the fine BA (Section 4.2.1). Sieved and washed BA were used for mixture BA20. This particular pre-treatment does not result in a significantly higher strength or resistance to expansion as opposed to the crushed BA. Nevertheless, sieving and washing normally leads to a somewhat lower workability than crushing. Still, the former pre-treatment was chosen since a crushing of the aggregates is more energy intensive and it does not improve the concrete workability that much. Instead, a suitable pre-wetting procedure for the BA was investigated in order to lower the required SP dosage for an adequate workability (see Section 4.5.1). Mixture BA20, still without pre-wetting of the BA aggregates, was manufactured for a further in-depth evaluation of the concrete mainly regarding its porosity (Section 4.3) and expected behavior in the long term (Section 4.4). Within Section 4.5.2, an additional reactive washing step was considered for the BA to overcome the expansion problem.

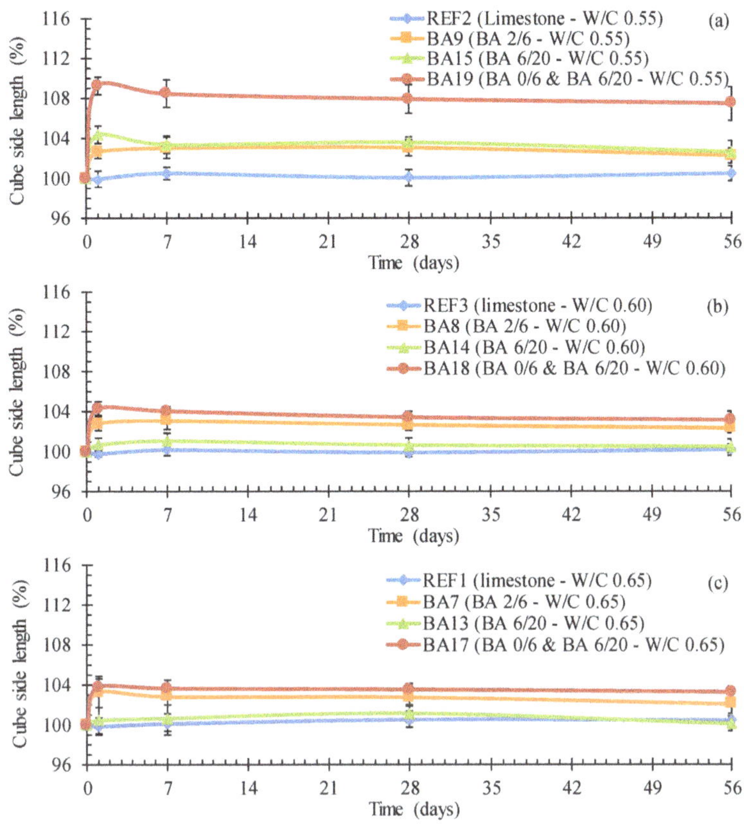

Figure 4. The higher susceptibility of BA concrete toward expansion in comparison with limestone concrete, an expansion which decreases with increasing W/C ratio 0.55 (**a**); 0.60 (**b**); 0.65 (**c**).

4.3. Porosity

From the vacuum saturation experiments, it can be concluded that the BA concrete is clearly more porous than the reference concrete (Table 6). The pronounced porosity difference between the BA and the reference concrete exists both after pre-drying at 40 °C and 105 °C.

Table 6. The higher open/permeable porosity of concrete BA20 *vs.* REF4 both after pre-drying at 40 °C (capillary porosity) and 105 °C (total porosity) and their similar calculated gel porosity.

Mixture	Capillary Porosity φ (%) after Pre-Drying at (40 ± 5) °C	Total Porosity φ (%) after Pre-Drying at (105 ± 5) °C	Gel Porosity φ (%)
REF4	11.1 ± 0.4	16.1 ± 0.6	5.1 ± 0.4
BA20	17.6 ± 0.4	22.5 ± 0.5	4.9 ± 0.1

In literature, the porosity obtained after pre-drying at 40 °C is said to correspond with the capillary porosity (pore diameter: 1–10 μm), while the difference in porosity obtained after pre-drying at 105 °C and 40 °C should represent the gel porosity (pore diameter: 10 nm) [25]. The capillary pores are the result of the fact that water and cement in unbound conditions take a larger volume than the hydrated cement. The gel pores relate to the gel water which is physically adsorbed by the hydration products and the monomolecular water film in between the flat crystallization products of interlayer water. When considering the theory of the capillary and gel porosity, one could state that the gel porosity is equal for both mixtures, while the capillary porosity is significantly different. However, since for both mixtures the same amount of water and cement was used, there is no reason to conclude that the capillary porosity should be different. Now, since the BA aggregates absorb a lot of water during the mixing process, it can be stated that the BA concrete has a lower effective water/cement ratio and thus also a lower capillary porosity. The increase in porosity after pre-drying at 40 ± 5 °C should be assigned to the existence of other voids, *i.e.*, the longitudinal expansion voids at the surface and the pores in the aggregates, if capillary pores reach these aggregates. The expansion voids will not directly attract water as their size (around 100 μm in width) does not really accommodate capillary suction. Still, these voids enlarge the possible water penetration surface from which more capillary pores can be reached. If a porous aggregate is in contact with them, an extra amount of water would be stored in the concrete as well.

Note that the capillary pores and the longitudinal expansion voids are influencing the strength and permeability of the concrete. The higher porosity after pre-drying at 40 ± 5 °C is a measure for the extra voids at the surface, but not of the inner structure. However, one could state that if these voids could be minimized, permeability will be diminished and strength will increase.

For a better insight on the inner porosity, a microscopic analysis was performed as well (Figure 5).

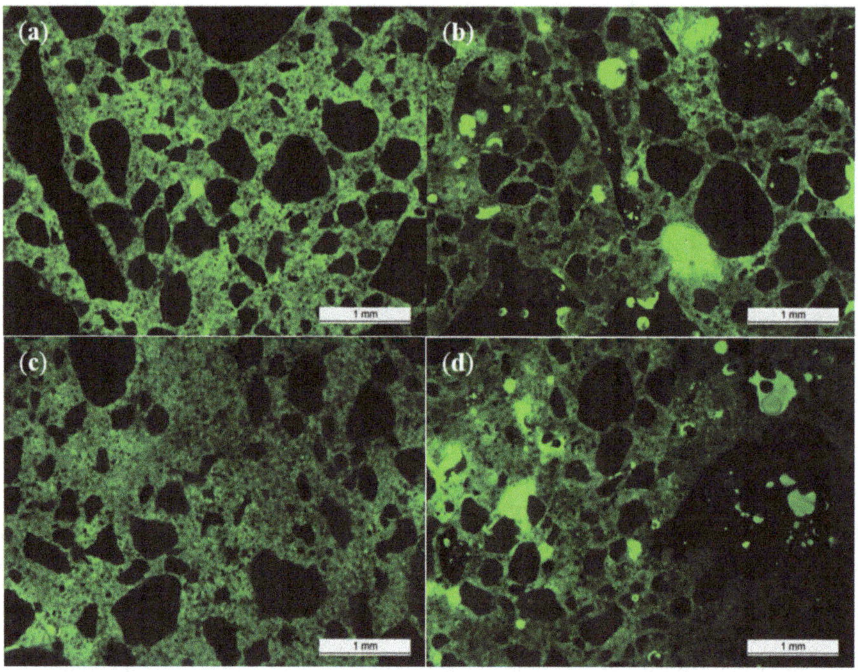

Figure 5. The darker green color of the thin sections of concrete BA20 *vs.* limestone concrete REF4 in fluorescent light mode, indicating a lower effective W/C ratio of the former. (**a**) REF4, thin section 1; (**b**) BA20, thin section 1; (**c**) REF4, thin section 2; (**d**) BA20, thin section 2.

When analyzing the thin sections in fluorescent light mode, it can be seen that the cement matrix of the BA concrete has a darker green color than the one of the reference concrete. This normally indicates that the effective W/C ratio of the BA concrete is lower. The concrete's effective water content was also calculated. In accordance with NBN EN 206-1, this is the reduced amount of mixing water available given the water absorption (WA_{24}) of the concrete's inert fraction. The calculation revealed that when aiming at a theoretical W/C ratio of 0.65, the effective W/C ratio due to the high water absorption of the BA would only be 0.56. It also means that the higher open/permeable porosity for the BA mixture after pre-drying at 40 °C cannot be assigned to the presence of more capillary pores. Hence, it must indeed be caused by the presence of the longitudinal expansion voids and the highly absorptive aggregates.

4.4. Long-Term Behavior

4.4.1. Susceptibility to Leaching

Leaching values stayed below the limiting values imposed by the Flemish VLAREMA standards for the normalized elements, except for bromine (Table 7). Copper was much more present in the leaching solution of the BA concrete as opposed to the one of the reference concrete (28 µg/L *vs.* <6 µg/L). Still, the measured concentration was acceptable (<80 µg/L). There is a two-fold explanation for the adequate leaching behavior. First of all, the BA itself already meets the VLAREMA criteria. Leaching components usually come from the bottle glass and metal fragments present in the BA, but in this study these fractions apparently had no problematic effects. Secondly, once in concrete they are immobilized in the cement matrix which reduces the leaching even more. Leaching data published by Vandecasteele *et al.* on the original non-processed BA 6/50 further confirm both effects [6]. The highest leaching value was recorded for Al (5302 µg/L), for which there are no VLAREMA criteria. It was about the 10-fold of the Al leaching value of the reference concrete.

Table 7. Leaching values for concrete BA20 and REF4 relative to demineralized water, indicating a fulfillment of most VLAREMA criteria, yet with a much higher Al leaching value for BA20.

Normalized	BA20	REF4	Demineralized Water	Vlarema Criterion	Non-Normalized	BA20	REF4	Demineralized Water
As (µg/L)	<12	<12	<12	80	Al (µg/L)	5302	620	<100
Ba (µg/L)	<4	<4	<4	200	B (µg/L)	480	182	49
Cd (µg/L)	<2	<2	<2	3	Mn (µg/L)	<6	<6	<6
Co (µg/L)	<2	<2	<2	50	Tl (µg/L)	<15	<15	<15
Cr (µg/L)	51	53	<4	260	Fe (mg/L)	<0.4	<0.4	<0.4
Cu (µg/L)	28	<6	<6	80	Ca (mg/L)	15	6.4	<1
Mo (µg/L)	49	31	<10	3000	K (mg/L)	169	136	<1
Ni (µg/L)	<2	<2	<2	75	Mg (mg/L)	<1	<1	<1
Pb (µg/L)	<20	<20	<20	130	Na (mg/L)	160	96	<1
Sb (µg/L)	<12	<12	<12	100	P (mg/L)	<0.02	0.06	<0.02
Se (µg/L)	<10	<10	<10	200	S (mg/L)	8.8	10.7	<0.02
Sn (µg/L)	<10	<10	<10	100	NO_2 (mg/L)	<1	<1	<1
V (µg/L)	165	169	<10	250	NO_3 (mg/L)	<1	<1	<1
Zn (µg/L)	<8	<8	<8	280	PO_4 (mg/L)	<5	<5	<5
F (mg/L)	<5	<5	<5	5.5	I (mg/L)	<5	<5	<5
Cl (mg/L)	<100	<100	<100	100				
Br (mg/L)	<4	<4	<4	2				
SO_4 (mg/L)	24	30	<10	220				
Hg (µg/L)	<1	<1	<1	2				

4.4.2. Resistance to ASR

After the modified Oberholster test, substantial expansion and cracking were observed for the BA concrete. The length increase was around twice the value of the reference concrete (Figure 6: 0.36% ± 0.16% *vs.* 0.18% ± 0.05%). Especially for the BA mixture, the recorded length increase exceeded 0.1% by far. This normally indicates a serious sensitivity to ASR. Surprisingly, this criterion was also not met for the reference limestone concrete. Note that the length change of the BA concrete is also characterized by a much larger standard deviation on the individual values. It demonstrates that the ASR reactivity of the BA depends on its specific composition and its often high heterogeneity.

The observed expansion during the modified Oberholster test for the BA concrete is rather in contrast with literature which normally reports an acceptable ASR resistance. Perhaps the accelerated testing conditions and the fact that the exposure solution is NaOH, which also reacts with the metallic Al present in the BA, may explain this unexpected behavior. To see whether the experiment indeed resulted in expansion and cracking due to ASR, additional microscopic thin section analysis on the test samples from the Oberholster test is needed. It would also be useful to perform an additional Oberholster test on concrete with BA that was subjected to the reactive washing operation. That investigation is, for the moment, still ongoing.

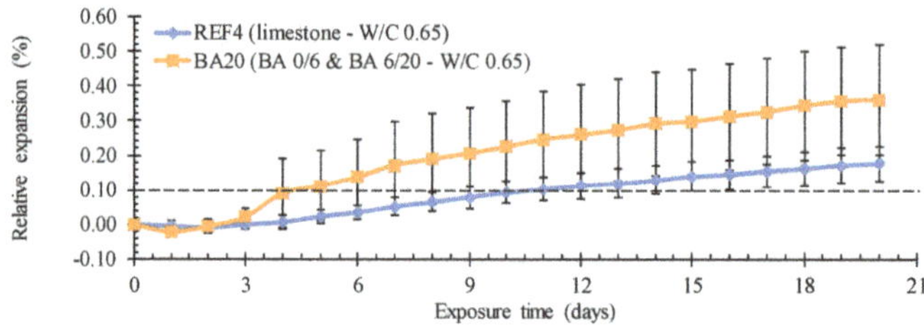

Figure 6. Higher mean relative expansion of concrete BA20 *vs.* limestone concrete REF4 after the modified Oberholster test.

4.4.3. Resistance to Acid Attack

From the recorded relative mass changes, concrete BA20 seems to be more resistant to cyclic lactic and acetic acid exposure than the reference limestone concrete REF4 (Figure 7). However, a visual inspection of the samples shows that one should state this conclusion rather carefully. Apparently, the reference concrete eroded rather homogenously and the sample surface remained smooth. The BA concrete showed a quite different behavior. There, only the cement matrix eroded in between the BA aggregates. This resulted in a rough sample surface and some loss of individual, mainly spherical and smooth-shaped BA grains (Figure 8). As the mass loss of individual aggregates can vary considerably per test sample, it is no surprise that the relative mass change of the BA concrete is characterized by a larger standard deviation on the individual values than the reference concrete (Figure 7).

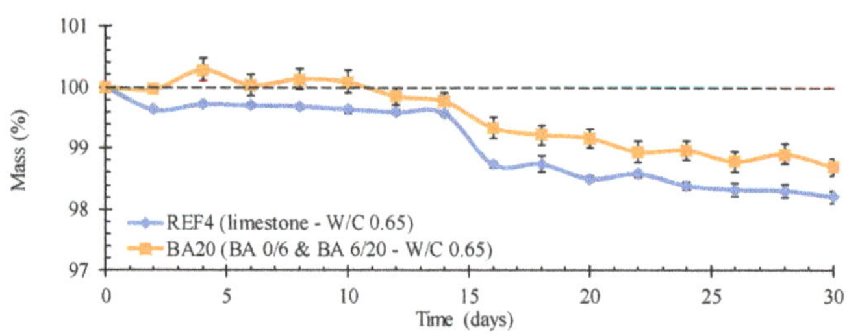

Figure 7. Lower mass loss of concrete BA20 *vs.* limestone concrete REF4 during cyclic exposure to a mixture of lactic and acetic acid.

Figure 8. Higher superficial deterioration of concrete BA20 (**a**) *vs.* limestone concrete REF4 (**b**) after cyclic exposure to lactic and acetic acid.

The difference in sample deterioration between the two concrete types can be explained as follows. The lactic and acetic acid react with both the calcium hydroxide present in the cement matrix and with the limestone aggregates which are, in essence, $CaCO_3$. In both cases, the reaction products are CO_2 and H_2O as well as either calcium acetate or lactate, depending on the acid. Thus, in the case of the reference concrete, both the cement matrix and the aggregate fraction deteriorated. It explains the still rather smooth appearance of the concrete surface after the acid exposure. As the calcium content of the BA-based aggregates is much less (around 10%–15%), it evidently contains much less of the vulnerable $CaCO_3$ in comparison with the limestone aggregates of the reference concrete. In other words, just the cement matrix of the BA concrete is susceptible to the acid attack. As this cement matrix is also more porous due to the presence of more longitudinal expansion voids (Section 4.3), it is also more accessible for the acidic exposure solution. As a consequence, it will deteriorate more than the cement matrix of the reference concrete. This, in contrast with the BA-based aggregates remaining more or less intact, resulted in a rather inhomogeneous deterioration of the BA concrete. The corresponding very distinctive roughening of the concrete surface is probably as undesirable as the higher mass loss of the reference concrete.

4.4.4. Resistance to Freeze-Thaw Attack

No problems were observed regarding the freeze-thaw resistance of the BA concrete. This conclusion follows from both the visual inspection and the ultrasonic testing. Visually, no damage phenomena such as cracks were observed. Secondly, the ultrasonic pulse velocity before freeze-thaw exposure (3951 ± 93 m/s) is more or less equal to the one after 14 freeze-thaw cycles (3994 ± 88 m/s). The ultrasonic pulse velocities recorded for the reference mixture did not change either (before: 4369 ± 82 m/s, after: 4320 ± 160 m/s). The difference in ultrasonic pulse velocity between the BA mixture and the reference mixture is due to the higher porosity and lower strength of the former concrete type. Note that the influence of de-icing salts on the freeze-thaw resistance was not investigated.

4.5. Optimization of the Bottom Ash Characteristics

4.5.1. Optimization regarding Workability and Water Absorption

Initially, it seemed that the large amount of fines was causing the high water absorption of the BA. Therefore, its water absorption with exclusion of the fraction <2 mm was determined and compared to the value of the non-sieved BA. Surprisingly, it was found that the water absorption was not really affected by this sieving operation (Table 8). Also, washing the BA with tap water instead of with industrial process water (containing more fine particles) did not result in a less absorptive behavior. Although the sieving operation influenced the concrete workability in a positive manner (Table 4), this is not attributable to the omission of the fine fraction.

Table 8. The limited influence of excluding the fraction <2 mm and washing with tap water instead of process water on the water absorption of the fine BA.

BA Treatment	BA 0/6 Sieved and Aged	BA 2/6 Sieved and Aged	BA 0/6 Sieved and Washed (Process Water)	BA 0/6 Sieved and Washed (Tap Water)
WA_{24} (%)	10.3 ± 1.6	11.8 ± 0.1	11.2 ± 0.3	11.0 ± 0.2

The high water absorption was believed to be caused by the high porosity of the BA rather than by the amount of fines. Attempts were made to separate the porous from the non-porous elements. However, this turned out rather difficult. The porous elements could also not be detected visually. For instance, quite some metal parts within the BA looked porous because they were covered by with porous-looking layers due to the sintering process during incineration. Floating tests based on the difference in density between the non-porous and porous elements were also unsuccessful.

However, it could be demonstrated that crushing the aggregates in a laboratory ball mill had a positive effect on the water absorption. As can be seen in Table 9, this crushing operation mainly influenced the smaller fractions of the bottom ash (BA 0/6 (⩾2 mm) and BA 6/20 (<4 mm)). In view of concrete manufacturing, it is thus advantageous to use the crushed material because the weaker porous elements are crushed. The slightly higher water absorption induced by having a somewhat greater fine fraction this way appears not to be comparable to the diminished absorption that can be obtained by removing the porous parts.

Table 9. The pronounced influence of a crushing operation on the reduction in water absorption of BA 0/6 (⩾2 mm) and BA 6/20 (<4 mm) *vs.* BA 6/20 (⩾4 mm).

BA Fraction	BA 0/6 (⩾2 mm)		BA 6/20 (⩾4 mm)		BA 6/20 (<4 mm)	
Crushing	Before	After	Before	After	Before	After
WA_{24} (%)	9.5 ± 0.9	6.2 ± 0.9	3.4 ± 0.2	3.0 ± 0.0	12.4 ± 1.1	7.1 ± 0.4

Also pre-wetting of the BA had a distinct positive influence. However, a lot depends on the method applied. The first technique described in Section 3.9.1 did not give satisfying results. Putting the BA in contact with water to achieve a 60% saturation resulted in a too-wet, non-surface-dry aggregate. The mixtures made with this material (Table 4: mixtures BA10–BA12) showed a good workability without SP addition. Still, mixture BA10 with the highest W/C ratio (=0.65) was on the verge of segregating. The overdose of water also affected the compressive strength in a negative way (Table 5). Its strength class was only C16/20, which is lower than the required C20/25.

The second and third pre-wetting methods were more promising in terms of gaining a sufficient workability without losing too much strength performance (Table 4, Table 5: mixtures BA21–BA22). It is recommended to pre-wet the BA in advance (third method) rather than to add the same surplus water content simultaneously with the mixing water (second method). The pre-wetting establishes a water-cement film around the grains in advance which prevents further absorption by the BA aggregates. As such, the aggregates do not get saturated, a higher effective water/cement ratio is reached and the workability is less of an issue. Thus, the sequence of adding the different components during concrete manufacturing is important. After testing several amounts of water, 1 m% of the coarse BA 6/20 and 2.5 m% of the fine BA 0/6 appeared to be the optimal pre-wetting content of water. This more or less corresponds with the amount of water absorbed in the first 2 min after the addition of the mixing water. Thus, it is not necessary to consider the whole 6 min mixing timeframe.

4.5.2. Optimization regarding the Susceptibility to Expansion

To diminish the expansion, LA cement could be used. From the length change measurements performed on mixture BA23, it could be concluded that the expansion was reduced from 3%–4% to 2.5%, but it was not completely eliminated (Figure 9: mixture BA23).

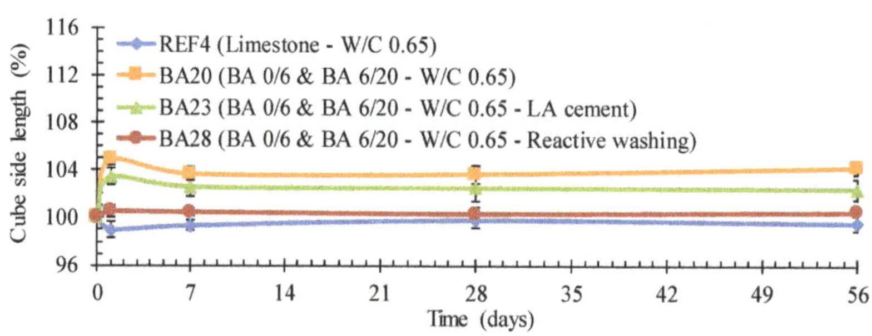

Figure 9. The achievable lower relative expansion of BA concrete with LA cement as binder or after reactive washing of the BA aggregates.

On the other hand, a reactive washing of the aggregates with 1 M NaOH was very effective and eliminated the problem completely (Figure 9: Mixture BA28). As shown in Table 10, reactive washing resulted in a significant decrease in Al reactivity after 14 days, especially in case of the fine BA. After only 4 h, the change in Al reactivity for both BA fractions was still non-significant.

Table 10. The pronounced effectiveness of a reactive washing operation on the Al reactivity of the fine and coarse BA, especially after 14 days.

Time	Al Reactivity (%)	
	BA 0/6	**BA 6/20**
Before reactive washing	2.22	0.16
After 4 h	1.93	0.21
After 14 days	0.73	0.02

Almost no expansion was recorded for concrete BA28 containing BA 0/6 and 6/20 washed with NaOH (Figure 9). There was a concern regarding the strength of this concrete because the surface showed signs of peeling. However, this appeared to be not an issue (Table 5: C25/30). Although this reactive washing can solve the expansion problem, one should remain aware of the fact that this chemical processing will increase the production cost of prefabricated Lego bricks significantly. Moreover, there will be some additional environmental issues. Further research is needed to see whether recycled NaOH obtained from other industrial processes can be used for this purpose.

4.6. Evaluation of the Lego Bricks

4.6.1. Applied Concrete Mixtures

Lego brick 1 was cast using a concrete with the same composition as mixture BA20. Lego brick 2 must be seen as the reference. Its composition was the same as mixtures REF 1 and REF 4. Finally, an optimized BA concrete composition was used for the production of Lego brick 3. It contained crushed BA 0/6 (with somewhat better workability properties) instead of the sieved and washed BA 0/6. For BA 6/20, there was no difference in water absorption between the sieved and washed and the crushed aggregates, so the former was used again. Both the BA 0/6 and 6/20 were sieved afterwards to exclude the fractions smaller than 1 mm and 4 mm, respectively. This enabled a better approximation of the particle size distribution of the reference limestone 2/6 and 6/20 aggregates. Since the BA 0/6 was the most reactive due to its higher metallic Al content, this fine aggregate was subjected to the reactive washing with 1 M NaOH for two to three weeks. Prior to concrete mixing, the aggregates were pre-wetted with 2.9 l water per 180 L of concrete (1 m% of BA 6/20 and 2.5 m% of BA 0/6). A LA cement CEM I 52.5 N HES LA HSR was used as binder. The predefined W/C ratio amounted to 0.65. The compositions of all Lego bricks were included in Table 4.

4.6.2. Observations Made in Fresh and Hardened State

The unadjusted BA concrete mixture (*cf.* BA20) that was used for Lego brick 1 was characterized by a poor workability. Even the addition of 10 mL SP per kg cement could not overcome the problem. This behavior can be explained by the fact that another delivery of coarse BA was used for this concrete which was characterized by a higher water absorption. Nevertheless, with some extra compaction efforts, it was possible to cast the Lego brick. All protrusions were filled in a proper manner and no difficulties were encountered during demolding. However, in hardened state, surface laitance, clear traces of the vibration needle, longitudinal expansion voids and compaction voids could be seen. The expansion itself was comparable to what was observed earlier for mixture BA20 on small concrete cubes. Despite the larger dimensions of the Lego brick, the expansion was not excessive. This can be explained by the fact that the ratio of the troweled surface over the total concrete surface available (troweled and cast surfaces) was lower for the Lego brick than for the cubes with a 150 mm side. The mechanical performance of the concrete used for Lego brick 1 was inadequate since its strength class was only C16/20. This was mainly due to the difficult compaction of the poorly workable concrete which resulted in more voids.

None of these issues were encountered when casting reference Lego brick 2. The only problem was that the concrete was too workable. Without SP addition the concrete was almost segregating. Soon after casting, a distinct water layer appeared on the surface and bleeding occurred. Apparently, applying a W/C ratio of 0.65 for a large concrete volume was too much. This is in contrast with the smaller batches of this concrete that were produced earlier on (REF1 and REF4).

Once Lego brick 3 was produced, the influence of the different optimization techniques could be evaluated. The use of crushed BA 2/6 and pre-wetting the aggregates solved the workability problem. No SP was needed anymore. The reactive washing of the fine BA 2/6 again appeared to be very effective in eliminating expansion and the longitudinal voids. The use of LA cement probably also helped to achieve this. The sieving operations performed on the BA to eliminate the excess in fine material and to approximate the particle size distributions of the traditional limestone aggregates resulted in a more evenly distributed use of sand and BA in accordance with the Fuller approach. This contributed to the better strength (C20/25) and workability (Table 4: slump: S4, flow: F5) performance as well. Figure 10 shows pictures of all three Lego bricks.

Figure 10. The three Lego bricks (3: the optimized BA-based Lego brick; 2: the reference limestone-based Lego brick; and 1: the BA-based unadjusted Lego brick).

5. Conclusions

It is possible to produce a proper concrete in which both the fine and coarse aggregates have been replaced with BA resulting from MSWI. The concrete seems suitable for casting prefabricated Lego bricks. The minimum required strength class (=C20/25) can be assured even with W/C ratios of no less than 0.65 and without using an expensive SP.

However, a dedicated processing of the BA prior to concrete manufacturing is imperative. Excluding the excess of fines to approximate the particle size distributions of the traditional limestone aggregates is advised. This results in a better packing and a higher strength of the concrete. A crushing operation before the sieving eliminates porous elements in the BA which absorb a lot of water. As such, a better workability of the concrete can be obtained. A controlled pre-wetting of the BA just before concrete mixing in relation to the time-dependent water absorption curve also helps to improve the concrete workability. Given these techniques, it will probably be possible to manufacture a sufficiently workable BA concrete with a design W/C ratio of only 0.60 in the future.

The significant presence of metallic Al in the BA makes the concrete susceptible to the formation of longitudinal voids and expansion. The use of a LA cement can somewhat reduce this problem. A reactive washing of the BA with 1 M NaOH eliminates the problem completely. Still, the latter procedure is not ideal from both an economic and an environmental point of view. Further research on more appropriate reactive washing techniques is still needed.

No problems regarding leaching and freeze-thaw were observed for the BA concrete. On the other hand, the concrete shows a pronounced surface roughening after acetic and lactic acid exposure and does not pass the modified Oberholster test for concrete. The expansion (>0.1%) after the latter test indicates ASR sensitivity. However, further microscopic analysis is needed to see whether ASR is indeed causing the problem. Moreover, the tested BA concrete was not fully optimized yet (e.g., with exclusion of the reactive washing stage). Additional tests on the finally optimized mixture used for Lego brick 3 are, for the moment, still ongoing.

Acknowledgments: The authors would like to thank Ghent University for the financial support and the technicians of the Magnel Laboratory for Concrete Research for their assistance during the experiments. Acknowledgments also go to Indaver nv for providing and treating the BA and performing the ICP, IC and FIMS analyses as well as the Al reactivity tests at their laboratories. A word of thanks also goes to CB Recycling BV for providing the mold for the manufacturing of the Lego bricks.

Author Contributions: Niels Ringoot and Arno Beirnaert worked on this research topic for their master dissertation. Together with their supervisors, Philip Van den Heede, Andres Van Brecht and Erwin Van den Brande, they conceived and designed the research plan and all experiments. They performed all tests and analyzed the data which they then compiled in their master dissertation. Philip Van den Heede wrote the paper which was revised and approved by all co-authors. Nele De Belie and Geert De Schutter were the promoters of the master dissertation and hence provided input at all stages of the research.

Conflicts of Interest: The authors declare no conflict of interest.

References

1. European Commission. *Being Wise with Waste: The EU's Approach to Waste Management*; Publications Office of the European Union: Luxembourg, 2010.
2. Chimenos, J.M.; Segarra, M.; Fernandez, M.A.; Espiell, F. Characterization of the bottom ash in municipal solid waste incinerator. *J. Hazard. Mater.* **1998**, *64*, 211–222. [CrossRef]
3. Siddique, R. Use of municipal waste ash in concrete. *Resour. Conserv. Recycl.* **2010**, *55*, 83–91. [CrossRef]
4. European Union. Directive 2008/98/EC of the European parliament and of the council of 19 November 2008 on waste and repealing certain directives. *Off. J. Eur. Union* **2008**, *312*, 3–20.
5. Müller, U.; Rübner, K. The microstructure of concrete made with municipal waste incinerator bottom ash as an aggregate component. *Cem. Concr. Res.* **2006**, *36*, 1434–1443. [CrossRef]
6. Vandecasteele, C.; Wauters, G.; Arickx, S.; Jaspers, M.; van Gerven, T. Integrated municipal solid waste treatment using a grate furnace incinerator: The Indaver case. *Waste Manag.* **2007**, *27*, 1366–1375. [CrossRef] [PubMed]
7. Van Gerven, T.; van Keer, E.; Arickx, S.; Jaspers, M.; Wauters, G.; Vandecasteele, C. Carbonation of MSWI bottom ash to decrease heavy metal leaching, in view of recycling. *Waste Manag.* **2005**, *25*, 291–300. [CrossRef] [PubMed]
8. Courard, L.; Degeimbre, R.; Darimont, A.; Laval, A.L.; Dupont, L.; Bertrand, L. Use of municipal solid waste incineration residues in the fabrication of concrete paving. *Mater. Struct.* **2002**, *35*, 365–372. (In French) [CrossRef]
9. Van Rooy, S. Feasibility Study of MSWI Bottom Ash Aggregate as Secondary Resource for Ground and Road Works. Master's Thesis, Hogeschool Antwerpen, Antwerpen, Belgium, 2007. (In Dutch)
10. Keulen, A.; Florea, M.; Brouwers, J. *Upgrading MSWI Bottom Ash as Building Material for Concrete Mixes*; Eindhoven University of Technology: Eindhoven, The Netherlands, 2012.
11. Pera, J.; Coutaz, L.; Ambroise, J.; Chababbet, M. Use of incinerator bottom ash in concrete. *Cem. Concr. Res.* **1997**, *27*, 1–5. [CrossRef]
12. Abbá, A.; Collivignarelli, M.C.; Sorlini, S.; Bruggi, M. On the reliability of reusing bottom ash from municipal solid waste incineration as aggregate in concrete. *Composites B* **2014**, *58*, 502–509. [CrossRef]
13. Song, G.J.; Kim, K.H.; Seo, Y.C.; Kim, S.C. Characteristics of ashes from different locations at the MSW incinerator equipped with various air pollution control devices. *Waste Manag.* **2003**, *24*, 99–106. [CrossRef]

14. Forteza, R.; Far, M.; Seguí, C.; Cerdá, V. Characterization of bottom ash in municipal solid waste incinerators for its use in road base. *Waste Manag.* **2004**, *24*, 899–909. [CrossRef] [PubMed]
15. Rübner, K.; Haamkens, F.; Linde, O. *Use of Municipal Solid Waste Incinerator Bottom Ash as Aggregate in Concrete*; European Geosciences Union General Assembly: Vienna, Austria, 2007.
16. Van der Wegen, G.; Hofstra, U.; Speerstra, J. Upgraded MSWI bottom ash as aggregate in concrete. *Waste Biomass Valoris.* **2013**, *4*, 737–743. [CrossRef]
17. Sorlini, S.; Abba, A.; Collivignarelli, C. Recovery of MSWI and soil washing residues as concrete aggregates. *Waste Manag.* **2010**, *31*, 289–297. [CrossRef] [PubMed]
18. Ginés, O.; Chimenos, J.M.; Vizcarro, A.; Formosa, J.; Rosell, J.R. Combined use of MSWI bottom ash and fly ash as aggregate in concrete formulation: Environmental and mechanical considerations. *J. Hazard. Mater.* **2009**, *169*, 643–650. [CrossRef] [PubMed]
19. Berg, E.R.; Neal, J.A. Municipal solid waste bottom ash as Portland cement concrete ingredient. *J. Mater. Civ. Eng.* **1998**, *10*, 168–173. [CrossRef]
20. Garcia-Gonzalez, J.; Rodriguez-Robles, D.; Juan-Valdes, A.; Moran-del Pozo, J.M.; Guerra-Romero, M.I. Pre-saturation technique of the recycled aggregates: Solution to the water absorption drawback in the recycled concrete manufacture. *Materials* **2014**, *7*, 6224–6236. [CrossRef]
21. Tegguer, D.A. Determining the water absorption of recycled aggregates utilizing hydrostatic weighing approach. *Constr. Build. Mater.* **2012**, *27*, 112–116. [CrossRef]
22. RILEM TC 106-AAR: Alkali-aggregate reaction. *Mater. Struct.* **2000**, *33*, 283–293.
23. Fabien, A.; de Boom, A.; Lenormand, T.; Lambrette, N.; Degrez, M.; Staquet, S. Reuse of municipal solid waste incinerator fly ash into cement-based materials. In Proceedings of the 1st Concrete Innovation Conference (CIC2014), Oslo, Norway, 11–13 June 2014; Justnes, H., Ed.; Sintef Building and Infrastructure: Trondheim, Norway, 2014.
24. CUR Bouw & Infra. *CUR Recommendation 116, Processed Municipal Incinerator Bottom Ash as Aggregate for Concrete*; CUR Bouw & Infra: Gouda, The Netherlands, 2012. (In Dutch)
25. Audenaert, K. Transport Mechanisms of Self-Compacting Concrete Related to Carbonation and Chloride Penetration. Ph.D. Thesis, Ghent University, Ghent, Belgium, 2006. (In Dutch)

5

Optical and Magneto-Optical Properties of $Gd_{22}Fe_{78}$ Thin Films in the Photon Energy Range From 1.5 to 5.5 eV

Eva Jesenská [1,2,*], Takahiro Hashinaka [1], Takayuki Ishibashi [1], Lukáš Beran [2], Ján Dušek [2], Roman Antoš [2], Kiyoshi Kuga [3], Ken-ichi Aoshima [3], Kenji Machida [3], Hidekazu Kinjo [3] and Martin Veis [2,*]

Academic Editor: Beatriz Noheda

1 Department of Materials Science and Technology, Nagaoka University of Technology, Nagaoka, Niigata 1603-1, Japan; takahiro_hashinaka@mst.nagaokaut.ac.jp (T.H.); t_bashi@mst.nagaokaut.ac.jp (T.I.)
2 Institute of Physics, Charles University, Prague 12116, Czech Republic; beranlu@gmail.com (L.B.); r.jan.dusek@seznam.cz (J.D.); antos@karlov.mff.cuni.cz (R.A.)
3 Science and Technology Research Laboratories, Japan Broadcasting Corporation (NHK), Tokyo 157-8510, Japan; kuga.k-lu@nhk.or.jp (K.K.); aoshima.k-ia@nhk.or.jp (K.-i.A.); machida.k-ge@nhk.or.jp (K.M.); kinjou.h-lk@nhk.or.jp (H.K.)
* Correspondance: jesenska@karlov.mff.cuni.cz (E.J.); veis@karlov.mff.cuni.cz (M.V.)

Abstract: Optical and magneto-optical properties of amorphous $Gd_{22}Fe_{78}$ (GdFe) thin films prepared by direct current (DC) sputtering on thermally oxidized substrates were characterized by the combination of spectroscopic ellipsometry and magneto-optical spectroscopy in the photon energy range from 1.5 to 5.5 eV. Thin SiN_x and Ru coatings were used to prevent the GdFe surface oxidation and contamination. Using advanced theoretical models spectral dependence of the complete permittivity tensor and spectral dependence of the absorption coefficient were deduced from experimental data. No significant changes in the optical properties upon different coatings were observed, indicating reliability of used analysis.

Keywords: GdFe; magneto-optics; spectroscopic ellipsometry

---

1. Introduction

Considerable attention has been paid to magnetic and magneto-optical (MO) properties of amorphous ferrimagnetic thin films composed of rare-earth and transition metals because of their useful technological applications [1–4]. As one of the important magneto-optical storage materials, GdFe has significant advantages, such as large magnetization density, and possibility to adjust its compensation temperature, coercive and saturation magnetization by changing the composition [5–7]. Because of these properties, GdFe has a substantial impact on modern micro- and nanoelectronic research, where it is often used in domain wall junctions or MO memories [1,3,4].

The GdFe shows perpendicular anisotropy when the Fe concentration is about the compensation concentration, which is for this material about 75% [8]. This composition is often used for MO applications such as MO disk storage or MO spatial light modulator driven by spin transfer torque (spin spatial light modulator (SLM)) [2]. It is very important to control the GdFe composition precisely, since it significantly affects the GdFe magnetic switching property. Coercivity shows maxima when the composition is the compensation one, and it gets smaller when the composition becomes Fe rich (compared to the compensation composition). Spin-torque switching current of the spin-SLM is significantly reduced with an increase in Fe concentration and it shows very small switching current when composition is slightly Fe richer (such as $Gd_{20-22}Fe_{78-80}$) than the compensation one [9,10]. Therefore it is meaningful to investigate optical properties of the GdFe material with the Fe concentration around 78%.

The main purpose of our investigation was to fully determine the dielectric permittivity tensor of the GdFe thin film. Knowledge of the permittivity tensor is crucial, since it allows theoretical prediction of complex physical properties of complicated multilayered nanostructures containing GdFe layers without necessity to manufacture multiple samples. This is especially useful for the design of advanced devices as holographic 3D displays based on structures consisting of more than 10 nanolayers [2,11].

Since GdFe is very easy to oxidize [12], it is usually covered by a protecting layer, which complicates its analysis. The main reason is that the optical properties of protecting layer materials (here Ru, SiN_x) may slightly differ in dependence on material they are deposited on. This is usually caused by the lattice mismatch between the film and the substrate, which induces strains of various kinds [13–15]. In this work, we have dealt with this problem by using two different coating layer materials which allowed more precise determination of GdFe permittivity tensor. Spectroscopic ellipsometry showed very similar optical properties of GdFe for both coatings, which allowed us to fit the optical constants from experimental data simultaneously.

We used spectroscopic ellipsometry at energies 1.2–6 eV and MO spectral measurements at energies 1.5–5.5 eV. From ellipsometric data we derived spectra of the real, ε_{1r}, and imaginary, ε_{1i}, part of diagonal permittivity tensor elements and the absorption coefficient spectra of the GdFe thin films. Magneto-optical properties were examined by polar magneto-optical Kerr effect (MOKE) rotation and ellipticity measurements. From these data we determined the spectral dependence of the real, ε_{2r}, and imaginary, ε_{2i}, part of off-diagonal GdFe permittivity tensor elements.

2. Results and Discussion

To obtain the GdFe dielectric permittivity tensor we analyzed optical and magneto-optical properties of two samples with structural compositions and nominal thicknesses listed in the Table 1.

Table 1. Structural compositions and nominal thicknesses of examined samples.

Sample	Substrate	Layer 1	Layer 2	Layer 3
Ru coated	Si	SiO_2 (300 nm)	$Gd_{22}Fe_{78}$ (100 nm)	Ru (3 nm)
SiN_x coated	Si	SiO_2 (300 nm)	$Gd_{22}Fe_{78}$ (100 nm)	SiN_x (20 nm)

2.1. Spectroscopic Ellipsometry

Figure 1 shows spectroscopic ellipsometry experimental data of studied samples. Since the measurements were performed under large incident angles, the difference between the ellipsometry data of two samples reflects not only the different optical parameters of coatings, but also the Fabry-Perot type resonance in much thicker SiN_x layer with respect to Ru. The GdFe optical constants and layer thicknesses were fitted to the theoretical models (based on structural compositions) and experimental data. We used the "Multi Sample Analysis" mode to derive the GdFe optical constants from the experimental data of both samples simultaneously (It is an advanced mode in CompleteEase software that allows multiple samples to be fitted simultaneously with some of the fit parameters common to all samples (GdFe optical constants) and other allowed to vary (coating layer material, thicknesses) [16] (p. 169)). This could be done because the GdFe optical constants obtained from the individual sample analysis were very similar. Therefore this mode allowed, suppressing the fit error and so more precise analysis. Obtained optical functions were afterwards parameterized to ensure Kramers-Kronig consistent results. Optical functions of the GdFe thin film were parameterized by the linear summation of two Lorentz, two Gaussian oscillators and one Drude term in the spectral range 1.5–6 eV. Gaussian oscillators have been used due to their ability to rapidly approach zero beyond the FWHM positions which makes them suitable for the parametrization of the steeper regions of the optical spectra. Drude term has been used because GdFe is an amorphous alloy; therefore, we also had to consider the free carrier effect on the optical response. Parameters of used oscillators are listed in the Table 2. Derived thicknesses are listed in the Table 3. Optical constants of Si, SiO_2, Ru and SiN_x were determined from the ellipsometry measurements on individual samples. As we can see from Figure 1, the experimental and theoretical values correspond well. Since the GdFe has large absorption coefficient, the average penetration depth of this material in measured spectral region is around 24 nm, which allows consider the GdFe layer as semi-infinite. Therefore, we also theoretically modeled optical constants for structures where GdFe was used as semi-infinite substrate. However, no significant changes in results given by full and semi-infinite model structures were observed, since the full structure model also includes the high absorption of GdFe. In this work we present results obtained with the full model structure.

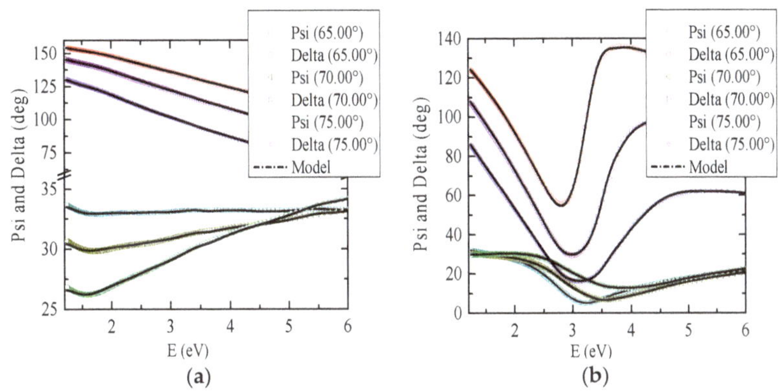

Figure 1. Measured variable angle spectroscopic ellipsometric data of (**a**) Ru coated sample and (**b**) SiN_x coated sample are compared with theoretical calculations (lines). Dark cyan, olive and green symbols correspond to Psi measurements at angles 65°, 70°, 75° respectively. Red, purple and violet symbols correspond to Delta measurements at angles 65°, 70°, 75° respectively.

Table 2. Parameters of oscillators for model of GdFe layer for 1.5–6 eV spectral range. In here, E stands for central energies of oscillators; Amp represents amplitudes of oscillators and Br broadenings. For Drude term, N represents carrier concentration, μ carrier mobility and m^* carrier effective mass.

Oscillator Type	*E* (eV)	Amp	Br (eV)
Lorentz	1.89	5.66	1.87
Lorentz	2.57	3.34	1.04
Gaussian	3.75	0.60	0.39
Gaussian	4.03	0.75	0.50
-	N (cm^{-3})	μ ($cm^2 \cdot V^{-1} s^{-1}$)	m^*
Drude	1.11×10^{23}	0.36	0.53

Table 3. Thicknesses used to model GdFe layer for 1.5–6 eV spectral range. In here, t stands for thickness and r for roughness on top.

Sample	t_{SiO_2} (nm)	t_{GdFe} (nm)	t_{Ru} (nm)	t_{SiN} (nm)	r (nm)
Ru coated	304	105	3.3	-	2
SiN_x coated	304	105	-	21	3

Figure 2 shows obtained spectra of the real, ε_{1r}, and imaginary part, ε_{1i}, of diagonal permittivity tensor elements. Figure 3 shows the calculated absorption coefficient spectra of the GdFe thin film. The ε_{1r} spectrum is characteristic by one global minimum at 2.9 eV while the ε_{1i} decreases its value with increasing energy in the whole measured spectral range. The spectra show similar behavior to Fe and Gd [17] (p. 394), [18] and also to previously reported results on GdFe films with different compositions [6,19]. The behavior in the spectral range 1.5–3 eV, where ε_{1r} decreases its value for higher energies is similar to the behavior of some transitions metals (including Cr, Ru, Ti, Gd [17] (p. 377), [20] (pp. 245, 259), [18]) and it was previously explained by intra-band transitions, which for some transition metals, are not negligible in a measured spectral region [17] (p. 375).

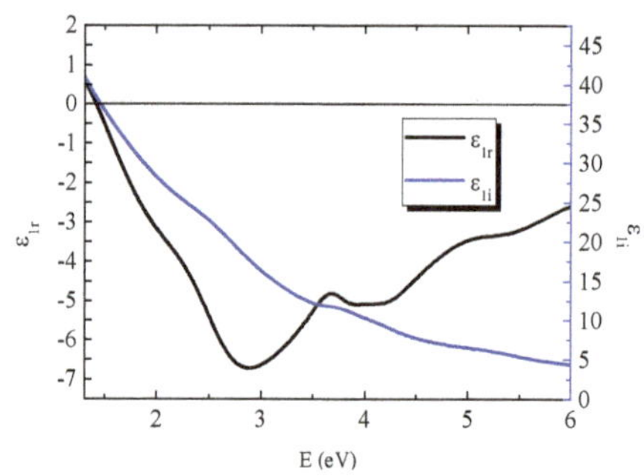

Figure 2. Real and imaginary part of diagonal permittivity tensor elements of GdFe. Black line corresponds to the real part ε_{1r} and blue line to the imaginary part ε_{1i} respectively.

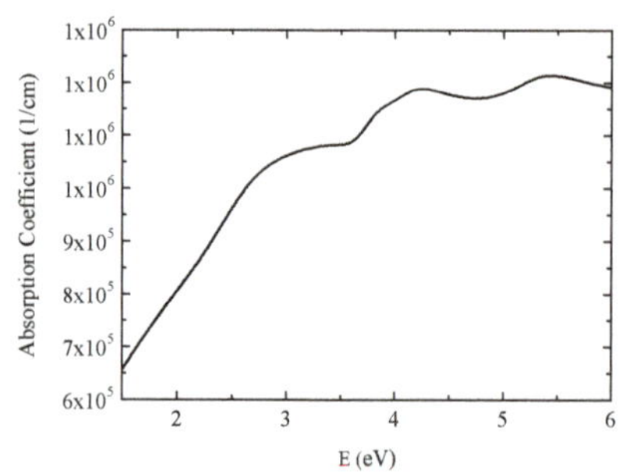

Figure 3. Calculated absorption coefficient of GdFe.

2.2. Magneto-Optical Kerr Effect (MOKE) Spectroscopy

Figure 4 shows experimental polar MOKE rotation and ellipticity spectra of the Ru and SiN_x coated samples. The MOKE spectra of the Ru coated sample are characteristic by increasing rotation and ellipticity amplitudes toward to smaller energies. The rotation spectrum of the SiN_x coated sample is characteristic by one global maximum at 2.6 eV and the ellipticity spectrum by one global maximum at 2 eV. We can also observe that the sample with SiN_x coating is, especially in the spectral range 1.5–3.5 eV, giving much higher MO signal than the sample with Ru coating. This is probably caused by the multiple reflections inside SiN_x coating layer, which results in Fabry-Perot like resonance causing the enhancement of the MOKE in this energy region. We used MOKE spectra to calculate the off-diagonal elements of the GdFe dielectric permittivity tensor.

For the off-diagonal elements calculations we used the diagonal elements of the GdFe dielectric permittivity tensor and thicknesses determined by the spectroscopic ellipsometry. The real and imaginary parts of off-diagonal elements, ε_{2r} and ε_{2i} were calculated from the MOKE spectra in the spectral range 1.5–5.5 eV. Figure 5 shows the spectra of the real and imaginary parts of off-diagonal permittivity tensor elements calculated from the MOKE spectra of the Ru and SiN_x coated samples. As one can see, the difference between the samples is rather small. Resulted values of the off-diagonal GdFe permittivity tensor elements ε_{2r} and ε_{2i} were obtained by averaging these two results and they are also shown in Figure 5. The ε_{2r} spectrum is characteristic by one global minimum around 2.5 eV while ε_{2i} is positive and decreases its amplitude with energy in the whole measured spectral range.

The obtained spectra revealed a similar shape as off-diagonal permittivity elements of iron and also previously reported spectra of GdFe with different composition [19,21]. Amplitudes of GdFe off-diagonal permittivity elements spectra are smaller than amplitudes for Fe, which is most probably caused by the presence of the Gd.

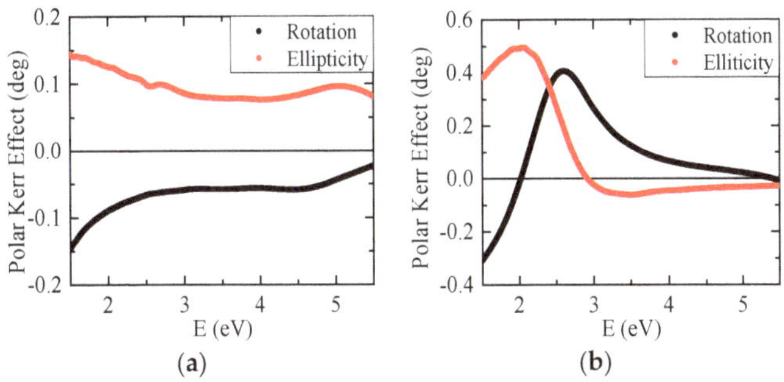

Figure 4. Polar magneto-optical Kerr effect (MOKE) rotation and ellipticity spectra of (**a**) Ru coated sample and (**b**) SiN_x coated sample. Black symbols correspond to Kerr rotation; red symbols correspond to Kerr ellipticity.

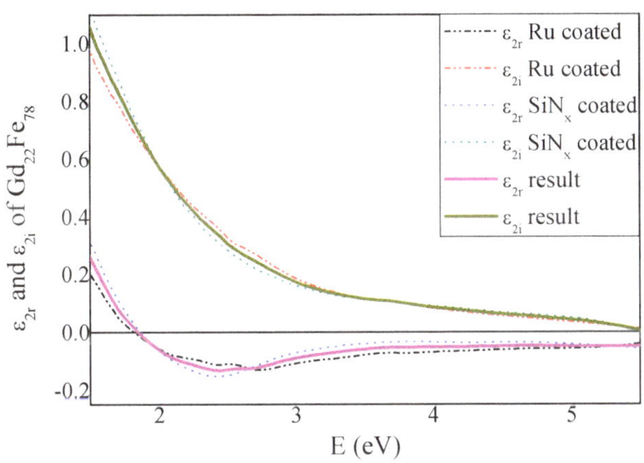

Figure 5. Real and imaginary part of off-diagonal permittivity tensor elements of GdFe. Dash-dot lines correspond to calculations from the Ru coated sample MOKE spectra, dotted lines correspond to calculations from the SiN_x coated sample MOKE spectra. Pink line corresponds to the averaged real part ε_{2r} and green line to the averaged imaginary part ε_{2i} respectively.

3. Materials and Methods

3.1. Theory

The material response on electromagnetic waves in the presence of an external magnetic field can be described by the dielectric permittivity tensor. If the magnetization vector is parallel to the z-axis of the Cartesian coordinate system (the magnetic film-ambient interface is normal to the z-axis, light is propagating along the z-axis) and if we restrict ourselves to linear MO effects, the dielectric permittivity tensor simplifies to the form [22]:

$$\begin{pmatrix} \varepsilon_1 & -i \cdot \varepsilon_2 & 0 \\ i \cdot \varepsilon_2 & \varepsilon_1 & 0 \\ 0 & 0 & \varepsilon_1 \end{pmatrix} \tag{1}$$

All elements of the tensor have real and imaginary parts:

$$\begin{aligned} \varepsilon_1 &= \varepsilon_{1r} - i \cdot \varepsilon_{1i} \\ \varepsilon_2 &= \varepsilon_{2r} - i \cdot \varepsilon_{2i} \end{aligned} \tag{2}$$

The optical behavior of the sample upon light reflection can be in Cartesian representation with the base of s and p polarizations described by the Jones matrix of reflection [23] (p. 170):

$$J_{sp}^R = \begin{pmatrix} r_{ss} & r_{sp} \\ r_{ps} & r_{pp} \end{pmatrix} J_{sp}^I \tag{3}$$

Matrix elements are amplitude reflection coefficients for the s and p waves. The change in the polarization state of the reflected beam can be then expressed by the ellipsometric parameters Psi (ψ) and Delta (Δ), which are defined as:

$$\tan\psi \cdot e^{i\Delta} = \rho = \frac{r_{pp}}{r_{ss}} \tag{4}$$

Where tanψ is the magnitude of the reflectivity ratio and Δ is the phase change between s and p polarized light. The r_{pp} and r_{ss} are measured from the AC signal (dual rotating compensator ellipsometer configuration) [16] (p. 34). Analyzing the experimental ellipsometric data, one can derive the diagonal elements of the permittivity tensor. The important step in the spectroscopic ellipsometry analysis is the proper parametrization of the dispersion of the unknown optical functions. In this work we used Kramers-Kronig (KK) consistent Lorentz, Gaussian and Drude models. Classic version of Lorentz oscillator model can be mathematically described as:

$$\varepsilon_{1\_Lorentz} = \frac{AmpBrE_0}{E_0^2 - E^2 - i \cdot EBr} \tag{5}$$

Parameters E_0, Amp, Br denote the center energy, amplitude and the broadening parameter respectively [16] (p. 343), [24]. Gaussian line shape in ε_{1i} is defined as:

$$\varepsilon_{1\_Gaussian} = Amp \left\{ \begin{array}{c} \left(\Gamma\left(\frac{E - E_0}{\sigma}\right) + \Gamma\left(\frac{E + E_0}{\sigma}\right) \right) + \\ i \cdot \left(\exp\left[-\left(\frac{E - E_0}{\sigma}\right)^2 \right] + \exp\left[-\left(\frac{E + E_0}{\sigma}\right)^2 \right] \right) \end{array} \right\} \tag{6}$$

where

$$\sigma = \frac{Br}{2\sqrt{\ln(2)}} \tag{7}$$

The function Γ is a convergence series that produces a Kramers-Kronig consistent line shape for ε_{1r} [16] (p. 344), [25]. In order to describe the free carrier effect on the dielectric response, Drude model is commonly used:

$$\varepsilon_{1\_Drude} = \frac{-\hbar^2 q^2 N\mu}{\varepsilon_0(\mu m * m_e E^2 + iq\hbar E)} \tag{8}$$

Parameters N, μ, m^* denote the carrier concentration, carrier mobility and carrier effective mass respectively. The physical constants are $\hbar$ (Planck constant/2π), q (electron charge), ε_0 (the vacuum dielectric constant) and m_e (the electron mass) [16] (p. 344) [26].

Spectroscopic MOKE can be used to derive off-diagonal elements of the permittivity tensor. In here we used the Yeh matrix formalism for anisotropic multilayers to theoretically calculate the MOKE effect in studied samples [23] (p. 344), [22,27]. The change in the polarization state of the reflected beam in the polar MOKE experiment can be expressed by the complex MO Kerr angle Φ_K, which is for p-polarization and small angles of incidence defined as follows:

$$\Phi_K = \theta_K - i \cdot e_K = \frac{r_{sp}}{r_{pp}} \tag{9}$$

In this equation θ_k is the Kerr rotation, e_k is the Kerr ellipticity. Let us consider the case of the three layered medium prepared on a bulk substrate. We will work in Cartesian coordinates where the sample interface is perpendicular to the z-axis, the wave vector of the incident light is perpendicular to the x-axis and each layer is characterized by the complex permittivity tensor and the thickness. In this case the Yeh Matrix Formalism allows to express the relationship between the electric field amplitudes on the substrate/film interface ($E_0{}^{(0)}(z)$) and the electric field amplitudes on the coating ambient interface ($E_0{}^{(4)}(z_3)$) as:

$$E_0^{(0)}(z) = \left\{ \begin{array}{c} [D^{(0)}]^{-1} D^{(1)} P^{(1)} [D^{(1)}]^{-1} D^{(2)} P^{(2)} [D^{(2)}]^{-1} D^{(3)} \cdot \\ P^{(3)} [D^{(3)}]^{-1} D^{(4)} E_0^{(4)}(z_3) \end{array} \right\} = M E_0^{(4)}(z_3) \tag{10}$$

Here M stands for transfer matrix (related to reflection coefficients) between substrate/film interface and coating ambient interface. Superscripts in brackets, n = 0, 1, 2, 3 and 4 are markers of the substrate (0), three layers and ambient half space (4). P stands for propagation matrix

$$P_{ij}^{(n)} = \delta_{ij} \exp(i \frac{\omega}{c} N_{zj}^{(n)} t_n) \tag{11}$$

and D for dynamical matrix defined as:

$$\begin{aligned} D_{1j}^{(n)} &= -\varepsilon_2^{(n)} (\varepsilon_1^{(n)} - N_y^2) \\ D_{2j}^{(n)} &= N_{zj}^{(n)} D_{1j}^{(n)} \\ D_{3j}^{(n)} &= (\varepsilon_1^{(n)} - N_y^2)(\varepsilon_1^{(n)} - N_y^2 - (N_{zj}^{(n)})^2) \\ D_{4j}^{(n)} &= -\varepsilon_1^{(n)} (\varepsilon_1^{(n)} - N_y^2 - (N_{zj}^{(n)})^2) \end{aligned} \tag{12}$$

where t_n, N_{zj} and N_y are the thickness of the n-th layer, z components of the reduced wave vector and y components of the reduced wave vector respectively [23] (p. 151).

The structural composition and nominal thicknesses used for the theoretical analysis of ellipsometric and MOKE experimental data are described in the Table 1. The model structure consisted of Si semi-infinite substrate followed by 300 nm thick buffer layer of SiO_2 and 100 nm thick layer of GdFe. Finally, it was followed by a coating which was 3 nm thick Ru layer for Ru coated sample and 20 nm thick SiN_x layer in the case of SiN_x coated sample. Surface roughness was also considered. We used the Bruggeman Effective Medium Approximation formula of the mixture of hosting material (ε) with void to simulate surface roughness as a thin layer with permittivity ε_{eff} defined as follows [28]:

$$0 = (1-f)\frac{\varepsilon - \varepsilon_{eff}}{\varepsilon + 2\varepsilon_{eff}} + \frac{1 - \varepsilon_{eff}}{1 + 2\varepsilon_{eff}} \tag{13}$$

In this formula f denotes the volume fraction of the void in the mixture (we fixed this value to 50%).

3.2. Experimental Details

In this work we analyzed two samples with structural compositions and nominal thicknesses listed in the Table 1. The GdFe and Ru layers were deposited by direct current sputtering technique in Kr gas of pressure 8.7×10^{-2} Pa with a deposition rate of 3.6 nm/min. The SiN_x film was prepared by RF ion beam sputtering technique in Kr gas of pressure 7.8×10^{-2} Pa with the deposition rate of 3.9 nm/min.

3.2.1. Spectroscopic Ellipsometry Measurements

Spectroscopic Ellipsometry measurements were performed by a Mueller matrix ellipsometer Woollam RC2 (J.A. Woollam Co. Inc., Lincoln, NE, USA). We measured ellipsometric Psi and Delta parameters of the reflected light in the spectral range from 1.3 to 6 eV for incident angles 65°, 70° and 75°. Obtained experimental data were analyzed using CompleteEase software in "Multi Sample Analysis" mode.

3.2.2. Magneto-Optical Measurements

Magneto-optical properties of samples were measured by the MOKE spectroscopy. The MOKE rotation and ellipticity spectra were measured in the polar configuration using a method of generalized magneto-optical ellipsometry with rotating analyzer, allowing the determination of the rotation angles with high accuracy. The spectra of polar Kerr rotation and ellipticity were acquired at the room temperature for nearly normal light incidence. Applied magnetic field was 1.2 T, which was enough for magnetic saturation of the samples. Incident light was p-polarized. The data were recorded in the photon energy range from 1.4 to 5.5 eV.

4. Conclusions

In this paper we presented systematic optical and magneto-optical study of the GdFe thin films prepared by DC sputtering technique. Since GdFe is easy to oxidize it was protected by the coating layers. Here Ru or SiN_x layers were chosen. We used the assumption that optical and MO properties of the GdFe material will not change regardless of whether Ru or SiN_x is deposited on its top. In experimental part, we performed spectroscopic ellipsometry and MOKE spectroscopy. Significant enhancement of MOKE was observed for the sample with SiN_x coating, which was ascribed to the multiple reflections in SiN_x layer. Combination of ellipsometric and MOKE spectroscopy allowed us to successfully determine the full dielectric permittivity tensor and the absorption coefficient spectra of the GdFe thin film prepared on the SiO_2 buffer layer. Optical constants of Si, SiO_2, Ru and SiN_x used in our analysis were determined from the ellipsometry measurements on individual samples. The knowledge of the permittivity tensor is crucial, since it allows theoretical prediction of complex physical properties of complicated multilayered nanostructures containing GdFe layers without necessity to manufacture multiple samples.

Acknowledgments: This research was supported by the National Institute of Information and Communications Technology (NICT). Financial support of Czech Grant Agency (grant No. 13-30397S) is also acknowledged.

Author Contributions: Takayuki Ishibashi, Martin Veis and Ken-ichi Aoshima conceived and designed the experiments; Kiyoshi Kuga and Ken-ichi Aoshima performed the composition analysis for the fabricated films and analysis equipment set up; Martin Veis co-wrote the paper; Kenji Machida set up sample deposition system; Hidekazu Kinjo and Takahiro Hashinaka fabricated samples; Lukáš Beran performed MOKE measurements; Ján Dušek performed spectroscopic ellipsometry measurements; Roman Antoš contributed to theoretical modeling; Eva Jesenská performed theoretical analysis of the data and wrote the paper.

Conflicts of Interest: The authors declare no conflict of interest.

References

1. Imamura, N.; Ota, C. Experimental study on magneto-optical disk exerciser with the laser diode and amorphous magnetic thin films. *Jpn. J. Appl. Phys.* **1980**, *19*, L731–L734. [CrossRef]
2. Aoshima, K.; Machida, K.; Kato, D.; Mishina, T.; Wada, K.; Cai, Y.-F.; Kinjo, H.; Kuga, K.; Kikuchi, H.; Ishibashi, T.; *et al.* A magneto-optical spatial light modulator driven by spin transfer switching for 3D holography applications. *J. Disp. Technol.* **2015**, *11*, 129–135. [CrossRef]

3. Mangin, S.; Marchal, G.; Bellouard, C.; Wernsdorfer, W.; Barbara, B. Magnetic behavior and resistivity of the domain-wall junction GdFe (1000 Å)/TbFe/GdFe (500 Å). *Phys. Rev. B* **1998**, *58*, 2748–2757. [CrossRef]
4. Nishimura, N.; Hirai, T.; Koganei, A.; Ikeda, T.; Okano, K.; Sekiguchi, Y.; Osada, Y. Magnetic tunnel junction device with perpendicular magnetization films for high-density magnetic random access memory. *J. Appl. Phys.* **2002**, *91*. [CrossRef]
5. Urner-Wille, M.; Witter, K. Compensation point switching in homogeneous amorphous GdFe-films. *J. Magn. Magn. Mater.* **1979**, *13*, 77–80. [CrossRef]
6. Shen, Z.; Li, J.; Wang, S.; Zhou, S.; Chen, L. Magneto-optical and optical properties of GdFe films. In Proceedings of the Fifth International Symposium on Optical Storage, Shanghai, China, 22–26 May 2000.
7. Sugano, S.; Kojima, N. *Magneto-Optics*; Springer: Tokyo, Japan, 2011; p. 282.
8. Boernstein, L. *Magnetic Properties of Metals*; Springer-Verlag: Tokyo, Japan, 1986; p. 172.
9. Aoshima, K.; Hashimoto, Y.; Funabashi, N.; Machida, K.; Kuga, K.; Kikuchi, H.; Shimidzu, N.; Ishibashi, T. Spin transfer switching of current-perpendicular-to-plane giant magnetoresistance with various Gd-Fe free-layer compositions. *J. Appl. Phys.* **2012**, *111*. [CrossRef]
10. Kinjo, H.; Machida, K.; Matsui, K.; Aoshima, K.-i.; Kato, D.; Kuga, K.; Kikuchi, H.; Shimidzu, N. Low-current-density spin-transfer switching in $Gd_{22}Fe_{78}$-MgO magnetic tunnel junction. *J. Appl. Phys.* **2014**, *115*. [CrossRef]
11. Goto, S.; Machida, K.; Aoshima, K.; Kuga, K.; Kikuchi, H.; Shimidzu, N.; Ishibashi, T. Magneto-optical Properties of (Pt/Co)/X/IZO (X = Ta, Au, Pt, Ru and Ag) Structures for Magneto-optical Spatial Light Modulators. *EPJ Web Conf.* **2013**, *40*. [CrossRef]
12. Terzieff, P.; Lee, K. Electron spectroscopy studies on amorphous GdFe and GdCo alloys. *J. Appl. Phys.* **1979**, *50*, 3565–3569. [CrossRef]
13. Henein, G.E.; Wagner, W.R. Stresses induced in GaAs by TiPt ohmic contacts. *J. Appl. Phys.* **1983**, *54*, 6395–6400. [CrossRef]
14. Greer, J.R.; Oliver, W.C.; Nix, W.D. Size dependence of mechanical properties of gold at the micron scale in the absence of strain gradients. *Acta Mater.* **2005**, *53*, 1821–1830. [CrossRef]
15. Nix, W.D. Mechanical properties of thin films. *Metall. Tran. A* **1989**, *20A*, 2217–2245. [CrossRef]
16. *CompleteEASE Data Analysis Manual*. Available online: https://crn2.3it.usherbrooke.ca/guide_sb/appareils/Ellipsometre/CompleteEASE%20Manual.pdf (accessed on 24 December 2015).
17. Palik, E.D. *Handbook of Optical Constants of Solids*, 2nd ed.; Academic Press: Waltham, MA, USA, 1991.
18. Quemerais, A.; Loisel, B.; Jezequel, G.; Thomas, J.; Lemonnier, J.C. Optical spectra of gadolinium and dysprosium: Study of the 5p thresholds. *J. Phys. F Met. Phys.* **1981**, *11*, 293–303.
19. Ferguson, P.E.; Romagnoli, R.J. Transverse Kerr magneto-optic effect and optical properties of transition-rare-earth *Alloys*. *J. Appl. Phys.* **1969**, *40*, 1236–1238. [CrossRef]
20. Palik, E.D. *Handbook of Optical Constants of Solids*, 3rd ed.; Academic Press: Waltham, MA, USA, 1998.
21. Visnovsky, S.; Lopusnik, R.; Nyvlt, M.; Das, A.; Krishnan, R.; Tessier, M.; Frait, Z.; Aitchison, P.; Chapman, J.N. Magneto-optic studies of Fe/Au multilayers. *J. Magn. Magn. Mater.* **1999**, *198*, 480–482. [CrossRef]
22. Višňovský, Š. Magneto-optical effects in crystals at the normal incidence. *Czech. J. Ph.* **1987**, *37*, 218–231.
23. Višňovský, Š. *Optics in Magnetic Multilayers and Nanostructures*; Taylor & Francis: Abingdon, VA, USA, 2006.
24. Wooten, F. *Optical Properties of Solids*; Academinc Press, Inc.: New York, NY, USA, 1972; p. 52.
25. Meneses, D.D.S.; Malki, M.; Echegut, P. Structure and lattice dynamics of binary lead silicate glasses investigated by infrared spectroscopy. *J. Non-Cryst. Solids* **2006**, *352*, 769–776. [CrossRef]
26. Tiwald, T.E.; Thompson, D.W.; Woollam, J.A.; Paulson, W.; Hance, R. Application of IR variable angle spectroscopic ellipsometry to the determination of free carrier concentration depth profiles. *Thin Solid Films* **1998**, *313–314*, 661–666. [CrossRef]
27. Yeh, P. Optics of anisotropic layered media: A new 4 × 4 matrix algebra. *Surf. Sci.* **1979**, *96*, 41–53. [CrossRef]
28. Halagačka, L.; Postava, K.; Vanwolleghem, M.; Vaurette, F.; Ben Youssef, J.; Dagens, B.; Pištora, J. Mueller matrix optical and magneto-optical characterization of Bi-substituted gadolinium iron garnet for application in magnetoplasmonic structures. *Opt. Mater. Express* **2014**, *4*, 1903–1919. [CrossRef]

6

Dependence of Liquid Supercooling on Liquid Overheating Levels of Al Small Particles

Qingsong Mei [1,*] and Juying Li [2]

Academic Editor: Andreas Taubert

1 Department of Materials Engineering, School of Power and Mechanical Engineering, Wuhan University, Wuhan 430072, China
2 School of Mechanical Engineering, Wuhan Polytechnic University, Wuhan 430023, China; jyli@whpu.edu.cn
* Correspondence: qsmei@whu.edu.cn

Abstract: The liquid thermal history effect on liquid supercooling behavior has been found in various metals and alloys; typically the degree of liquid supercooling (ΔT^-) increases with the increase of liquid overheating (ΔT^+) up to several to tens of degrees above the equilibrium melting point (T_0). Here we report quantitative experimental measurements on the ΔT^--ΔT^+ relationship of Al small particles encapsulated in Al_2O_3 shells by using a differential scanning calorimeter. We find a remarkable dependence of ΔT^- on ΔT^+ of Al small particles, extending to at least 340 °C above T_0 of Al (~1.36T_0), which indicates the existence of temperature-dependent crystallization centers in liquid Al up to very high liquid overheating levels. Our results demonstrate quantitatively the significant effect of liquid thermal history on the supercooling behavior of Al and its alloys, and raise new considerations about the dependence of ΔT^- on ΔT^+ at very high ΔT^+ levels.

Keywords: liquid overheating; liquid supercooling; solidification; thermal history; Al particles

---

1. Introduction

Solidification and melting are transformations basic to technological applications such as casting, crystal growth, and glass formation. Historically, the liquid thermal history was discovered to have a significant effect on the liquid supercooling behavior and thus, the nucleation and growth of crystals and their qualities [1–13]. The thermal history effect can be quantified by the relationship between liquid overheating (ΔT^+), which is measured by the difference between liquid temperature and the equilibrium melting point (T_0), and liquid supercooling (ΔT^-), which is measured by the difference between T_0 and the temperature of solidification. Thus far, the dependence of ΔT^- on ΔT^+ has been investigated in metals/semimetals (Bi, Sn, Ga) [2–5,11,14] and alloys [6–11]; typically ΔT^- increases with an increase of ΔT^+ up to several to tens of degrees above T_0. Despite the fundamental and technical importance, the underlying mechanism of this phenomenon in relation to the liquid structure and the kinetics of heterogeneous nucleation of solidification is not well understood. Turnbull [14] suggested that crystals in microcavities, which are on the container or surface of impurity particles inside a melt, can be retained above T_0, *i.e.*, they have elevated melting points. These retained crystals can serve as nuclei for solidification at a certain supercooling level, which results in an increase of ΔT^- with the increase of ΔT^+ [14].

The cavity theory has been quantitatively validated by various experimental results [2–5], but it was considered to be not general, e.g., the dependence of ΔT^- on ΔT^+ in Bi, Sn, SnSb and SnPb was found to be either continuous or discontinuous [10,11], which points to the evolution of transient short range order structures in the liquid state. Obviously, more investigations of the ΔT^+-ΔT^- relationship (thermal history effect) in metals and alloys are needed. In the casting of Al alloys, liquid thermal history effects on the microstructure of solidification have been reported [15,16]. Hereby, a quantitative study of the dependence of ΔT^- on ΔT^+ in the simple metallic system Al is considered to be interesting. For experimental measurements of the ΔT^+-ΔT^- relationship, a precise control of temperature and heating/cooling rate is needed, and extraneous effects (e.g., oxidation) should be excluded [3,4]. Meanwhile, in bulk continuous systems, the factors affecting the solidification process are rather accidental, and the crystallization centers in one region may have an effect on the freezing behavior of the whole. In this study, we prepared samples of Al particles encapsulated in Al_2O_3 shells to eliminate possible effects of oxidation and coalescence at high temperatures. By using differential scanning calorimeter (DSC), we quantitatively measured the ΔT^+-ΔT^- relationship of encapsulated Al particles and found a remarkable dependence of ΔT^- on ΔT^+ that extends to ultrahigh ΔT^+ levels.

2. Materials and Methods

Al small particles were prepared by means of active H_2 plasma evaporation and condensation, using bulk Al (with a purity of 99.8%) as the starting material. The particles were *in situ* passivated at room temperature before exposure to air. The passivated Al particles were further oxidized in air to produce a thick surface oxide shell. Samples with different oxidation extents were produced by changing the temperature/duration of oxidation.

The morphology and size of Al particles was investigated by a transmission electron microscope (TEM, Philips, Amsterdam, The Netherlands) and a high-resolution transmission electron microscope (HRTEM, JEOL, Tokyo, Japan), which was conducted on a Philips EM 420 microscope (Philips) with an accelerating voltage of 100 kV and a JEM 2010 high-resolution microscope (JEOL) with an accelerating voltage of 200 kV, respectively. Samples for TEM and HRTEM observations were prepared by dispersing the particles in ethanol by sonication and then, dropping on a carbon-coated copper grid. Formation of an oxide shell on the Al particle surface was confirmed by TEM and HRTEM observations. The surface oxide shell was found to be able to retain the shape of Al particles effectively upon heating and cooling so that the particle can melt and freeze independently.

Thermal analysis was performed on a Netzsch high-temperature differential scanning calorimeter (DSC 404C, Netzsch, Selb, Germany). The sample in Al_2O_3 crucible was measured in a dynamic Ar atmosphere with a gas flow rate of 50 mL/min and a heating/cooling rate of 20 °C/min. The temperature and enthalpy were calibrated by the melting endotherms of pure In, Al and Ag.

3. Results and Discussion

As shown in Figure 1a, the original Al particles are nearly spherical in shape with sizes mostly ranged from ~30 to ~500 nm. Al particles were encapsulated in Al_2O_3 shells after thermal oxidation at 500 °C for 90 min in air (Figure 1b,c). X-ray diffraction (XRD) analysis also confirmed the formation of oxide shells (Figure 1d). The Al_2O_3 shells are able to prevent the coalescence of Al nanoparticles upon heating and retain the particles shape even at temperatures well above T_0 [17,18], allowing for the independent melting and solidification of encapsulated Al particles.

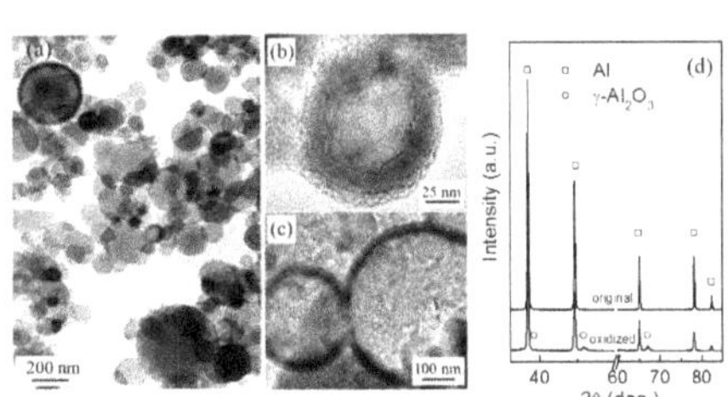

Figure 1. (**a**) A bright-field transmission electron microscope (TEM) image of the as-evaporated (original) Al particles; (**b**,**c**) high-resolution transmission electron microscope (HRTEM) images of small and large Al particles oxidized at 500 °C for 90 min, showing existence of the surface oxide shell; (**d**) XRD profiles of the original and oxidized (at 500 °C for 90 min) Al particles.

The melting and solidification behaviors of encapsulated Al particles were measured by DSC. As shown in Figure 2a, melting endotherms of encapsulated Al particles are quite similar after several heating/cooling cycles, showing a reduced melting temperature (onset temperature of the melting peak: ~650 °C) due to the size-dependent melting point depression [19]. It is noted that previous studies [20–23] indicated the important effects of stress relaxation and damage of the oxide shell on the melting of Al nanoparticles coated by oxide shells, while in the present study the good repeatability of melting endotherms suggests the good thermal stability of encapsulated Al particles after several DSC thermal cycles, which may be attributed to the thicker oxide shells that were formed by thermal oxidation in the present study. Figure 2b demonstrates DSC curves of the Al particles cooled from different liquid temperatures/overheating levels. All DSC curves consist of two main exothermic peaks, one with small supercooling (P1) and another with larger supercooling (P2). Most importantly, the solidification behavior shows an evident dependence on the liquid temperature. For a quantitative illustration, we take ΔT^- as the difference between the onset temperature of P1, and T_0 of Al and ΔT^+, as that between the liquid temperature and T_0. As shown in Figure 2c, an evident dependence of ΔT^- on ΔT^+ is identified: ΔT^- increases with the increasing of ΔT^+, most evidently in the ΔT^+ range of 140–340 °C. In Figure 2c, the relative area ratio of the two peaks ($A1/A2$) also shows a strong dependence on ΔT^+, indicating that when cooled from higher liquid temperatures, more Al particles solidify at larger supercoolings. In our experiments, the holding time at which the liquid was kept at ΔT^+ (up to 1 h) was found to have little effect on the ΔT^+-ΔT^- relationship, and the supercooling behavior was only related to the highest temperature to which the liquid was heated, regardless of the intermediate temperatures at which it was held before solidification, in agreement with predictions by the cavity theory [14]. Compared to previous experiments using fast scanning calorimetry with a ultrahigh cooling rate up to the order of 10^5 °C/min [2,14,24,25], the cooling rate in the present study is very small (20 °C/min), which can exclude the cooling rate effect on the solidification of Al particles cooled from different liquid temperatures and rationalize the independence of the holding time at ΔT^+. To confirm the reproducibility of the results and clarify possible effects of the morphologies changes on the oxidized Al particles induced by the high-temperature annealing/heating on the freezing behavior, we carried out several heating-cooling cycles for different heating temperatures of the liquid and obtained DSC curves immediately after the 1200 °C heating-cooling cycle. As shown in Figure 3, ΔT^- still shows an evident dependence on ΔT^+ regardless of the previous 1200 °C heating-cooling cycle, indicating that the freezing behavior is only dependent on the maximum liquid heating temperature before solidification. We also carried out DSC measurements on samples with different oxidation degrees prepared by changing oxidation temperature/duration, all showing a similar ΔT^+-ΔT^- relationship, *i.e.*, ΔT^- increases with increasing ΔT^+ (Figure 4a–c). To our knowledge, such ΔT^+-ΔT^- relationship has not been reported in Al before. Moreover, the ΔT^+ level (up to 340 °C) to which the dependence of ΔT^- on ΔT^+ extends in the present study is significantly larger than that observed in other metals and alloys (typically several to tens of degrees) [14].

For comparison, bulk Al samples with different impurity levels were also tested, as shown in Figure 4d–f. For bulk Al sample with a purity of 99.999%, the solidification behavior varies randomly for each heating-cooling cycle, which is due to the accidental nature of catalysts for nucleation of solidification. For bulk Al sample with a purity of 99.8%, the DSC cooling curves are similar for different ΔT^+s, with a smaller supercooling than bulk Al sample with a higher purity level (99.999%), which simply can be attributed to the higher content of catalysts for nucleation. The supercooling behavior of the consolidated sample, made by cold-pressing the oxidized Al particles, remains almost unchanged for different ΔT^+s although two exothermic peaks are also observed, *i.e.*, no dependence of ΔT^- on ΔT^+ as that in the powder sample is observed. This is possible because, as mentioned above, in bulk continuous systems, crystallization centers in one region may affect the solidification behavior of the whole. It is also shown in Figure 4 that the solidification temperatures of P1 of the powder sample, the consolidated sample, and bulk Al sample with a purity of 99.8% are similar but different from that of the bulk Al sample with a purity of 99.999%, indicating the important effect of impurities.

For a particle/droplet, different nucleation mechanisms, namely bulk and surface nucleation, can have significant effects on the solidification behavior [24,25]. To identify the two solidification processes of Al particles, the sample was heated to different temperatures followed by cooling down so that only a part of Al particles in the sample were melted and solidified. As shown in Figure 5, Al particles melted at lower temperatures tend to solidify at larger supercoolings (P2), and those with higher melting temperatures at smaller supercoolings (P1). Since smaller particles have lower melting points than larger ones due to the size effect [19], it is believed that P1 corresponds to the solidification of larger Al particles while P2 corresponds to smaller ones. Previous studies [26,27] showed that the multi-stage solidification behaviors in various embedded particle/matrix systems can be related to the different catalyzing effects of impurities and matrix/coating. Hereby, we can further conclude that P1 corresponds to the solidification of larger particles (with more impurities) related to the catalyzing effect of impurities while P2 corresponds to smaller particles (with less impurities) catalyzed by the coating. It is noted that previous phase field simulations [28] indicated that Al melt can completely wet alumina with a contact angle of zero. However, interfacial energy between Al and Al_2O_3 is also dependent on factors such as crystallography, temperature, impurity, *etc*. Here, it is believed that upon solidification, heterogeneous nucleation on the Al/Al_2O_3 interface would be more likely than homogenous nucleation, as long as upon nucleation the Al/Al_2O_3 interface is formed so that the contact angle of solid Al on Al_2O_3 is between 0° and 180°, by e.g., forming a favored local crystallography. The relative change in peak areas shown in Figure 2c where the increasing ΔT^+ is due to the modification of the two nucleation sites upon elevated temperature exposure: bulk heterogeneous nucleation (P1) and surface/interface heterogeneous nucleation (P2). The above analysis indicates that the dependence of ΔT^- on ΔT^+ of encapsulated Al particles observed in the present experiment is related to effect of impurity (alloying) atoms despite the minute amount. Hereby, a critical problem is how the liquid overheating can lead to a transfer between the two nucleation mechanisms, *i.e.*, how bulk heterogeneous nucleation is suppressed by liquid overheating.

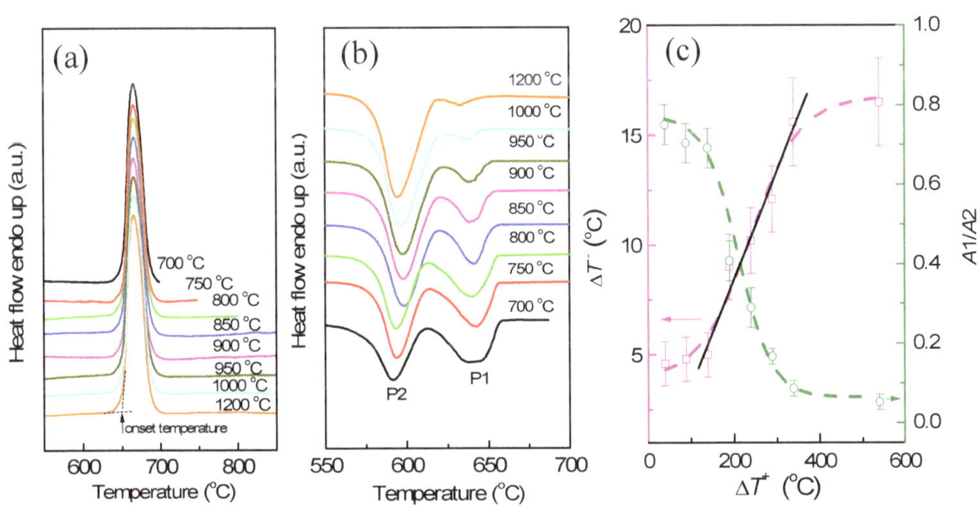

Figure 2. (**a**) Differential scanning calorimeter (DSC) melting curves of the encapsulated Al particles (oxidized at 500 °C for 90 min) heated to different temperatures as indicated, showing the good repeatability of melting endotherms after several thermal cycles. The definition of the onset temperature of DSC peak is indicated in (a); (**b**) DSC curves of the encapsulated Al particles (oxidized at 500 °C for 90 min) cooled from different temperatures as indicated, showing variations of the two freezing exotherms P1 and P2 with liquid temperature; (**c**) Variations of ΔT^- of P1 and relative area ratio of P1 to P2 ($A1/A2$) as functions of ΔT^+. In (c), the dashed lines are guides for eyes, and the solid line is a linear fitting of the data in the ΔT^+ range of 140–340 °C.

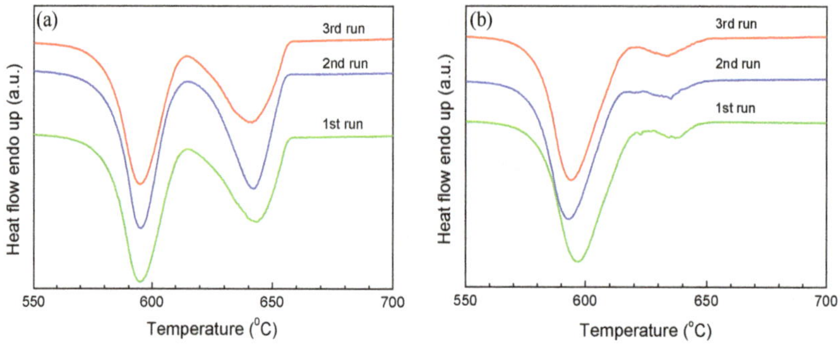

Figure 3. DSC cooling curves of the encapsulated Al particles (oxidized at 500 °C for 90 min) for different liquid overheating temperatures (**a**) 800 °C and (**b**) 1000 °C. Each run was performed immediately after a previous 1200 °C heating-cooling cycle.

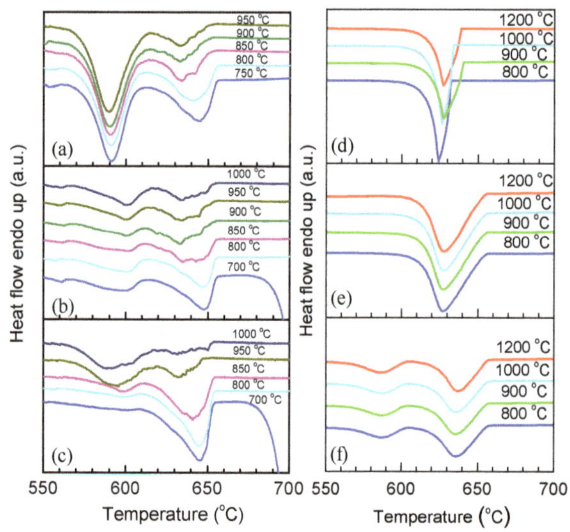

Figure 4. DSC curves corresponding to the cooling from different temperatures as indicated for different samples. (**a**) Al particles oxidized at 700 °C for 90 min; (**b**) Al particles oxidized at 800 °C for 180 min; (**c**) Al particles oxidized at 900 °C for 180 min; (**d**) 99.999% bulk Al; (**e**) 99.8% bulk Al; (**f**) Consolidated samples of oxidized Al particles (at 500 °C for 90 min) made by cold pressing.

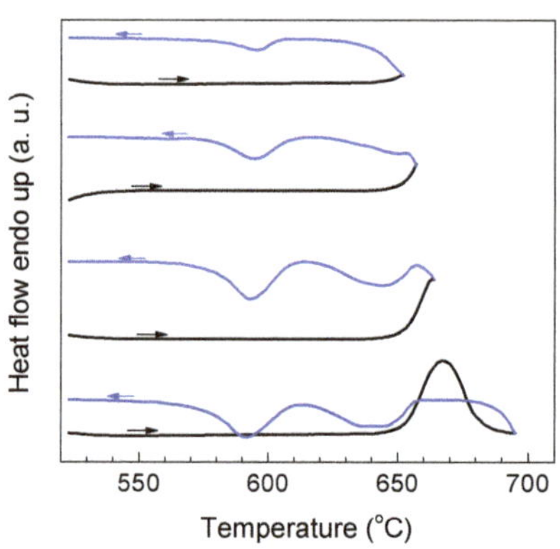

Figure 5. DSC partial melting-cooling curves for the encapsulated Al particles (oxidized at 500 °C for 90 min). Melting of the encapsulated Al particles was interrupted at different temperatures (as indicated) for 60 s before the cooling step.

The dependence of ΔT^- on ΔT^+ obtained in our study indicates the existence of temperature-dependent crystals in liquid Al up to a very high ΔT^+ (~340 °C), unlike the transient short range orders as in the cases of semimetals and semiconductors [11,12]. Previous studies indicated that the crystals confined by a coating/matrix can be superheated metastably providing epitaxial interfaces are formed [19,29,30]. However, in our study, the Al-Al_2O_3 interface is non-epitaxial. Although a small pressure-induced superheating up to ~15 °C was observed [31], it cannot obviously account for the very large superheating (~340 °C) that is close to the inverse Kauzmann point (highest superheating limit) [32] in the present study. As mentioned above, the cavity theory [14] provided a possible explanation for the dependence of ΔT^- on ΔT^+. According to the cavity theory [14], a crystal contained in a cavity with a radius of r can remain unmelted at a certain liquid overheating level (ΔT^+), given by:

$$\Delta T^+ = \frac{2T_0\sigma_{sl}\cos\theta}{\Delta H_m r} \tag{1}$$

where σ_{sl} is the solid-liquid interfacial energy of the host metal, θ is the contact angle between the cavity and the host liquid, and ΔH_m is the melting enthalpy of the host metal. The cavity theory predicted a linear relationship between ΔT^+ and ΔT^- [14]:

$$\Delta T^-/\Delta T^+ = \tan\theta \tag{2}$$

In the present experiment, the ΔT^+-ΔT^- relationship in the ΔT^+ range of 140 to 340 °C is approximately linear, as shown in Figure 2. By linear fitting of the experimental data according to Equation (2), we get $\theta = 2.6°$. Meanwhile, for ΔT^+_{max}= 340 °C and ΔT^+_{min}= 140 °C with available data of ΔH_m and σ_{sl} for Al [33], we get r_{min} = 0.6 nm and r_{max} = 1.5 nm from Equation (1), corresponding to the radius of the smallest and largest cavities, respectively. However, the existence of cavities with such extremely small sizes and contact angles is hardly reasonable, indicating the insufficiency of the cavity theory for interpreting the dependence of ΔT^- on ΔT^+ extending to ultrahigh ΔT^+ in the present experiment. In a Ni-base superalloy, existence of Ni_3(Al,Ti,Nb)-like clusters and residual MC (M = Nb, Ti) carbide in the alloy melt superheated to 1500 °C (170 °C above the liquidus temperature) was detected by high temperature X-ray diffraction, resulting in a liquid thermal history effect on solidification [6]. Hereby, a possible consideration is that the persisted crystals are not likely pure Al but alloyed clusters containing Al and other (impurity) atoms, as indicated by the impurity-related evidence of the dependence of ΔT^- on ΔT^+ in the present study. However, for a better interpretation of the present results, direct experimental evidence for the size, surface energy, and melting point of these clusters is needed.

4. Conclusions

In summary, through quantitative experimental measurements, we found a remarkable dependence of ΔT^- on ΔT^+ extending to ultrahigh ΔT^+ levels in Al small particles encapsulated in Al_2O_3 shells. While the analysis pointed to the existence of temperature-dependent crystallization centers in liquid Al at very high liquid temperatures, the cavity theory was found to be insufficient to interpret the present results. The present study highlighted quantitatively the significant effect of liquid overheating levels on the supercooling behavior of Al and Al alloys, and raised new considerations for the dependence of liquid supercooling on liquid overheating at very high liquid overheating levels.

Acknowledgments: Financial supports from the National Natural Science Foundation of China (Grant 51371128) and Wuhan University (Grant 410100018) are gratefully acknowledged.

Author Contributions: Qingsong Mei conducted the experiments. Qingsong Mei and Juying Li contributed to the analyses and discussion of the results and prepared the manuscript.

Conflicts of Interest: The authors declare no conflict of interest.

References

1. Fine, M.E. *Introduction to Phase Transformations in Condensed Systems*; Macmillan: New York, NY, USA, 1964.
2. Yang, B.; Perepezko, J.H.; Schmelzer, J.W.P.; Gao, Y.L.; Schick, C. Dependence of crystal nucleation on prior liquid overheating by differential fast scanning calorimeter. *J. Chem. Phys.* **2014**, *140*, 104513. [CrossRef] [PubMed]
3. Webster, W.L. Phenomena occurring in the melting of metals. *Proc. R. Soc. Lond. Ser. A* **1933**, *140*, 653–660. [CrossRef]
4. Richard, W.T. The persistence and development of crystal nuclei above the melting temperature. *J. Am. Chem. Soc.* **1932**, *54*, 479–495. [CrossRef]
5. Luzny, W. Kinetics of heterogeneous nucleation and the "memory effect" in liquid gallium. *J. Phys. Condens. Matter* **1990**, *2*, 10183–10190. [CrossRef]
6. Yin, F.S.; Sun, X.F.; Guan, H.R.; Hu, Z.Q. Effect of thermal history on the liquid structure of a cast nickel-base superalloy M963. *J. Alloy. Compd.* **2004**, *364*, 225–228. [CrossRef]
7. Koh, H.J.; Rudolph, P.; Schafer, N.; Umetsu, K.; Fukuda, T. The effect of various thermal treatments on supercooling of PbTe melts. *Mater. Sci. Eng. B* **1995**, *34*, 199–203. [CrossRef]
8. Muhlbug, M.; Rudolph, P.; Laasch, M. The correlation between superheating and supercooling in CdTe melts during unseeded bridgman growth. *J. Cryst. Growth* **1993**, *128*, 571–575. [CrossRef]
9. Rudolph, P.; Koh, H.J.; Schafer, N.; Fukuda, T. The crystal perfection depends on the superheating of the mother phase too—Experimental facts and speculations on the "melt structure" of semiconductor compounds. *J. Cryst. Growth* **1996**, *166*, 578–582. [CrossRef]
10. Tong, H.Y.; Shi, F.G. Dependence of supercooling of a liquid on its overheating. *J. Chem. Phys.* **1997**, *107*, 7964–7966. [CrossRef]
11. Tong, H.Y.; Shi, F.G. Abrupt discontinuous relationships between supercooling and melt overheating. *Appl. Phys. Lett.* **1997**, *70*, 841–843. [CrossRef]
12. Zhou, Z.; Wang, W.; Sun, L. Undercooling and metastable phase formation in a Bi95Sb5 melt. *Appl. Phys. A* **2000**, *71*, 261–265. [CrossRef]
13. Zhao, B.; Li, L.; Lu, F.; Zhai, Q.; Yang, B.; Schick, C.; Gao, Y. Phase transitions and nucleation mechanisms in metals studied by nanocalorimetry: A review. *Thermochim. Acta* **2015**, *603*, 2–23. [CrossRef]
14. Turnbull, D. Kinetics of heterogeneous nucleation. *J. Chem. Phys.* **1950**, *18*, 198–203. [CrossRef]
15. Li, P.; Nikitin, V.I.; Kandalova, E.G.; Nikitin, K.V. Effect of melt overheating, cooling and solidification rates on Al–16wt.%Si alloys. *Mater. Sci. Eng. A* **2002**, *332*, 371–374. [CrossRef]
16. Xu, C.L.; Jiang, Q.C. Morphologies of primary silicon in hypereutectic Al–Si alloys with melt overheating temperature and cooling rate. *Mater. Sci. Eng. A* **2006**, *437*, 451–455. [CrossRef]
17. Révész, Á.; Lendvai, J.; Ungár, T. Melting point depression and microstructure in ball-milled nanocrystalline aluminium powders. *J. Metastab. Nanocryst. Mater.* **2000**, *8*, 326–331. [CrossRef]
18. Echert, J.; Holzer, J.C.; Ahn, C.C.; Fu, Z.; Johnson, W.L. Melting behavior of nanocrystalline aluminum powders. *Nanostruct. Mater.* **1993**, *2*, 407–413. [CrossRef]
19. Mei, Q.S.; Lu, K. Melting and superheating of crystalline solids: From bulk to nanocrystals. *Prog. Mater. Sci.* **2007**, *52*, 1175–1262. [CrossRef]
20. Sun, J.; Simon, S.L. The melting behavior of aluminum nanoparticles. *Thermochim. Acta* **2007**, *463*, 32–40. [CrossRef]
21. Levitas, V.I.; Pantoya, M.L.; Chauhan, G.; Rivero, I. Effect of the alumina shell on the melting temperature depression for aluminum nanoparticles. *J. Phys. Chem. C* **2009**, *113*, 14088–14096. [CrossRef]
22. Ruffino, B.; Coulet, M.V.; Bouchet, R.; Isnard, O.; Denoyel, R. Structural changes and thermal properties of aluminium micro- and nano-powders. *Acta Mater.* **2010**, *58*, 4224–4232. [CrossRef]
23. Firmansyah, D.A.; Sullivan, K.; Lee, K.S.; Kim, Y.H.; Zahaf, R.; Zachariah, M.R.; Lee, D. Microstructural behavior of the alumina shell and aluminum core before and after melting of aluminum nanoparticles. *J. Phys. Chem. C* **2012**, *116*, 404–411. [CrossRef]
24. Yang, B.; Abyzov, A.S.; Zhuravlev, E.; Gao, Y.L.; Schmelzer, J.W.P.; Schick, C. Size and rate dependence of crystal nucleation in single tin drops by fast scanning calorimetry. *J. Chem. Phys.* **2013**, *138*, 054501. [CrossRef] [PubMed]

25. Yang, B.; Gao, Y.L.; Zou, C.D.; Zhai, Q.J.; Abyzov, A.S.; Schmelzer, J.W.P.; SchickSize, C. Cooling rate dependence of undercooling of pure Sn single drop by fast scanning calorimetry. *Appl. Phys. A* **2011**, *104*, 189–196. [CrossRef]
26. Moore, K.I.; Zhang, D.L.; Cantor, B. Solidification of Pb particles embedded in Al. *Acta Metall. Mater.* **1990**, *38*, 1327–1342. [CrossRef]
27. Goswami, R.; Chattopadhyay, K.; Kim, W.T.; Cantor, B. Heterogeneous nucleation of Pb particles embedded in a Zn matrix. *Metall. Trans. A* **1992**, *23*, 3207–3218. [CrossRef]
28. Levitas, V.I.; Samani, K. Melting and solidification of nanoparticles: Scale effects, thermally activated surface nucleation, and bistable states. *Phys. Rev. B* **2014**, *89*, 075427. [CrossRef]
29. Cahn, R. Melting and the surface. *Nature* **1986**, *323*, 668–669. [CrossRef]
30. Sheng, H.W.; Ren, G.; Peng, L.M.; Hu, Z.Q.; Lu, K. Superheating and melting-point depression of Pb nanoparticles embedded in Al matrices. *Philos. Mag. Lett.* **1996**, *73*, 179–186. [CrossRef]
31. Mei, Q.S.; Wang, S.C.; Cong, H.T.; Jin, Z.H.; Lu, K. Pressure-induced superheating of Al nanoparticles encapsulated in Al_2O_3 shells without epitaxial interface. *Acta Mater.* **2005**, *53*, 1059–1066. [CrossRef]
32. Fecht, H.J.; Johnson, W.L. Entropy and enthalpy catastrophe as a stability limit for crystalline material. *Nature* **1988**, *334*, 50–51. [CrossRef]
33. Pluis, B.; Frenkel, D.; van der Veen, J.F. Surface-induced melting and freezing II. A semi-empirical Landau-type model. *Surf. Sci.* **1990**, *239*, 282–300.

7

Enhanced Physicochemical and Biological Properties of Ion-Implanted Titanium Using Electron Cyclotron Resonance Ion Sources

Csaba Hegedűs [1], Chia-Che Ho [2], Attila Csik [3], Sándor Biri [3,*] and Shinn-Jyh Ding [2,4,*]

Academic Editor: Jung Ho Je

1 Department of Biomaterials and Prosthetic Dentistry, University of Debrecen, Nagyerdei krt. 98, H-4032 Debrecen, Hungary; chegedus@edu.unideb.hu
2 Institute of Oral Science, Chung Shan Medical University, Taichung City 402, Taiwan; sfox1223@gmail.com
3 Institute for Nuclear Research (ATOMKI), Hungarian Academy of Sciences, Bem tér 18/c, H-4026 Debrecen, Hungary; csik.attila@atomki.mta.hu
4 Department of Dentistry, Chung Shan Medical University Hospital, Taichung City 402, Taiwan
* Correspondence: biri@atomki.mta.hu (S.B.); sjding@csmu.edu.tw (S.-J.D.)

Abstract: The surface properties of metallic implants play an important role in their clinical success. Improving upon the inherent shortcomings of Ti implants, such as poor bioactivity, is imperative for achieving clinical use. In this study, we have developed a Ti implant modified with Ca or dual Ca + Si ions on the surface using an electron cyclotron resonance ion source (ECRIS). The physicochemical and biological properties of ion-implanted Ti surfaces were analyzed using various analytical techniques, such as surface analyses, potentiodynamic polarization and cell culture. Experimental results indicated that a rough morphology was observed on the Ti substrate surface modified by ECRIS plasma ions. The *in vitro* electrochemical measurement results also indicated that the Ca + Si ion-implanted surface had a more beneficial and desired behavior than the pristine Ti substrate. Compared to the pristine Ti substrate, all ion-implanted samples had a lower hemolysis ratio. MG63 cells cultured on the high Ca and dual Ca + Si ion-implanted surfaces revealed significantly greater cell viability in comparison to the pristine Ti substrate. In conclusion, surface modification by electron cyclotron resonance Ca and Si ion sources could be an effective method for Ti implants.

Keywords: titanium; ion implantation; calcium; silicon; implant

---

1. Introduction

Titanium and titanium alloys are the metallic materials most commonly used in load-bearing applications because of their superior mechanical properties. Poor bioactivity is a significant shortcoming of metallic implants, which results in the failure of bonding directly to bone tissues. Many studies have been done for improving and optimizing metallic implant surfaces by creating bioactive coatings or modifying the surfaces, because the biomaterial surface plays a crucial role in biological interaction. To ensure successful medical use, the key physical properties of a metal implant material should be retained, while the outermost surface is modified to affect the interaction between the implant and its human host. Surface modification methods applied to the inert implants include texturing/passivation, anodization, ion implantation, silanization and deposition processing [1–7]. Surface texturing, such as sandblasting and acid etching, could increase the surface area of the implant, thereby enhancing osseointegration by increasing the area to which bone can bond [1]. In ion implantation, high-energy ions bombard the implant surface and enhance the corrosion resistance and/or biocompatibility of the titanium through the formation of a ceramic surface layer [2].

The incorporation of osteoconductive or osteoinductive molecules on Ti surfaces is one of the effective modifications [7–11]. It has been documented that extracellular Ca is a potent regulator of cell behavior and has significant effects on the proliferation and differentiation of osteoblasts [12]. Sawada *et al.* found that $Ca(OH)_2$-treated titanium disks induced osteogenic differentiation in human mesenchymal stem cells [10]. Titanium implant surfaces modified with calcium ions using sonication dipping stimulate platelet adhesion and activation and provisional matrix formation [11]. Hanawa and co-workers found that Ca ion-implanted titanium in an amount of 10^{17} ions/cm^2 was superior to titanium alone for bone conduction after implantation in rat tibia [13]. In addition to Ca ion, Si species can enhance attachment, proliferation, differentiation and mineralization of osteoblast-like cells and mesenchymal stem cells [7,14], as well as osteogenesis *in vivo* [15].

It is not easy to find a suitable facility that is able to deliver calcium, silicon and other ion beams with the required charge, intensity and energy. During the process of finding the most optimal solution, finally, a special device is chosen, which has been used mainly in physical research and up-to-date has not spread in medical research. The electron cyclotron resonance (ECR) ion source (ECRIS) is used world-wide to deliver highly charged ions for particle accelerators to get very high energy beams at the end. Their many unique features make them one of the most promising devices to produce a wide variety of low energy ion beams for direct surface treatments. ECRIS can produce ion beams from many gaseous or solid materials. By changing the ion extraction voltage of the source and/or the charge state distribution of the plasma, the implantation depth of the ions into the materials can be varied easily. Usually, it should be between 1 and 100 nm, which is available applying standard terminal voltages. Thus, the surface properties of the materials only are modified, the bulk mechanical properties remaining unchanged. The Atomki-ECRIS [16,17] seems to be a suitable device for this task, because it is an independent ion source (there is no post-acceleration) offering beam time freely. Furthermore, the modularity of this source opens the possibility to change its configuration within a reasonable time. In this study, ECRIS consisting of Ca or Ca + Si ions was used to modify the Ti surfaces. Calcium and silicon ion vapors are ionized by electrons heated by microwave radiation. The microstructure, corrosion resistance and biocompatibility of ion-implanted Ti substrates were characterized.

2. Experimental Section

2.1. Preparation of Substrate

Grade 2 commercially available 1 mm-thick titanium plates (99.6 at%, Grade 2, Spemet Co., Taipei, Taiwan) of 10×10 mm^2 were selected as the substrate materials. Prior to ion implantation, the substrate surface was mechanically polished to #2000 grit level, followed by 1-μm Al_2O_3 powder to produce a mirror-like surface. The substrates were then etched in 30% HNO_3 for 30 min at room temperature, followed by ultrasonic cleaning in ethanol for 30 min, rinsing with distilled water for 30 min and air drying.

2.2. Ion Implantation

The Atomki-ECRIS (Room-temperature, 14.3 GHz; Atomki, Debrecen, Hungary) system can produce a variety of charged ion beams from H, He, CO, CH4, N, O, Ne, Ar, Kr and Xe gases and C, C_{60}, Fe, Ni, Zn and Au ions from solids. However, in the present study, new methods for calcium and silicon ion beam production were necessary to develop. A calcium plasma and ion beam was obtained using a commercial filament oven (PK10-0107-DT01, Pantechnik, Bayeux, France). The oven head was placed close to the plasma chamber of the ECRIS system and filled by pure calcium. Due to the vapor pressure of the calcium, the optimal operation temperature was found between 500 and 700 °C. Helium was used as the support gas. Multiply-charged calcium plasma was generated with some He background. A typical calcium ion beam spectrum is shown in Figure 1 [17]. The plasma was optimized Ca^{3+} with the maximum extracted Faraday cup current (FCU) of about 20 eμA.

To obtain a silicon ion beam, we used SiH_4 (silane) gas. Special care had to be applied because of its highly flammable feature. A special gas handling system was designed to transfer the silane gas from a high volume and high pressure (2 dm^3 and 50 bar, respectively) gas bottle to a smaller (50 cm^3 and 1.5 bar, respectively) one. The small bottle was connected to the ion source through a gas dosing valve, and a silicon-hydrogen mixture ion beam was produced. Due to the safety arrangements, the risk was reduced to the normal operation level. For low and middle charge states (Q = 1 to 8), silicon beam currents between 25 and 100 eμA were easily obtained (Figure 2), which gave us enough freedom to vary the kinetic energy of the beam and, thus, to regulate the implantation depth [17].

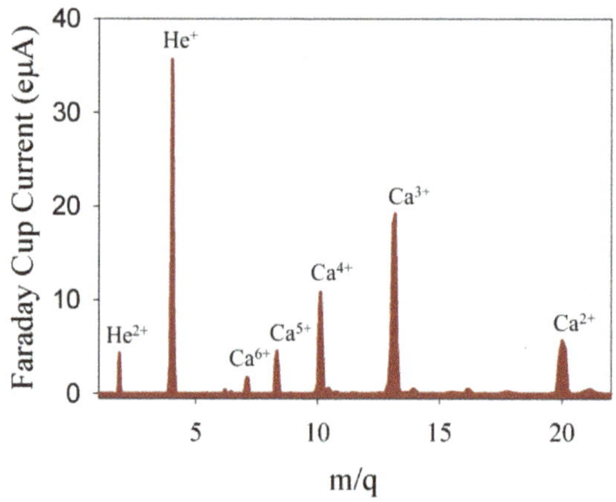

Figure 1. Calcium ion beam spectra. The extraction voltage and the temperature of the oven were 5 kV and 700 °C, respectively. The plasma was optimized for Ca^{3+}.

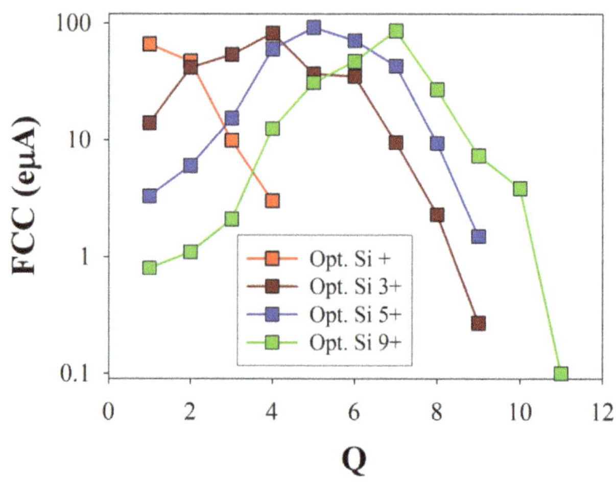

Figure 2. Charge state distributions of the analyzed silicon ion beam spectra as a function of the optimized charge states.

Besides the development of the necessary ion beams (Ca and Si), a new irradiation chamber was also designed and manufactured. Thus, the exchange of the materials to be irradiated became fast and easy without breaking the vacuum. The size of the ion beam was about 50 mm in diameter and circular. In order to reach as high Ca and Si dose as possible, the irradiation facility was built very close to the ECRIS. By this arrangement, all of the extracted components of the plasma reach and hit the samples, of course. We believe, however, that the effect of the unnecessary, but unavoidable components of the beam (mainly H and He) is negligible. The extraction voltage was 10 kV, and the ECRIS system was tuned for Ca^{2+} or Si^{2+} production. This meant that ions with charges from 1 up to 4 or 5 hit the samples, resulting in different implantation depths. Ti samples were implanted by 5×10^{16} ion/cm^2 Ca, 1×10^{17} ion/cm^2 Ca and 5×10^{16} ion/cm^2 Ca + 5×10^{16} ion/cm^2 Si dose, respectively, which were also denoted as 5Ca, 10Ca and 5CaSi, respectively.

2.3. *Composition and Morphology*

A secondary neutral mass spectrometry (SNMS) system (INA-X, SPECS GmbH, Berlin, Germany) was used for the depth profiling of the samples. Phase analysis of sample surfaces with and without ion implantation was performed using a grazing incidence X-ray diffractometer (GIXRD; Bruker D8 SSS, Bruker Corporation, Karlsruhe, Germany) with Ni-filtered CuKα radiation operating at 40 kV and 100 mA with a glancing incident angle at 0.5° and a scanning speed of 1 °/min. The chemical structure of the samples was analyzed using Fourier transform infrared spectroscopy (FTIR; Bruker Vertex 80v, Bruker Corporation) with a spectral resolution of 1 cm^{-1} in reflection at grazing incidence $\theta = 80°$ and a wavenumber range of 400 to 4000 cm^{-1}. Surface morphologies of various samples were coated with gold and observed under a field emission scanning electron microscope (SEM; JEOL JSM-6700F, JEOL Ltd., Tokyo, Japan). The Surfcorder SE-40G profilometer (Kosaka Laboratory, Tokyo, Japan) was used to measure the surface roughness (Ra; arithmetic mean roughness) with a cutoff value of 0.25 mm, a measurement length of 1 mm and a drive speed of 0.1 mm/s. The Ra data provided for each group were the mean of five independent measurements.

2.4. *Corrosion Measurement*

The corrosion measurements included open circuit potential (OCP) time methods and potentiodynamic polarization in a non-deaerated simulated body fluid (SBF) solution, using a CHI 660A electrochemical system (CH Instrument, Austin, TX, USA). The SBF solution, the ionic composition of which is similar to that of human blood plasma, consisted of 7.9949 g NaCl, 0.3528 g $NaHCO_3$, 0.2235 g KCl, 0.147 g K_2HPO_4, 0.305 g $MgCl_2 \cdot 6H_2O$, 0.2775 g $CaCl_2$ and 0.071 g Na_2SO_4 in 1000 mL distilled H_2O and was buffered to pH 7.4 with hydrochloric acid (HCl) and tris-hydroxymethyl aminomethane (Tris, $CH_2OH)_3CNH_2$) [18]. All chemicals used were of reagent grade and used as obtained. For OCP measurement, only two electrodes (working electrode and reference electrode) were involved, whereas for the potentiodynamic polarization method, a conventional three-electrode cell was used. A saturated calomel reference electrode (SCE) and a platinum counter electrode were employed. The sample surface was cleaned by distilled water. The evaluation of potentiodynamic polarization was started after immersion in SBF for 1 h. The scanned potential range varied from −1 to 1 V toward the anodic direction at a sweep rate of 1 mV/s in the Tafel mode. The current was recorded in the absence of stirring or gas bubbling into the electrolyte. The corrosion potential and corrosion current were provided after being analyzed by the software. The results were obtained from five separate experiments.

2.5. *Hemocompatibility*

A 5-mL whole-blood sample obtained from a healthy human donor was added to 10 mL of phosphate-buffered saline (PBS). The red blood cells (RBCs) were harvested by centrifugation of the diluted blood solution at 10,400 rpm for 10 min and were then washed with PBS 5 times. A total of 5×10^7 RBCs in 1 mL of PBS was incubated with different samples in 24-well plates for 6 h at 37 °C. Then, the solution was centrifuged to obtain the supernatant; the absorbance of the supernatant at 570 nm was measured by a Sunrise microtiter plate reader (Tecan Austria Gesellschaft, Salzburg, Austria) with a reference of 655 nm. The absorbance of RBCs in deionized water and PBS served as positive (Cp) and negative controls (Cn), respectively. The hemolysis ratio was calculated using the following equation: hemolysis (%) = $(A - Cn)/(Cp - Cn) \times 100\%$, where A represents the absorbance of the RBC solution incubated with the samples [19]. The presented data are representative of three independent experiments.

2.6. Cytoskeleton Observation

The implant biocompatibility was evaluated by incubating the samples with MG63 human osteoblast-like cells (BCRC 60279, Hsinchu, Taiwan). Before cell incubation, the samples were sterilized by soaking in a 75% ethanol solution and exposure to ultraviolet light for 2 h. MG63 cells were suspended in Dulbecco's Modified Eagle's Medium (DMEM; Gibco, Langley, OK, USA) containing 10% fetal bovine serum (FBS; GeneDireX, Las Vegas, NV, USA) and 1% penicillin/streptomycin solution (Caisson, North Logan, UT, USA). The MG63 suspensions (2×10^4 cells per well) were directly seeded over each sample and placed in a 24-well plate. To clarify the effects of the various ion-implanted surfaces on cell attachment, the actin cytoskeleton was analyzed via fluorescence microscopy. After 6 h of incubation, the unbound cells were washed off with cold PBS, and the adherent cells were fixed in 4% p-formaldehyde (Sigma-Aldrich, St. Louis, MO, USA) for 30 min at room temperature and permeabilized with 0.1% Triton X-100 (Sigma-Aldrich) in PBS for 10 min. The cells were then incubated with phalloidin conjugated to Alexa Fluor 594 (Invitrogen, Grand Island, NY, USA) to stain the cytoskeleton for 1 h. Next, the nuclei was stained with 300 nM 4',6-diamidino-2-phenylindole (DAPI) (Invitrogen) for 1 h. After washing three times with PBS, the cells were viewed under indirect fluorescence using a Zeiss Axioskop2 microscope (Carl Zeiss, Thornwood, NY, USA) at 200× magnification.

2.7. Cell Viability

Cell viability was performed with an initial cell density of 5000 cells/well on Days 1, 3 and 7. Fresh culture medium was replaced every 2 days. At the end of the culture time, the medium was removed, and the samples were washed twice with PBS. A total of 650 μL of 0.5 mg/mL MTT ((3-(4,5-dimethylthiazol-2-yl)-2,5-diphenyltetrazolium bromide; Sigma-Aldrich) in culture medium was added and incubated at 37 °C for 2.5 h. Subsequently, 500 μL of dimethylsulfoxide (Sigma-Aldrich) was added, and the absorbance at 570 nm was determined using a Sunrise Microtiter Reader. The data provided for each group were the mean of three independent samples.

2.8. Statistical Analysis

One-way analysis of variance (ANOVA) was used to evaluate significant differences between means in the measured data. In the event of a significant difference between test groups, this necessitated testing all of the possible differences via multiple comparisons that were characterized by considering any significant differences between all possible pairs of groups. Scheffé's multiple comparison testing was used to determine the significance of the standard deviations in the measured data from each specimen under different experimental conditions. In all cases, the results were considered statistically significant with a p-value of less than 0.05.

3. Results and Discussion

3.1. Chemical Composition

Figure 3A shows the GIXRD patterns of Ti surfaces before and after ion implantation. Similar GIXRD patterns were obtained for all of the samples. Three characteristic peaks located at around 35.1°, 38.4° and 40.1° can be attributed to the (100), (002) and (101) crystal faces of the Ti substrate phase [9]. Interestingly, the reduction of (100) reflection was found on the ion implanted surface. It seems that there were two broad peaks at around 35.9° and 42.4° visible in the two Ca-implanted samples (the 5Ca and 10Ca groups). This was possibly because of an ultrafine crystal size (amorphous phase) associated with the surface prepared by the implantation [5]. As also observed in Figure 3B, the FTIR profile of the ion-implanted samples was almost featureless, indicating a high degree of amorphousness. Krupa *et al.* also reported amorphous calcium-ion implanted Ti samples using a dose of 10^{17} Ca/cm^2 [20].

In Figure 4, SNMS shows the depth profiles of three elements (Ti, Ca and Si) as a function of depth. Surface contaminants, such as carbon, oxygen and nitrogen, were also identified during the SNMS analysis in both control and ion-implanted samples. However, for the sake of clarity, we did not show the depth distribution of these elements here. It is reasonable to consider that the concentration of Ti on the surface of all groups increased and reached the saturation slowly, due to the formation of a titanium-based mixture on the surface. Depth profiling of the samples showed the presence of Ca in the surface layer of all ion-implanted Ti, while the Si signal was appreciable on the 5CaSi surface (Figure 4C), as expected. The presence of Si was identified by SNMS, indicating a 25-nm average depth.

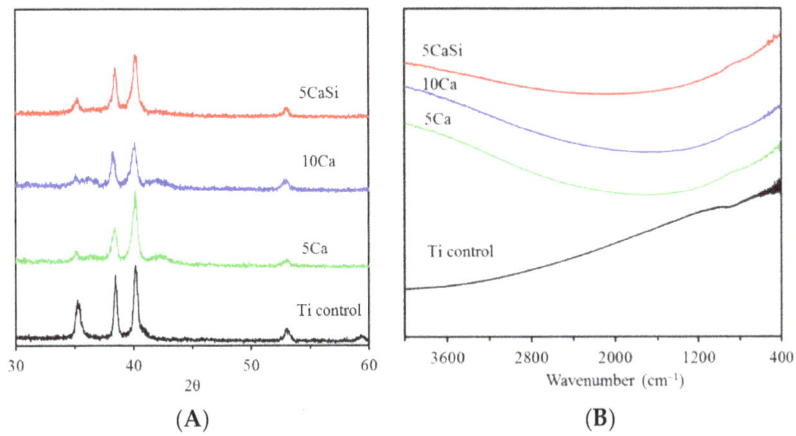

Figure 3. (**A**) Grazing incidence X-ray diffractometer (GIXRD) patterns and (**B**) FTIR spectra of the Ti samples with and without ion implantation.

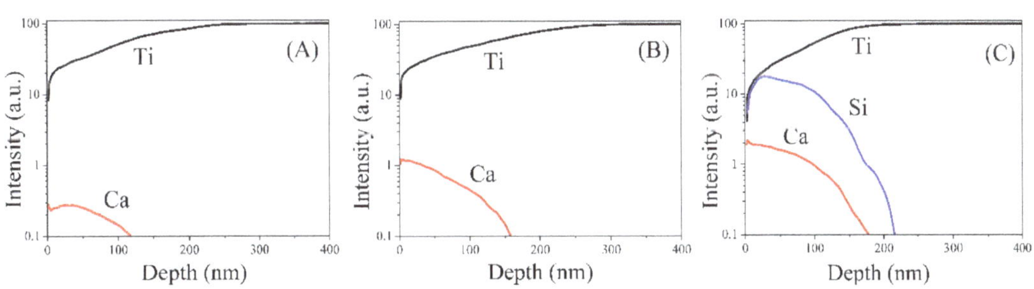

Figure 4. Secondary neutral mass spectrometry (SNMS) depth profiles of (**A**) 5Ca; (**B**) 10Ca and (**C**) 5CaSi ion-implanted surfaces.

3.2. Morphology

In general, the surface of Ti implants is pre-treated by sandblasting with a large grit and acid etching (SLA) for biomechanical interlocking that allows the bone ongrowth. However, the aim of the present study was to examine the ion implantation effect; relatively smooth sample surfaces were used throughout the study to reduce the interference of the roughened substrate and to facilitate the evaluation of the currently-used modification. The changed morphologies of the Ti substrate before and after ion implantation are shown in Figure 5. Compared to the pristine Ti substrate (Figure 5A), uniform distributed Ca granules with a size of about 300 nm built up the structure on the pristine Ti surface in a well-dispersed pattern (Figure 5B,C). In contrast, the incorporation of Si into the Ca-implanted surface induced numerous smaller granules over the Ti surface (Figure 5D), which could be distinguished from the Ca-implanted surfaces alone. Using this ion implantation, the smooth surface of the pristine Ti substrate was changed to a rougher morphology due to the formation of granules with a convex texture. As expected, the surface roughness results indicated a significantly ($p < 0.05$) higher Ra value ranging from 68 to 86 nm for ion-implanted groups compared to that obtained for the Ti control of 48 nm (Table 1). However, there were no significant differences ($p > 0.05$) in Ra roughness between the ion-implanted groups. The implantation process created surface changes that could result in altered topography and possibly even altered surface charge due to micro- or nano-scale pores or pits [21].

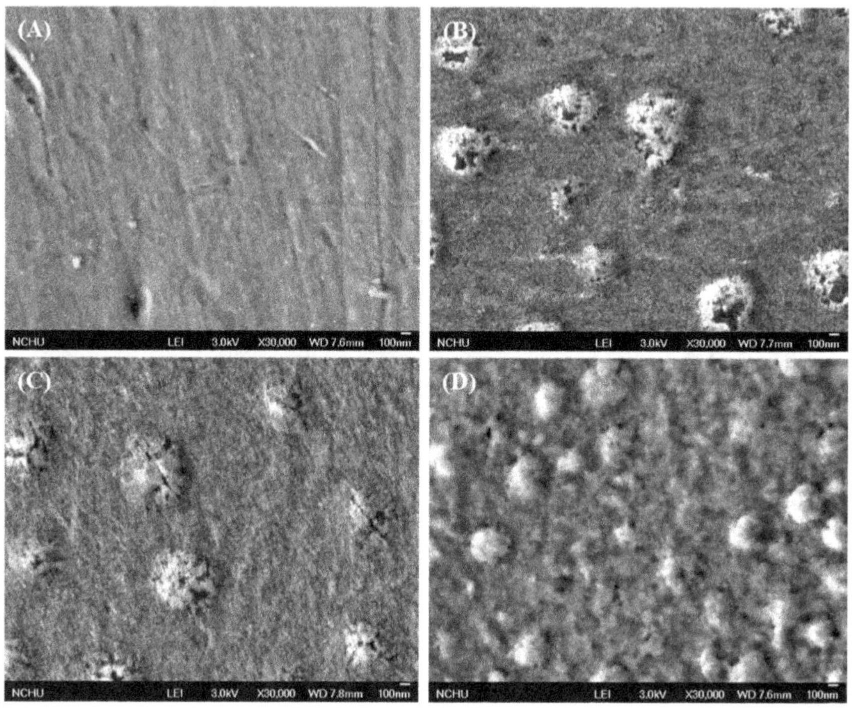

Figure 5. Surface SEM images of (**A**) Ti control; (**B**) 5Ca; (**C**) 10Ca and (**D**) 5CaSi groups.

Table 1. Roughness (Ra), electrochemical parameters and hemolysis ratio of all samples.

Sample Code	Ra (nm)	Potentiodynamic Polarization		Hemolysis (%)
		E_{corr} (mV)	I_{corr} (nA)	
Control	48 ± 8 [a]	-699 ± 114 [a]	54 ± 27 [a,b]	1.32 ± 0.16 [a]
5Ca	68 ± 8 [b]	-603 ± 98 [a,b]	98 ± 22 [b,c]	0.10 ± 0.23 [b]
10Ca	86 ± 11 [b]	-471 ± 84 [b]	131 ± 55 [c]	0.05 ± 0.38 [b]
5CaSi	76 ± 11 [b]	-575 ± 37 [a,b]	30 ± 3 [a]	0.16 ± 0.09 [b]

Values are the mean ± standard deviation. Mean values followed by the same superscript letter (e.g., a, b, c) were not significantly different ($p > 0.05$) according to Scheffé's *post hoc* multiple comparison.

3.3. Corrosion Behavior

In order to evaluate the corrosion behavior of the Ti implants with and without ion modifications, the OCP over time of the samples in SBF is demonstrated in Figure 6. The OCP of all samples shifted towards the active direction after soaking in SBF, possibly due to the dissolution that occurred at the sample surfaces [4], followed by reaching a steady state. Compared to the Ti control, the ion-implanted samples showed a higher initial potential and attained more noble potential values, indicating a superior corrosion behavior. Among the samples, the 10Ca surface might exhibit a better corrosion resistance during the OCP measurements than the other sample surfaces.

Typical potentiodynamic polarization curves are presented in Figure 7. The measured corrosion potential (E_{corr}) and corrosion current (I_{corr}) were obtained using Tafel extrapolation methods, as summed in Table 1. For the corrosion potential, the E_{corr} value of the pristine Ti substrate was −699 mV, while the E_{corr} values of the ion-implanted samples ranged from −471 to −603 mV. Similar to the findings of OCP examination, the 10Ca group had a corrosion potential (−471 mV) significantly ($p < 0.05$) higher than that of Ti (−699 mV). However, no significant differences ($p > 0.05$) were observed among the three ion-implanted samples. Concerning the effect of Si on polarization, it seems not to affect the corrosion potential in comparison with the other two ion-implanted samples. Concerning I_{corr}, the comparative study displayed that the average values between 30 and 131 nA were dependent on the types of the samples. The 10Ca group had higher I_{corr} values compared to the pristine Ti substrate and the 5CaSi group, possibly due to the dissolution of Ca ions when soaked in SBF. The released Ca ion may facilitate the cell growth. According to the literature [22], in tissue culture models, elevation in Ca concentration increases osteoblast chemotaxis and proliferation and alters the levels of expression of some differentiation markers. Compared to the pristine Ti substrate, the increased corrosion potential and the decreased corrosion current observed in 5CaSi samples can be explained by the structure of the surface layer being altered [20].

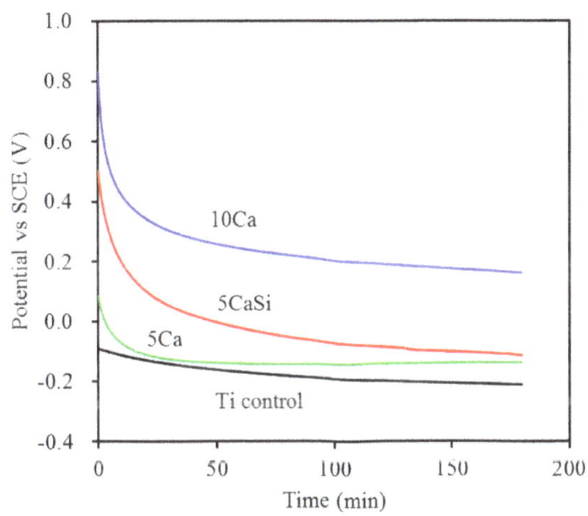

Figure 6. Open circuit potential (OCP) of Ti substrates with and without ion implantation. SCE, saturated calomel reference electrode.

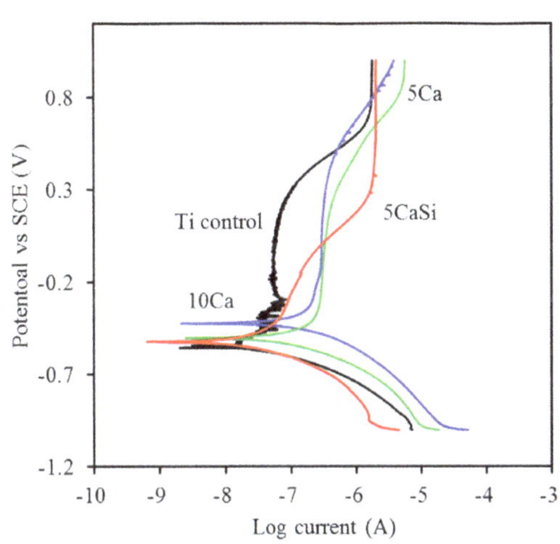

Figure 7. Typical potentiodynamic polarization curves of all samples in simulated body fluid (SBF).

3.4. Hemocompatibility

The hemolytic activities of all samples were evaluated by measuring the amount of hemoglobin released by ruptured RBCs. The results in Table 1 demonstrate that the hemolysis ratios of all samples are less than 2%, thus showing that these materials meet this criterion for biomedical materials (a hemolysis ratio lower than 5%) [19]. More importantly, all ion-implanted samples had a significantly lower ($p < 0.05$) hemolysis ratio than the Ti control. The hemolysis is linked to the surface characteristics of the test samples [19].

3.5. Biocompatibility

The biocompatibility of biomaterials is a prerequisite factor in the modulation of cellular function [23]. Surface chemistry is the most direct way to influence cell behavior [14]. To elucidate the effects of ion-implanted surfaces on biocompatibility, the responses of MG63 cells cultured to various samples were evaluated using fluorescent staining to actin. Actin stress fibers have been shown to act as a critical component contributing to cell migration [7]. The images revealed that after 6 h, MG63 cells cultured on all of the samples displayed highly tensioned actin stress fibers (red) (Figure 8). Cells were found growing in close proximity to all sample surfaces. Nayab *et al.* demonstrated that attachment and spreading of MG63 cells depended on the level of implanted Ca ions [24], which indicated a higher degree of cell spreading with increasing doses of Ca ions.

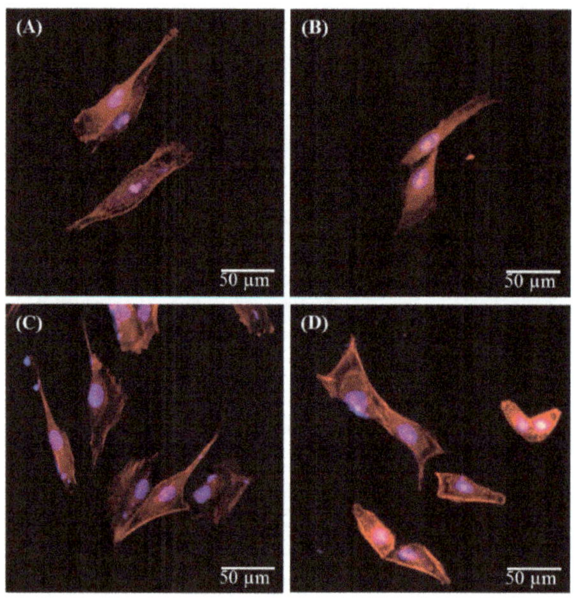

Figure 8. Fluorescent cytoskeleton staining of MG63 cells cultured on (**A**) Ti control; (**B**) 5Ca; (**C**) 10Ca and (**D**) 5CaSi after 6 h of seeding. The cells were stained for nuclei (blue) and actin cytoskeleton (red). Original magnification: 200×.

The quantification of adherent MG63 cells on different sample surfaces was performed using an MTT assay. The absorbance value steadily increased for all of the samples on Days 1 to 7, indicating increasing numbers of viable cells (Figure 9). There was no significant difference in the absorbance between all samples following culture for one and three days. However, cells rapidly attached to the 10Ca and 5CaSi surfaces after seven days compared to the pristine Ti and 5Ca groups. For example, the 5CaSi surface elicited a significant increment of 34% in comparison with the pristine Ti substrate after seven days of cell seeding. The 10Ca and 5CaSi implant surfaces could provide an optimal environment for osteoblasts to function normally. The cells seeded on the high Ca surface (10Ca group) for seven days had elevated metabolic activity, while cells incubated with a lower dose Ca surface (5Ca group) were not similarly activated. Krupa *et al.* also indicated that the viability and alkaline phosphatase bioactivity of the osteoblasts cultured in direct contact with non-implanted titanium and Ca-ion-implanted titanium (at a dose of 1×10^{17} ion/cm^2) were the same [20]. Nevertheless, an appropriate Si concentration implanted onto the Ti surface was effective in supporting the proliferation of osteoblasts, in agreement with a recent study [7]. Last, but not least, the present results highlighted the beneficial effects of the incorporation of Si in the Ca surface layer on cell viability, which is worthy of further evaluation.

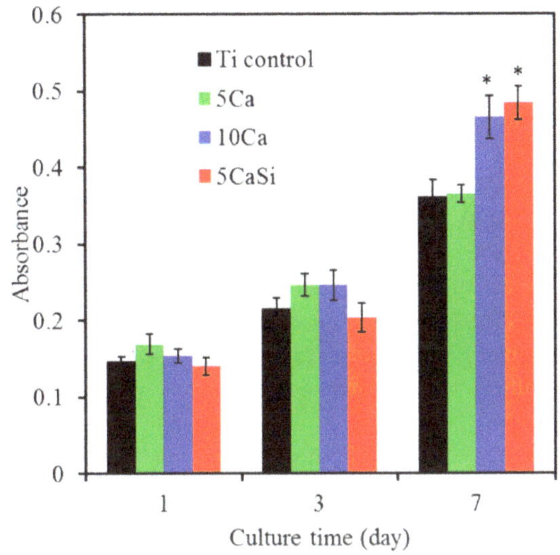

Figure 9. MTT assay for cell viability of MG63 cultured on the various samples after 1, 3 and 7 days of incubation. * Statistically significant difference ($p < 0.05$) from the Ti control.

4. Conclusions

The nature of the biomaterial surfaces may affect cell-material interactions, which in turn affect tissue development. Surface Ca and Ca/Si components were implanted onto the Ti metal by ECRIS. In light of the results obtained in this study, this kind of ion implantation technique did influence the surface topography of the pristine Ti substrate. The ion-implanted surface seemed to possess a better corrosion-resistant ability compared to the pristine Ti substrate. The improved surfaces did actively promote the viability of human osteoblast-like cells. The optimum protection on the Ti substrates was provided by a 10Ca or 5CaSi ion implantation in terms of the corrosion resistance and biocompatibility. Further studies, such as *in vitro* proliferation, differentiation and mineralization, are currently underway to evaluate the biological properties of the potential implant materials.

Acknowledgments: This work was partly supported by a Hungarian-Taiwan bilateral research program, No. NKM-92/2014, and MOST 103-2911-I-040-502. Work was also supported by the Hungarian TAMOP 4.2.2.A-11/1/KONV-2012-0036 project, which is co-financed by the European Union and the European Social Fund.

Author Contributions: Csaba Hegedűs was responsible for analyzing the results and writing the paper. Chia-Che Ho, Attila Csik and Sándor Biri performed the experiments. Shinn-Jyh Ding coordinated all tasks in the paper, planned the experiment and developed the results.

Conflicts of Interest: The authors declare no conflict of interest.

References

1. Perrin, D.; Szmukler-Moncler, S.; Echikou, C.; Pointaire, P.; Bernard, J. Bone response to alteration of surface topography and surface composition of sandblasted and acid etched (SLA) implants. *Clin. Oral Implant. Res.* **2002**, *13*, 465–469. [CrossRef]

2. Tsyganov, I.; Maitz, M.F.; Wieser, E. Blood compatibility of titanium-based coatings prepared by metal plasma immersion ion implantation and deposition. *Appl. Surf. Sci.* **2004**, *235*, 156–163. [CrossRef]
3. Ding, S.J. Properties and immersion behavior of magnetron-sputtered multi-layered hydroxyapatite/titanium composite coatings. *Biomaterials* **2003**, *24*, 4233–4238. [CrossRef]
4. Chen, C.C.; Huang, T.H.; Kao, C.T.; Ding, S.J. Characterization of functionally graded hydroxyapatite/titanium composite coatings plasma-sprayed on Ti alloys. *J. Biomed. Mater. Res. B* **2006**, *78*, 146–152. [CrossRef] [PubMed]
5. Rautray, T.R.; Narayanan, R.; Kwon, T.Y.; Kim, K.H. Surface modification of titanium and titanium alloys by ion implantation. *J. Biomed. Mater. Res. B* **2010**, *93*, 581–591. [CrossRef] [PubMed]
6. Milovanović, D.S.; Radak, B.B.; Gaković, B.M.; Batanic, D.; Momčilović, M.D.; Trtica, M.S. Surface morphology modifications of titanium based implant induced by 40 picosecond laser pulses at 266 nm. *J. Alloys Compd.* **2010**, *501*, 89–92. [CrossRef]
7. Ho, C.C.; Ding, S.J. Novel SiO_2/PDA hybrid coatings to promote osteoblast-like cell expression on titanium implants. *J. Mater. Chem. B* **2015**, *3*, 2698–2707. [CrossRef]
8. Hanawa, T.; Ukai, H.; Murakami, K.; Asaoka, K. Structure of surface-modified layers of calcium-ion-implanted Ti-6Al-4V and Ti-56Ni. *Mater. Trans. JIM* **1995**, *36*, 438–444. [CrossRef]
9. Weng, H.A.; Wu, C.C.; Chen, C.C.; Ho, C.C.; Ding, S.J. Preparation and properties of gold nanoparticle-electrodeposited titanium implant metals with Arg-Gly-Asp-Cys peptides. *J. Mater. Sci. Mater. Med.* **2010**, *21*, 1511–1519. [CrossRef] [PubMed]
10. Sawada, R.; Kono, K.; Isama, K.; Haishima, Y.; Matsuoka, A. Calcium-incorporated titanium surfaces influence the osteogenic differentiation of human mesenchymal stem cells. *J. Biomed. Mater. Res. A* **2013**, *101*, 2573–2585. [CrossRef] [PubMed]
11. Anitua, E.; Prado, R.; Orive, G.; Tejero, R. Effects of calcium-modified titanium implant surfaces on platelet activation, clot formation, and osseointegration. *J. Biomed. Mater. Res. A* **2015**, *103*, 969–980. [CrossRef] [PubMed]
12. Dvorak, M.M.; Siddiqua, A.; Ward, D.T.; Carter, D.H.; Dallas, S.L.; Nemeth, E.F.; Riccardi, D. Physiological changes in extracellular calcium concentration directly control osteoblast function in the absence of calciotropic hormones. *Proc. Natl. Acad. Sci. USA* **2004**, *101*, 5140–5145. [CrossRef] [PubMed]
13. Hanawa, T.; Kamiura, Y.; Yamamoto, S.; Kohgo, T.; Amemiya, A.; Ukai, H.; Murakami, K.; Asaoka, K. Early bone formation around calcium-ion-implanted titanium inserted into rat tibia. *J. Biomed. Mater. Res.* **1997**, *36*, 131–136. [CrossRef]
14. Huang, S.C.; Wu, B.C.; Ding, S.J. Stem cell differentiation-induced calcium silicate cement with bacteriostatic activity. *J. Mater. Chem. B* **2015**, *3*, 570–580. [CrossRef]
15. Xu, S.; Lin, K.; Wang, Z.; Chang, J.; Wang, L.; Lu, J.; Ning, C. Reconstruction of calvarial defect of rabbits using porous calcium silicate bioactive ceramics. *Biomaterials* **2008**, *29*, 2588–2596. [CrossRef] [PubMed]
16. Biri, S.; Rácz, R.; Pálinkás, J. Status and special features of the Atomki ECR ion source. *Rev. Sci. Instrum.* **2012**, *83*, 02A341. [CrossRef] [PubMed]
17. Rácz, R.; Biri, S.; Hajdu, P.; Csik, A.; Vad, K.; Kökényesi, S.; Csarnovich, I.; Hegedűs, C.; Radics, T.; Bakó, J.; Hegedűs, V.; Pálinkás, J. Application of an ecr ion source for ionic fuctionalization of implant materials on the nanoscale. In Proceedings of the 21st International Workshop on ECR Ion Sources (ECRIS2014), Nizhny Novgorod, Russia, 24–28 August 2014; pp. 135–139.
18. Chen, C.C.; Wang, C.W.; Hsueh, N.S.; Ding, S.J. Improvement of *in vitro* physicochemical properties and osteogenic activity of calcium sulfate cement for bone repair by dicalcium silicate. *J. Alloys Compd.* **2014**, *585*, 25–31. [CrossRef]
19. Lee, P.R.; Ho, C.C.; Hwang, C.S.; Ding, S.J. Improved physicochemical properties and biocompatibility of stainless steel implants by PVA/ZrO_2-based composite coatings. *Surf. Coat. Technol.* **2014**, *258*, 374–380. [CrossRef]
20. Krupa, D.; Baszkiewicz, J.; Kozubowski, J.A.; Barcz, A.; Sobczak, J.W.; Biliński, A.; Lewandowska-Szumie, M.; Rajchel, B. Effect of calcium-ion implantation on the corrosion resistance and biocompatibility of titanium. *Biomaterials* **2001**, *22*, 2139–2151. [CrossRef]
21. Pham, M.T.; Matz, W.; Grambole, D.; Herrmann, F.; Reuther, H.; Richter, E.; Steiner, G. Solution deposition of hydroxyapatite on titanium pre-treated with a sodium ion implantation. *J. Biomed. Mater. Res.* **2002**, *59*, 716–724. [CrossRef] [PubMed]

22. Eklou-Kalonji, E.; Denis, I.; Lieberherr, M.; Pointillart, A. Effects of extracellular calcium on the proliferation and differentiation of porcine osteoblasts *in vitro*. *Cell Tissue Res.* **1998**, *292*, 163–171. [CrossRef] [PubMed]
23. Wu, C.C.; Wei, C.K.; Ho, C.C.; Ding, S.J. Enhanced hydrophilicity and biocompatibility of dental zirconia ceramics by oxygen plasma treatment. *Materials* **2015**, *8*, 684–699. [CrossRef]
24. Nayab, S.N.; Jones, F.H.; Olsen, I. Effects of calcium ion implantation on human bone cell interaction with titanium. *Biomaterials* **2005**, *26*, 4717–4727. [CrossRef] [PubMed]

8

Preparation and Characterization of Microencapsulated Phase Change Materials for Use in Building Applications

Jessica Giro-Paloma [1], Refat Al-Shannaq [2], Ana Inés Fernández [1] and Mohammed M. Farid [2,*]

Academic Editor: Luisa F. Cabeza

[1] Department of Materials Science and Metallurgical Engineering, Faculty of Chemistry, Universitat de Barcelona, C/Martí i Franquès, Barcelona 1. 08028, Spain; jessicagiro@ub.edu (J.G.-P.); ana_inesfernandez@ub.edu (A.I.F.)

[2] Department of Chemical and Materials Engineering, University of Auckland, Private Bag 92019, 20 Symonds Street, Auckland 1142, New Zealand; rals187@aucklanduni.ac.nz

* Correspondence: m.farid@auckland.ac.nz

Abstract: A method for preparing and characterizing microencapsulated phase change materials (MPCM) was developed. A comparison with a commercial MPCM is also presented. Both MPCM contained paraffin wax as PCM with acrylic shell. The melting temperature of the PCM was around 21 °C, suitable for building applications. The M-2 (our laboratory made sample) and Micronal® DS 5008 X (BASF) samples were characterized using SEM, DSC, nano-indentation technique, and Gas Chromatography/Mass spectrometry (GC-MS). Both samples presented a 6 μm average size and a spherical shape. Thermal energy storage (TES) capacities were 111.73 J$\cdot$ g^{-1} and 99.3 J$\cdot$ g^{-1} for M-2 and Micronal® DS 5008 X, respectively. Mechanical characterization of the samples was performed by nano-indentation technique in order to determine the elastic modulus (E), load at maximum displacement (P_m), and displacement at maximum load (h_m), concluding that M-2 presented slightly better mechanical properties. Finally, an important parameter for considering use in buildings is the release of volatile organic compounds (VOC's). This characteristic was studied at 65 °C by CG-MS. Both samples showed VOC's emission after 10 min of heating, however peaks intensity of VOC's generated from M-2 microcapsules showed a lower concentration than Micronal® DS 5008 X.

Keywords: microencapsulated phase change material; volatile organic compounds; nano-indentation; differential scanning calorimetry

---

1. Introduction

Thermal energy storage (TES) using phase change materials (PCM) has shown a significant increased attention because of its important role on energy conservation in buildings [1–4]. PCM can be used for TES in buildings [5] either in passive [6] or active systems [7–9], aiming to improve the thermal managements of these buildings. In most of the applications, PCM were used either in macroencapsulated [10,11] or microencapsulated [12–14] forms, for heating [15], air-conditioning [16], ventilation [17], refrigeration [18], and heat exchangers [19] for building applications [3,20–22].

Microencapsulation process is defined as a technique in which small particles or droplet are enclosed by a coating, or surrounded in a homogeneous or heterogeneous matrix, to provide microcapsules (1–100 μm). For this reason, the microencapsulated phase change materials (MPCM) are composed of PCM as a core and a polymer as a shell used to preserve the spherical shape of the microcapsule and avoid PCM leakage during phase change [12,23]. The use of MPCM in buildings [24–27] can decrease daily inner temperature fluctuation during summer and winter [28].

The suitability of the shell used in encapsulating specific core PCM is a key issue in order to ensure proper thermal performance of the MPCM [13,29,30], especially in preventing PCM leakage when it melts. Additionally, MPCM can be easily incorporated in gypsum board [31,32], plaster [33], and concrete [34] used in buildings.

The complete characterization of materials used in indoor environments like MPCM is very important. The exposure to chemical compounds could cause health problems (nausea; dry skin; eye, nose or throat irritations; headache; irritated eyes; dizziness; difficulty in concentrating; psychological stress) in indoor environments [35–38] (buildings, for example [39–47]) or outdoor environments [48]. These problems are known as sick building syndrome (SBS) [37,49–51]. Volatile organic compounds (VOC's) are defined as any organic compound having an initial boiling point less than or equal to 250 °C at a standard pressure of 101.3 kPa [52]. VOC's are one of the most important groups of trace contaminants in the atmosphere known for its photochemical, toxic, and radioactive effects. For this reason there are some studies, guides [53,54], and database [55,56] related to this effect. Formaldehyde [39,46,57] and benzene [58] are some of the most studied pollutants since they are classified in Group 1 of human carcinogens by the IARC 2004 (International Agency for Research on Cancer). Other chemicals known for their health hazard are acetaldehyde, toluene, and xylenes [59]. By this way, VOC's evaluation of the outdoor and indoor air quality has been evaluated [28–31] in materials for buildings like gypsum base PCM composites [60] but it has not been reported for building materials containing MPCM. For this reason, the characterization of VOC's of MPCM is an important contribution to the state of the art of the environmental properties of MPCM.

The main purpose of this research is to develop, prepare, characterize, study, and compare thermal and mechanical properties of microcapsules containing organic PCM in order to assess their suitability for use in buildings and other applications. The samples under study are commercial MPCM (Micronal® DS 5008 manufactured by BASF, Berlin, Germany) and a laboratory prepared one by us (M-2). Micronal® DS 5008 sample has been used extensively in concrete, gypsum, lime plaster, and gypsum plaster, without being fully characterized for fire hazards. The comparison includes fire retardancy and gas emission released to environment from upon fire. It is important to establish a characterization methodology, which will include both volatile emission measurements and nano-indentation technique to measure the shell mechanical strength of the microcapsules. This is very important issue in the selection of PCM products, especially for use in building application. PCM microcapsules should have high phase change enthalpy, uniform spherical shape, acceptable thermal stability, good mechanical properties, and low release of hazardous gases in the form of volatile organic compounds.

2. Materials and Methods

2.1. Materials

The chemical preparation of microcapsules required the following reagents:

- Shell: Methyl methacrylate (MMA) (99%, contains $\leqslant$ 30 ppm monomethyl ether hydroquinone (MEHQ) as inhibitor, Sigma Aldrich, Auckland, New Zealand) and pentaerythritol tetraacrylate (PETRA) (contains 350 ppm (MEHQ), Sigma Aldrich, Auckland, New Zealand) were used as a monomer and cross-linking agent respectively in order to obtain proper shells for MPCM.
- Free radical thermal initiator: Luperox® A75, Benzoyl peroxide (BPO) (75%, contains 25% water, Sigma Aldrich, Auckland, New Zealand) was used as free radical thermal initiator.
- Surfactants: Polyvinyl alcohol (PVA) (M_w 85,000–124,000, Sigma Aldrich, Auckland, New Zealand) and sodium dodecyl sulfate (SDS) (BioXtra, 99%, Sigma Aldrich, Auckland, New Zealand) were used as a non-ionic and ionic surfactant, respectively.
- PCM: a commercial paraffinic PCM, Rubitherm® RT 21 (T_m = 21 °C, ΔH_m = 135 J·g^{-1}, Rubitherm® Technologies GmbH, Berlin, Germany) was used.

The bulk density of M-2 microcapsules is 0.496 g· mL^{-1}. The commercial MPCM, Micronal® DS 5008 X (BASF®), was also selected for characterization and was compared with the microcapsules produced in this work. This sample is also composed by an acrylate shell and organic PCM in the core [13], and its bulk density is 0.445 g· mL^{-1}.

2.2. Synthesis of PCMs Microcapsules

2.2.1. Emulsification

A standard procedure was used as reported elsewhere [61]. Wherein, an aqueous solution of surface-active agent (called aqueous phase) and a mixture of MMA, PETRA, BPO, and PCM (called organic phase) were prepared separately. The organic phase was added to the aqueous phase and emulsified mechanically using a high shear mixer (Silverson L5M-A laboratory Mixer, Silverson LTD, East Longmeadow, MA, USA). A stirring rate of 3000 rpm for 5 min was chosen to achieve the required emulsification.

2.2.2. Polymerization

The produced emulsion was transferred to a 2-L four-neck glass reactor (LR-2.ST laboratory reactor-IKA-Werke, Gmbh@Co.KG, Staufen, Germany) consisting of EUROSTAR 200 control P4, Anchor stirrer LR 2000.1, HBR 4 digital heating bath. The agitation speed was set at approximately 300 rpm, and the temperature of the water bath was maintained at 70 °C for 2 h, and then adjusted to 85 °C for another 4 h. The water bath was then switched off and allowed to cool down naturally to room temperature. After cooling, the suspension of PCM microcapsules was transferred to a clean glass beaker and washed three times with distilled water to remove the unreacted monomers and the non-encapsulated PCM. The separated microcapsules were spread on a tray and placed in an oven at 50 °C for 48 h for drying. The dried microcapsules were then collected for testing.

3. Characterization of Microcapsules

3.1. Scanning Electron Microscopy (SEM)

To study the shape and size of microcapsules SEM was used (a FEI Quanta 200 FEG-Field Emission Gun with an EDS Detector SiLi (Lithium drifted) with a Super Ultra-Thin window, FEI Company, Hillsboro, OR, USA). The sputter coater used was a Quorum Q150RS (FEI Company, Hillsboro, OR, USA), and it is designed to give an appropriate thin, slight metal coating proper for SEM observation, using Pt as a target. The coating thickness and uniformity of the sample depends on different factors: distance between sample and target, topography of the sample, and affinity of the material with the metal coating.

3.2. Differential Scanning Calorimetry (DSC)

Phase change properties of the fabricated PCM microcapsules and the pure PCM (such as melting and solidification temperatures and their phase change enthalpies) were determined using a SHIMADZU DSC-60 differential scanning calorimeter (Shimadzu Company, Kyoto, Japan). The measurements were performed by varying the temperature between −15 °C and 40 °C with heating and cooling rate of 3 °C· min^{-1}. Each sample was analyzed for three times and the average was taken. Consequently, the percentage PCM encapsulated can be determined using DSC results and the following Equation (1) [62,63]. The mass content obtained by DSC measurements does not provide accurate measure of the core mass content. Equation (1), which was used to estimate the mass content from DSC measurements, does not take in account the sensible heat of coating materials. The TGA method provides more accurate measure of the core material mass content than DSC. In our previous publication [63] the core material mass content of sample M-2 obtained by TGA is 77.5 wt%, which is less than the one obtained by DSC.

$$\% \text{ PCM in microcapsules by mass} = \Delta H_{\text{microcapsules}}/\Delta H_{\text{Pure PCM}} \times 100\% \quad (1)$$

where $\Delta H_{microcapsules}$ ($J \cdot g^{-1}$) is the latent heat of the microcapsule containing PCM; and $\Delta H_{purePCM}$ ($J \cdot g^{-1}$) is the latent heat of pure PCM (before encapsulation). In Equation (1), it is assumed that the latent heat of the microcapsule without PCM is zero, which is true if phase change does not occur in the shell does.

3.3. Nano-Indentation Technique

To characterize the mechanical properties of M-2 and commercial Micronal® DS 5008 X samples, a nano-indentation technique was used. Nano-indentation is identified as an appropriate technique to test the strength of individual microcapsules [64]. MTS Nano Indenter XP (MTS Company, Eden Prairie, MN, USA) was the instrument used. Aluminium stubs of 20 mm height and 30 mm diameter were needed to stick the samples at the top to characterize them using a red glue to stick the samples as shown in Figure 1. The instrument parameters were set the same for the two studied samples for a more accurate comparison.

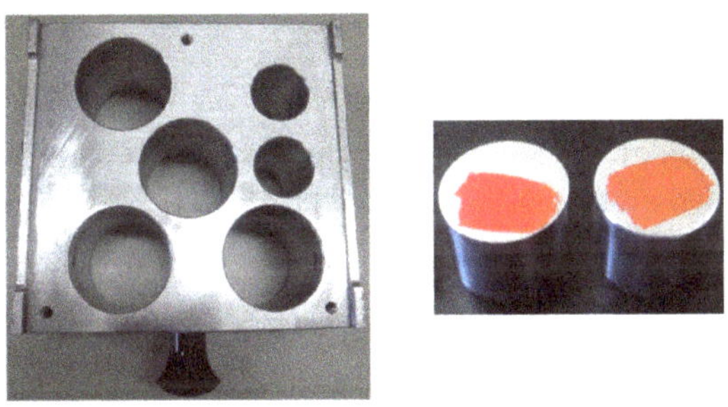

Figure 1. Holder and aluminum stubs with the sample over the red glue.

There are some required inputs to set before starting the experiments: strain rate target of 0.05 s^{-1}, allowable drift rate of 0.05 $nm \cdot s^{-1}$, a Poissons' ratio of 0.18 for the tip indenter [65,66] a peak hold time of 10 s, a surface speed of 10 $nm \cdot s^{-1}$, 25% of surface approach sensitivity, 90% to unload, an approach distance to store of 1000 nm, a surface approach distance of 1000 nm, and finally, a depth limit of 5000 nm.

To determine the elastic modulus (E) of the studied samples, a Berkovich tip TB-13288 (Micro Star Technologies, Huntsville, TX, USA) was used. The use of nano-indentation for the characterization of mechanical properties of materials has been extensively studied by several authors. Oliver and Pharr developed extensively the methodology for characterizing ceramic materials [66,67]. They described the typical load *vs.* displacement curve, where increasing the load (P) increases the displacement (h) until reaching a maximum load (P_{max}) and a maximum displacement (h_{max}). Following that, the indenter is removed out of the material (unloading section), the load will be zero, and the final displacement (h_f) will be measured. After that, the E value for each sample can be calculated. Hochstetter *et al.* [68] presented results for glassy polymers and Giro-Paloma *et al.* [69] compared both methodologies using continuous stiffness measurement (CSM) by applying a small oscillation to the quasi-static component of loading using different thermoplastics suitable as containers for PCM. They concluded that Loubet's method produce lower values of H and E because it uses a dynamic approach for stiffness measurements and the contact depth is larger due to the contribution of the apparent tip effect. In the light of these findings, it was concluded that Loubet's method should be used only with polymeric materials having a low viscous character ($T_g > T_{measurement}$). Additionally, in depth-sensing indentation (DSI), which is the mode used in this paper, load is applied as a function of penetration depth during the loading and unloading cycle, as described by Fischer-Cripps [70]:

- Hardness (H) is defined as the maximum indentation load divided by the cross-sectional area of the indenter specified at the maximum indentation depth ($A(h_m)$).
- Load (P_m) at maximum displacement (mN): It is the load recorded at the maximum load, which occurs when sample fails.
- Elastic modulus (E) is evaluated following Equations (2) and (3) from the nano-indentation test using the maximum indentation load (P_m) and the depth sensing indentation. Hardness (H), elastic work (W_e), and total work (W) can be calculated by integrating the areas under the indentation unloading. W_e and W are the elastic work and total work, which are equal to the areas under the unloading and loading curves, respectively which is correlated with E and H through the function ψ described in [71]. W_e/W value is independent of the degree of work-hardening behavior [66].

$$H = P_m/A(h_m) \tag{2}$$

$$E = H/\psi\left(\frac{W_e}{W}\right) \tag{3}$$

- Displacement (h_m) at maximum load (nm) is a measure of the extent the tip penetrates into the material.

3.4. *Emission of Volatile Organic Compounds (VOC's)*

A GC-17A Gas chromatograph Shimadzu (Shimadzu Corporation, Kyoto, Japan) coupled to GCMS-QP5000 Gas chromatograph/Mass Spectrometer Shimadzu (Shimadzu Corporation, Kyoto, Japan) was used to characterize the VOC's emissions from each sample. A calibration was performed for each pure compound: $C_{14}H_{30}$, $C_{15}H_{32}$, $C_{16}H_{34}$, $C_{17}H_{36}$, $C_{18}H_{38}$, $C_{20}H_{42}$, $C_{22}H_{46}$, and $C_{24}H_{50}$, at different temperatures: 25 °C, 35 °C, 45 °C, 55 °C, and 65 °C. A total of 40 experiments for calibration were executed. When the calibration was completed, the same procedure was performed for the two studied samples: M-2 and Micronal® DS 5008 X.

Each sample was independently located inside a crystal vial HS of 50 mL capacity. The vials were submerged during 30 min in a water bath until reaching the required temperature. Later on, a fibber solid-phase microextraction (SPME) holder with lot number P268618D 57330-U, provided by Supelco (Sigma-Aldrich Corporation, St. Louis, MO, USA), was punctured on the top of the silicone cap. Following that, 10 min desorption was applied. The temperature inside the device was 60 °C during 2 min. After that, a ramp of 15 $°C\cdot min^{-1}$ was programmed.

A HP-5MS (1553434H) (Agilent Technologies, Santa Clara, CA, USA) was the column used. Its thickness, length and diameter were 0.5 μm, 30 m, and 0.32 cm, respectively. Additionally, the injection and interface temperature were 200 °C and 280 °C, respectively inside the gas chromatographer. There were more parameters to take into account, such as inlet pressure: 1 kPa, flow: 1.1 $mL\cdot min^{-1}$, lineal velocity: 38.7 $cm\cdot s^{-1}$, split ratio between peaks of 20, and finally total flow for the He gas of 23.1 $mL\cdot min^{-1}$. On the other hand, the mass spectrometer m/z values are from 35 to 350. Moreover, the solvent cut time was 0.5 min.

4. Results and Discussion

4.1. *Characterization of MPCM Shape, Size, and Morphology*

SEM images for the two studied samples are shown in Figure 2. M-2 microcapsules morphology appears to be compact and with smooth surface as shown in Figure 2a. Furthermore, their size is around 6 μm and has a spherical shape. On the other hand, commercial Micronal® DS 5008 X looks made of a big sphere (around 150 μm) composed of hundreds of other small microcapsules (the ones which contain the PCM) with 6 μm in size, approximately as shown in Figure 2b. As Figure 2 shows, the 6 μm microcapsules of Micronal® DS 5008 X samples are deformed, which is probably due to the process of agglomerating of these microcapsules to form larger particle of 150 μm, which probably has been made for ease of handling. Accelerated thermal cycling experiments of PCM microcapsules (M-2 containing RT-58 sample) were performed in a controlled heating/cooling water bath at temperatures cycling between 2 and 40 °C in our previous publication [63]. The results showed that slight changes in phase transition temperatures of the PCM microcapsules (M-2 containing RT58 sample) after 2000 cycles. The latent heat of M-2 sample (based on DSC measurements) showed only a minor change of 2% after 2000 cycles. Furthermore, SEM images showed that the capsule shape remained spherical and no shell cracks were observed at the end of 2000 cycles.

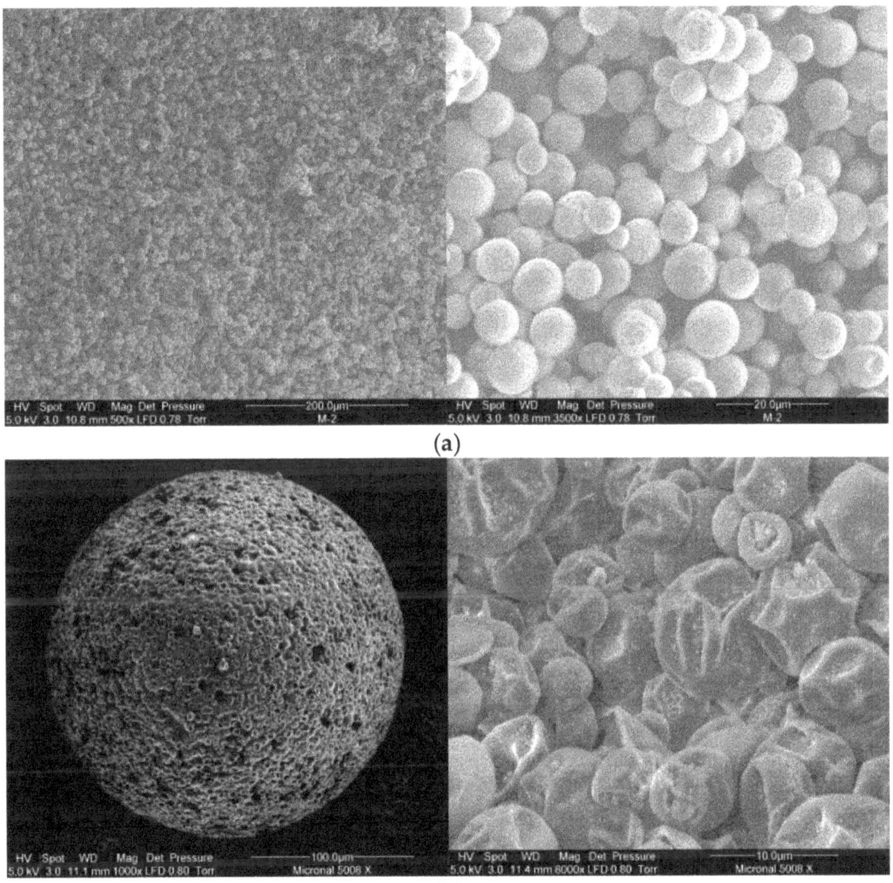

Figure 2. SEM images for the studied samples: (**a**) M-2 (magnification: ×500 left and ×3500 right); (**b**) Micronal® DS 5008 X (magnification: ×1000 left and ×8000 right).

4.2. *Thermophysical Properties of MPCM*

Thermal properties of M-2 and Micronal® DS 5008 X in terms of phase change enthalpies and phase change temperatures were investigated using DSC as shown in Figure 3. Based on DSC measurements, the % mass of the PCM for M-2 sample is 81.7 wt% and for Micronal® DS 5008 X is 78.1 wt%. The DSC results show that the thermal energy storage capacity of the M-2 is 113.8 $J \cdot g^{-1}$ which corresponds to 85 wt% of RT-21 encapsulation. The melting temperature of the RT-21 in M-2 microcapsules is similar to that of the bulk RT-21. In contrast, the solidification temperature of the PCM microcapsules was about 14 °C lower than that of the bulk RT-21 (super-cooling phenomena) as previously reported [72] (see Table 1). The supercooling of PCM in microcapsules has also been reported by Qiu *et al.* [73]. The increase of the degree of super-cooling of RT-21 microcapsules could be attributed either to the decrease in the amount of RT-21 nuclei inside each microcapsule compared to the bulk RT-21 [74] or due to formation of vacuum pockets space inside the microcapsules [75]. To reduce the supercooling of PCM microcapsules, additives were mixed with the PCM prior encapsulation to act as a nucleating agent. These nucleating agents are usually materials with a similar crystal structure as the solid PCM which allow nucleation at their surface but have a higher melting temperature. Figure 4 shows the DSC curve for the microcapsules containing nucleating agent. Commercial RT-58 (paraffin) with peak melting temperatures of 58 °C was used as nucleating agent. The degree of supercooling has been reduced dramatically as reported in Table 1 and shown in Figure 4.

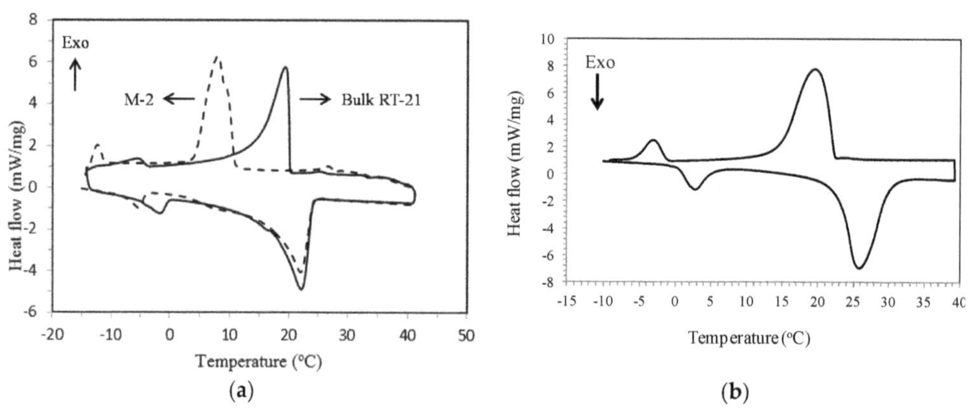

Figure 3. DSC results for: (**a**) M-2; and (**b**) Micronal® DS 5008 X.

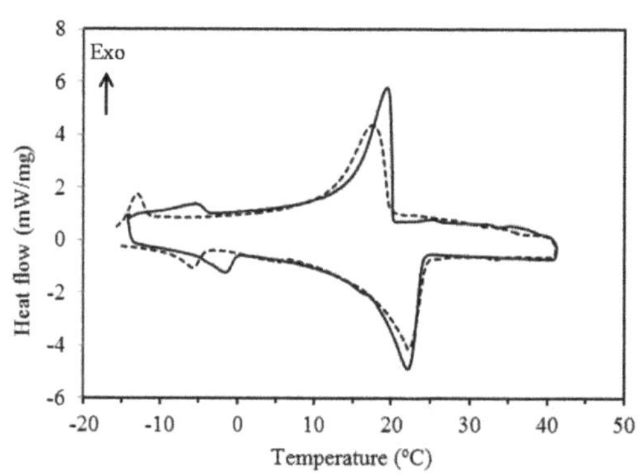

Figure 4. DSC curves of the bulk RT-21 (solid line) and M-2 microcapsules containing nucleating agent, RT-58 (dotted line).

Table 1. Thermophysical properties of fabricated M-2 and commercial Micronal® DS 5008 X.

Status	Transition Temperatures and Heat of Fusion	M-2	RT-21	Micronal® DS 5008 X	M-2 Containing RT-58
Heating	T_{onset} (°C)	16.7	16.5	22.1	16.1
	T_{peak} (°C)	22.0	22.1	20.3	22.3
	T_{endset} (°C)	24.0	23.9	26.3	24.3
	ΔH ($J \cdot g^{-1}$)	113.9	132.0	99.9	110.4
Cooling	T_{onset} (°C)	10.9	20.2	22.5	19.8
	T_{peak} (°C)	7.9	19.4	24.2	17.5
	T_{endset} (°C)	4.2	14.5	17.5	11.4
	ΔH ($J \cdot g^{-1}$)	111.9	132.5	103.5	108.3

4.3. *Mechanical Properties of MPCM*

Results of nano-indentation technique by DSI for the two samples were summarized in Table 2. Ten tests were performed for each sample, but some results were discarded because of the dispersion attributed to indentations in the edge of the microcapsules. For this reason, six results were finally selected. From these measurements the mean value and standard deviation of the elastic modulus (E) were calculated for each sample. These results show that M-2 microcapsules are stiffer than Micronal® DS 5008 X ones.

Table 2. Elastic modulus results of M-2 and Micronal® DS 5008 X.

Mechanical Property	M-2	Micronal® DS 5008 X
E (GPa)	1.89	0.15
	1.04	0.19
	1.16	0.17
	1.68	0.24
	1.38	0.22
	1.9	0.28
Mean	1.51	0.21
Standard Deviation	0.37	0.05

Figure 5 is a representation of the P-h curves for the M-2 and Micronal® respectively. The measured typical drift behavior can be observed on the plateau at the end of the unloaded curve. It can be concluded that the mechanical response for M-2 is better than Micronal® DS 5008 X as M-2 microcapsules are more rigid and show higher strength. This fact can be attributed to various factors such as the differences in the encapsulation ratio between samples, shell thickness, type of polymers, as well as degree of polymerization of the shells. These results should be compared when mixing these MPCM in a real system, such us mixing them with a cement-based matrix or gypsum.

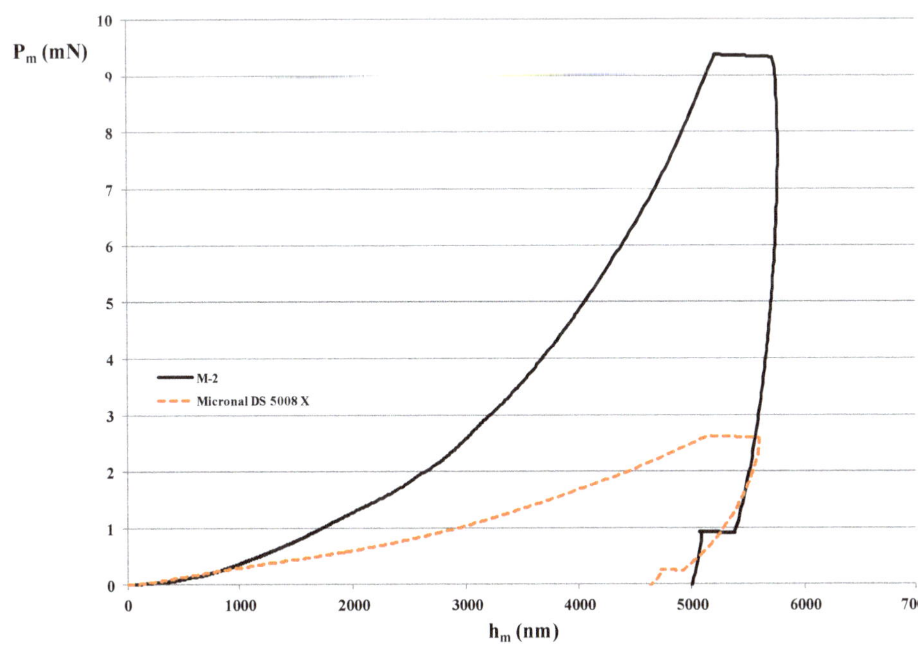

Figure 5. h_m *vs.* P_m, comparison of the samples under study.

4.4. Characterization of Volatile Organic Compounds (VOCs)

The results of VOCs' short-term release from the two studied samples are shown in Table 3. Each peak in the GC spectra was given at a certain retention time (t_r, in minutes), which was related to a specific compound, according to the calibration conducted in this work.

Table 3. VOC emission results for M-2 and Micronal® DS 5008 X samples.

Temperature	M-2		Micronal® DS 5008 X	
	t_r (min)	Compound	t_r (min)	Compound
25 °C	No signal		No signal	
35 °C	No signal		No signal	
45 °C	No signal		11.88 12.66	$C_{17}H_{36}$ $C_{18}H_{38}$
55 °C	No signal		11.88 12.65	$C_{17}H_{36}$ $C_{18}H_{38}$
65 °C	9.25 10.18 11.06 11.91 12.68	$C_{14}H_{30}$ $C_{15}H_{32}$ $C_{16}H_{34}$ $C_{17}H_{36}$ $C_{18}H_{38}$	- - - 11.85 12.63	- - - $C_{17}H_{36}$ $C_{18}H_{38}$

It may be observed from the results that some emissions from Micronal® DS 5008 X sample after 10 min exposure at 45 °C and 55 °C were detected while nothing was detected from M-2 microcapsules. After 10 min exposure at 65 °C, both samples release volatile compounds, but with much lower level for M-2 microcapsules in comparison with Micronal® DS 5008 X. This is confirmed by the high intensity peak for Micronal® DS 5008 X comparing both figures in Figure 6. The multiple peaks for M-2 show the presence of tetradecane, pentacosane, hexadecane, heptadecane, and octadecane in the original PCM, which is RT-21, while only two peaks were observed for the PCM used in Micronal® DS 5008 X, indicating that the latter is of higher purity. Although the results for the short-term emissions are relevant, the long term time-dependent release should be studied in a system after prolonged exhibition to evaluate the VOCs' emissions in real conditions.

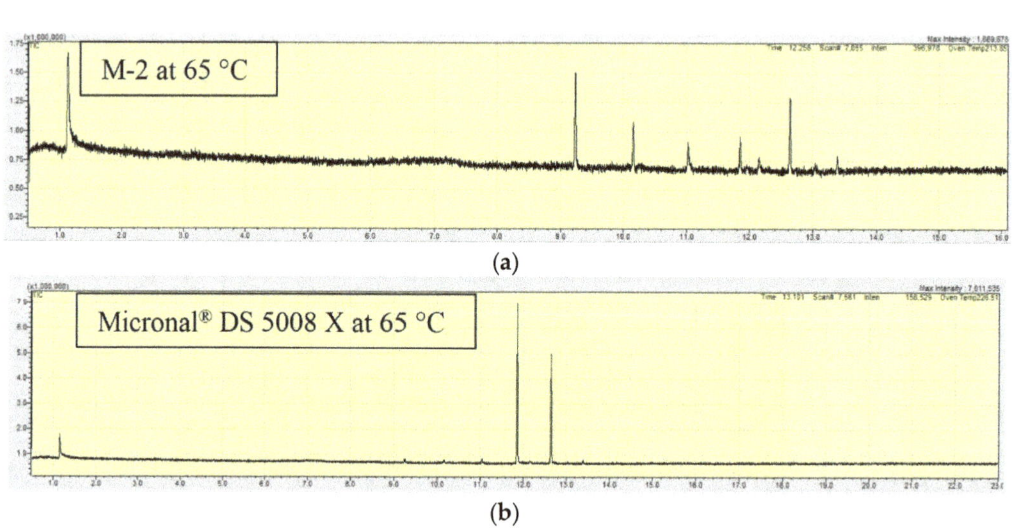

Figure 6. Revealed peaks for the GC/MS analysis for: (**a**) M-2 and; (**b**) Micronal® DS 5008 X samples at 65 °C.

5. Conclusions

A comparison for the characteristic of two MPCM samples was conducted: one sample was Micronal® DS5008 X while the other was made in our laboratory (M-2). The shell material for both samples is similar in terms of chemical composition. Following SEM observation, it can be concluded that the two samples have similar shape and diameter of about 6 μm, but they have different morphology since Micronal® DS 5008 X capsules were produced as aggregates of many microcapsules. Regarding the thermophysical properties for both samples, their thermal energy storage (TES) capacity were 111.7 $J \cdot g^{-1}$ and 99.3 $J \cdot g^{-1}$ for M-2 and Micronal® DS 5008 X, respectively. The mechanical testing was performed by measuring elastic modulus (E), load at displacement (P_m), and displacement at maximum load (h_m) using nano-indentation technique. Different results were obtained for both samples, showing that evaluating the isolated M-2 sample has better mechanical resistance and stiffness. Finally, a comparative evaluation of the VOC's release at 25 °C, 35 °C, 45 °C, 55 °C, and 65 °C was performed in order to study the volatiles emission from these microcapsules. M-2 microcapsules show better stability with less short-term emission of VOC's than Micronal® DS 5008 X.

Acknowledgments: Authors acknowledge the support received from the European Union's Seventh Framework Programme (FP7/2007–2013) under grant agreement No: PIRSES-GA-2013-610692 (INNOSTORAGE). Also, we would like to acknowledge the support we have received form Qatar Foundation in supporting the project No: NPRP 5-093-2-034.

Author Contributions: Mohammed M. Farid and Ana Inés Fernández conceived and designed the experiments; Jessica Giro-Paloma and Refat Al-Shannaq performed the experiments; all the authors analyzed the data; Mohammed M. Farid contributed reagents/materials/analysis tools; Jessica Giro-Paloma and Refat Al-Shannaq wrote the paper.

Conflicts of Interest: The authors declare no conflict of interest. The founding sponsors had no role in the design of the study; in the collection, analyses, or interpretation of data; in the writing of the manuscript, and in the decision to publish the results.

References

1. Zhou, D.; Zhao, C.Y.; Tian, Y. Review on thermal energy storage with phase change materials (PCMs) in building applications. *Appl. Energy* **2012**, *92*, 593–605. [CrossRef]
2. Sharma, A.; Tyagi, V.V.; Chen, C.R.; Buddhi, D. Review on thermal energy storage with phase change materials and applications. *Renew. Sustain. Energy Rev.* **2009**, *13*, 318–345. [CrossRef]
3. Kuznik, F.; David, D.; Johannes, K.; Roux, J.-J. A review on phase change materials integrated in building walls. *Renew. Sustain. Energy Rev.* **2011**, *15*, 379–391. [CrossRef]
4. Farid, M.M.; Khudhair, A.M.; Razack, S.A.K.; Al-Hallaj, S. A review on phase change energy storage: Materials and applications. *Energy Convers. Manag.* **2004**, *45*, 1597–1615. [CrossRef]
5. Cabeza, L.F.; Castell, A.; Barreneche, C.; de Gracia, A.; Fernández, A.I. Materials used as PCM in thermal energy storage in buildings: A review. *Renew. Sustain. Energy Rev.* **2011**, *15*, 1675–1695. [CrossRef]
6. Soares, N.; Costa, J.J.; Gaspar, A.R.; Santos, P. Review of passive PCM latent heat thermal energy storage systems towards buildings' energy efficiency. *Energy Build.* **2013**, *59*, 82–103. [CrossRef]
7. Griffiths, P.W.; Eames, P.C. Performance of chilled ceiling panels using phase change material slurries as the heat transport medium. *Appl. Therm. Eng.* **2007**, *27*, 1756–1760. [CrossRef]
8. Huang, L.; Petermann, M.; Doetsch, C. Evaluation of paraffin/water emulsion as a phase change slurry for cooling applications. *Energy* **2009**, *34*, 1145–1155. [CrossRef]
9. Youssef, Z.; Delahaye, A.; Huang, L.; Trinquet, F.; Fournaison, L.; Pollerberg, C.; Doetsch, C. State of the art on phase change material slurries. *Energy Convers. Manag.* **2013**, *65*, 120–132. [CrossRef]
10. Pendyala, S. Macroencapsulation of Phase Change Materials for Thermal Energy Storage. Master Thesis, University of South Florida, Tampa, FL, USA, 22 June 2012.
11. Calvet, N.; Py, X.; Olivès, R.; Bédécarrats, J.-P.; Dumas, J.-P.; Jay, F. Enhanced performances of macro-encapsulated phase change materials (PCMs) by intensification of the internal effective thermal conductivity. *Energy* **2013**, *55*, 956–964. [CrossRef]

12. Zhao, C.Y.; Zhang, G.H. Review on microencapsulated phase change materials (MEPCMs): Fabrication, characterization and applications. *Renew. Sustain. Energy Rev.* **2011**, *15*, 3813–3832. [CrossRef]
13. Giro-Paloma, J.; Oncins, G.; Barreneche, C.; Martínez, M.; Fernández, A.I.; Cabeza, L.F. Physico-chemical and mechanical properties of microencapsulated phase change material. *Appl. Energy* **2013**, *109*, 441–448. [CrossRef]
14. Al Shannaq, R.; Farid, M.M. Microencapsulation of phase change materials (PCMs) for thermal energy storage systems. In *Advances in Thermal Energy Storage Systems*; Cabeza, L.F., Ed.; Elsevier: Amsterdam, The Netherlands, 2015; pp. 247–284.
15. Alvarado, J.L.; Marsh, C.; Sohn, C.; Phetteplace, G.; Newell, T. Thermal performance of microencapsulated phase change material slurry in turbulent flow under constant heat flux. *Int. J. Heat Mass Transf.* **2007**, *50*, 1938–1952. [CrossRef]
16. Diaconu, B.M.; Varga, S.; Oliveira, A.C. Experimental study of natural convection heat transfer in a microencapsulated phase change material slurry. *Energy* **2010**, *35*, 2688–2693. [CrossRef]
17. Zhang, P.; Ma, Z.W.; Wang, R.Z. An overview of phase change material slurries: MPCS and CHS. *Renew. Sustain. Energy Rev.* **2010**, *14*, 598–614. [CrossRef]
18. Zhang, P.; Ma, Z.W. An overview of fundamental studies and applications of phase change material slurries to secondary loop refrigeration and air conditioning systems. *Renew. Sustain. Energy Rev.* **2012**, *16*, 5021–5058. [CrossRef]
19. Medrano, M.; Yilmaz, M.O.; Nogués, M.; Martorell, I.; Roca, J.; Cabeza, L.F. Experimental evaluation of commercial heat exchangers for use as PCM thermal storage systems. *Appl. Energy* **2009**, *86*, 2047–2055. [CrossRef]
20. Su, J.-F.; Wang, L.-X.; Ren, L. Preparation and characterization of double-MF shell microPCMs used in building materials. *J. Appl. Polym. Sci.* **2005**, *97*, 1755–1762. [CrossRef]
21. Khudhair, A.M.; Farid, M.M. A review on energy conservation in building applications with thermal storage by latent heat using phase change materials. *Energy Convers. Manag.* **2004**, *45*, 263–275. [CrossRef]
22. Alawadhi, E.M. Thermal analysis of a building brick containing phase change material. *Energy Build.* **2008**, *40*, 351–357. [CrossRef]
23. Sarı, A.; Alkan, C.; Karaipekli, A.; Uzun, O. Microencapsulated n-octacosane as phase change material for thermal energy storage. *Sol. Energy* **2009**, *83*, 1757–1763. [CrossRef]
24. Cabeza, L.F.; Castellón, C.; Nogués, M.; Medrano, M.; Leppers, R.; Zubillaga, O. Use of microencapsulated PCM in concrete walls for energy savings. *Energy Build.* **2007**, *39*, 113–119. [CrossRef]
25. Castellón, C.; Medrano, M.; Roca, J.; Cabeza, L.F.; Navarro, M.E.; Fernández, A.I.; Lázaro, A.; Zalba, B. Effect of microencapsulated phase change material in sandwich panels. *Renew. Energy* **2010**, *35*, 2370–2374. [CrossRef]
26. Toppi, T.; Mazzarella, L. Gypsum based composite materials with micro-encapsulated PCM: Experimental correlations for thermal properties estimation on the basis of the composition. *Energy Build.* **2013**, *57*, 227–236. [CrossRef]
27. Entrop, A.G.; Brouwers, H.J.H.; Reinders, A.H.M.E. Experimental research on the use of micro-encapsulated Phase Change Materials to store solar energy in concrete floors and to save energy in Dutch houses. *Sol. Energy* **2011**, *85*, 1007–1020. [CrossRef]
28. Tyagi, V.V.; Kaushik, S.C.; Tyagi, S.K.; Akiyama, T. Development of phase change materials based microencapsulated technology for buildings: A review. *Renew. Sustain. Energy Rev.* **2011**, *15*, 1373–1391. [CrossRef]
29. Zalba, B.; Marín, J.M.; Cabeza, L.F.; Mehling, H. Review on thermal energy storage with phase change: Materials, heat transfer analysis and applications. *Appl. Therm. Eng.* **2003**, *23*, 251–283. [CrossRef]
30. Alkan, C.; Sari, A. Fatty acid/poly(methyl methacrylate) (PMMA) blends as form-stable phase change materials for latent heat thermal energy storage. *Sol. Energy* **2008**, *82*, 118–124. [CrossRef]
31. Zhang, H.; Xu, Q.; Zhao, Z.; Zhang, J.; Sun, Y.; Sun, L.; Xu, F.; Sawada, Y. Preparation and thermal performance of gypsum boards incorporated with microencapsulated phase change materials for thermal regulation. *Sol. Energy Mater. Sol. Cells* **2012**, *102*, 93–102. [CrossRef]
32. Borreguero, A.M.; Garrido, I.; Valverde, J.L.; Rodríguez, J.F.; Carmona, M. Development of smart gypsum composites by incorporating thermoregulating microcapsules. *Energy Build.* **2014**, *76*, 631–639. [CrossRef]

33. Karkri, M.; Lachheb, M.; Albouchi, F.; Nasrallah, S.B.; Krupa, I. Thermal properties of smart microencapsulated paraffin/plaster composites for the thermal regulation of buildings. *Energy Build.* **2015**, *88*, 183–192. [CrossRef]
34. Eddhahak-Ouni, A.; Drissi, S.; Colin, J.; Neji, J.; Care, S. Experimental and multi-scale analysis of the thermal properties of Portland cement concretes embedded with microencapsulated Phase Change Materials (PCMs). *Appl. Therm. Eng.* **2014**, *64*, 32–39. [CrossRef]
35. Schlink, U.; Rehwagen, M.; Damm, M.; Richter, M.; Borte, M.; Herbarth, O. Seasonal cycle of indoor-VOCs: Comparison of apartments and cities. *Atmos. Environ.* **2004**, *38*, 1181–1190. [CrossRef]
36. Ayoko, G.A. Volatile organic compounds in indoor environments. In *The Handbook of Environmental Chemistry*; Pluschke, P., Ed.; Springer-Verlag: Heidelberg, Germany, 2004; pp. 1–35.
37. Gallego, E.; Roca, X.; Perales, J.F.; Guardino, X. Determining indoor air quality and identifying the origin of odour episodes in indoor environments. *J. Environ. Sci.* **2009**, *21*, 333–339. [CrossRef]
38. Missia, D.A.; Demetriou, E.; Michael, N.; Tolis, E.I.; Bartzis, J.G. Indoor exposure from building materials: A field study. *Atmos. Environ.* **2010**, *44*, 4388–4395. [CrossRef]
39. Shin, S.H.; Jo, W.K. Volatile organic compound concentrations, emission rates, and source apportionment in newly-built apartments at pre-occupancy stage. *Chemosphere* **2012**, *89*, 569–578. [CrossRef] [PubMed]
40. Huang, H.; Haghighat, F. Building materials VOC emissions—A systematic parametric study. *Build. Environ.* **2003**, *38*, 995–1005. [CrossRef]
41. Zhang, Y.; Xu, Y. Characteristics and correlations of VOC emissions from building materials. *Int. J. Heat Mass Transf.* **2003**, *46*, 4877–4883. [CrossRef]
42. Kim, S.; Choi, Y.-K.; Park, K.-W.; Kim, J.T. Test methods and reduction of organic pollutant compound emissions from wood-based building and furniture materials. *Bioresour. Technol.* **2010**, *101*, 6562–6568. [CrossRef] [PubMed]
43. Lee, C.-S.; Haghighat, F.; Ghaly, W. Conjugate mass transfer modeling for VOC source and sink behavior of porous building materials: When to apply it? *J. Build. Phys.* **2006**, *30*, 91–111. [CrossRef]
44. Haghighat, F.; Huang, H. Integrated IAQ model for prediction of VOC emissions from building material. *Build. Environ.* **2003**, *38*, 1007–1017. [CrossRef]
45. Lee, C.-S.; Haghighat, F.; Ghaly, W.S. A study on VOC source and sink behavior in porous building materials—Analytical model development and assessment. *Indoor Air* **2005**, *15*, 183–96. [CrossRef] [PubMed]
46. Kim, S.; Kim, H.-J.; Moon, S.-J. Evaluation of VOC emissions from building finishing materials using a small chamber and VOC analyser. *Indoor Built Environ.* **2006**, *15*, 511–523. [CrossRef]
47. Magee, R.J.; Bodalal, A.; Biesenthal, T.A.; Lusztyk, E.; Brouzes, M.; Shaw, C.Y. Prediction of VOC Concentration Profiles in a Newly Constructed House Using Small Chamber Data and an IAQ Simulation Program. In Proceedings of the 9th International Conference on IAQ and Climate, Monterey, CA, USA, 30 June–5 July 2002; pp. 298–303.
48. Jia, C.; Batterman, S.A.; Relyea, G.E. Variability of indoor and outdoor VOC measurements: An analysis using variance components. *Environ. Pollut.* **2012**, *169*, 152–159. [CrossRef] [PubMed]
49. Norbäck, D.; Torgén, M.; Edling, C. Volatile organic compounds, respirable dust, and personal factors related to prevalence and incidence of sick building syndrome in primary schools. *Br. J. Ind. Med.* **1990**, *47*, 733–741. [CrossRef] [PubMed]
50. Redlich, C.A.; Sparer, J.; Cullen, M.R. Sick-building syndrome. *Lancet* **1997**, *349*, 1013–1016. [CrossRef]
51. Apte, M.G.; Daisey, J.M. VOCs and "Sick Building Syndrome": Application of a New Statistical Approach for SBS Research to US EPA BASE Study Data. In Proceedings of the Indoor Air 99, Edinburgh, Scotland, 30 June–5 July 1999; pp. 2–7.
52. Rufford, T.E.; Zhu, J.; Hulicova-Jurcakova, D. *Green Carbon Materials: Advances and Applications*; Pan Stanford: Boca Raton, FL, USA, 2014; pp. 1–288.
53. *Standard Guide for Small-Scale Environmental Chamber Determinations of Organic Emissions from Indoor Material/Products*; ASTM Standard D5116-97; ASTM International: West Conshohocken, PA, USA, 1997.
54. *Standard Test Method for Determining Formaldehyde Concentrations in Air from Wood Products Using a Small Scale Chamber*; ASTM Standard D6007-96; ASTM International: West Conshohocken, PA, USA, 1996.
55. Won, D.; Magee, R.J.; Lusztyk, E.; Nong, G.; Zhu, J.P.; Zhang, J.S.; Reardon, J.T.; Shaw, C.Y. A Comprehensive VOC Emission Database for Commonly-Used Building Materials. In Proceedings of the 7th International Conference of Healthy Buildings, Singapore, 7–11 December 2003; pp. 1–6.

56. Ouazia, B.; Reardon, J.; Sander, D. Making the Case for Reducing Ventilation Requirements through Selection of Low-Emission Materials. In Proceedings of the 10th International Conference on Indoor Air Quality and Climate, Beijing, China, 4–9 September 2005.
57. Xiong, J.; Zhang, Y.; Huang, S. Characterization of VOC and formaldehyde emission from building materials in static environmental chamber: Model development and application. *Indoor Built Environ.* **2011**, *20*, 217–225. [CrossRef]
58. Knöppel, H.; Schauenburg, H. Screening of household products for the emission of volatile organic compounds. *Environ. Int.* **1989**, *15*, 413–418. [CrossRef]
59. Lin, C.-C.; Yu, K.-P.; Zhao, P.; Lee, G. Evaluation of impact factors on VOC emissions and concentrations from wooden flooring based on chamber tests. *Build. Environ.* **2009**, *44*, 525–533. [CrossRef]
60. Safari, V.; Barreneche, C.; Castell, A.; Basatni, A.; Navarro, L.; Cabeza, L.; Haghighat, F. Volatile Organic Emission from PCM Building Materials. In Proceedings of the Innostock 2012—The 12th International Conference on Energy Storage, Lleida, Spain, 16–18 May 2012.
61. Sánchez, L.; Sánchez, P.; de Lucas, A.; Carmona, M.; Rodríguez, J.F. Microencapsulation of PCMs with a polystyrene shell. *Colloid Polym. Sci.* **2007**, *285*, 1377–1385. [CrossRef]
62. Sánchez-Silva, L.; Rodríguez, J.F.; Romero, A.; Borreguero, A.M.; Carmona, M.; Sánchez, P. Microencapsulation of PCMs with a styrene-methyl methacrylate copolymer shell by suspension-like polymerisation. *Chem. Eng. J.* **2010**, *157*, 216–222. [CrossRef]
63. Al-Shannaq, R.; Kurdi, J.; Al-Muhtaseb, S.; Dickinson, M.; Mohammed, F. Supercooling elimination of phase change materials (PCMs) microcapsules. *Energy* **2015**, *87*, 654–662. [CrossRef]
64. Rahman, A.; Dickinson, M.E.; Farid, M.M. Microencapsulation of a PCM through membrane emulsification and nanocompression-based determination of microcapsule strength. *Mater. Renew. Sustain. Energy* **2012**, *1*. [CrossRef]
65. Troyon, M.; Huang, L. Correction factor for contact area in nanoindentation measurements. *J. Mater. Res.* **2011**, *20*, 610–617. [CrossRef]
66. Oliver, W.C.; Pharr, G.M. Measurement of hardness and elastic modulus by instrumented indentation: Advances in understanding and refinements to methodology. *J. Mater. Res.* **2004**, *19*, 3–20. [CrossRef]
67. Oliver, W.C.; Pharr, G.M. An improved technique for determining hardness and elastic modulus using load and displacement sensing indentation experiments. *J. Mater. Res.* **1992**, *7*, 1564–1580. [CrossRef]
68. Hochstetter, G.; Jimenez, A.; Loubet, J.L. Strain-rate effects on hardness of glassy polymers in the nanoscale range. Comparison between quasi-static and continuous stiffness measurements. *J. Macromol. Sci. Part B* **2006**, *38*, 681–692. [CrossRef]
69. Giro-Paloma, J.; Roa, J.J.; Díez-Pascual, A.M.; Rayón, E.; Flores, A.; Martínez, M.; Chimenos, J.M.; Fernández, A.I. Depth-sensing indentation applied to polymers: A comparison between standard methods of analysis in relation to the nature of the materials. *Eur. Polym. J.* **2013**, *49*, 4047–4053. [CrossRef]
70. Fischer-Cripps, A.C. *Introduction to Contact Mechanics*, 2nd ed.; Springer: New York, NY, USA, 2007; p. 221.
71. Ma, D. New relationship between Young's modulus and nonideally sharp indentation parameters. *J. Mater. Res.* **2004**, *19*, 2144–2151. [CrossRef]
72. Al-Shannaq, R.; Farid, M.; Al-Muhtaseb, S.; Kurdi, J. Emulsion stability and cross-linking of PMMA microcapsules containing phase change materials. *Sol. Energy Mater. Sol. Cells* **2015**, *132*, 311–318. [CrossRef]
73. Qiu, X.; Li, W.; Song, G.; Chu, X.; Tang, G. Fabrication and characterization of microencapsulated n-octadecane with different crosslinked methylmethacrylate-based polymer shells. *Sol. Energy Mater. Sol. Cells* **2012**, *98*, 283–293. [CrossRef]
74. Zhang, X.; Fan, Y.; Tao, X.; Yick, K. Crystallization and prevention of supercooling of microencapsulated n-alkanes. *J. Colloid Interface Sci.* **2005**, *281*, 299–306. [CrossRef] [PubMed]
75. Chaiyasat, P.; Ogino, Y.; Suzuki, T.; Okubo, M. Influence of water domain formed in hexadecane core inside cross-linked capsule particle on thermal properties for heat storage application. *Colloid Polym. Sci.* **2008**, *286*, 753–759. [CrossRef]

9

Buckling Behavior of Substrate Supported Graphene Sheets

Kuijian Yang [1], Yuli Chen [1,*], Fei Pan [1], Shengtao Wang [1], Yong Ma [1] and Qijun Liu [2]

Academic Editor: Owen J. Guy

1 Institute of Solid Mechanics, International Research Institute for Multidisciplinary Science, Beihang University, Beijing 100191, China; yangkj@buaa.edu.cn (K.Y.); f_pan@buaa.edu.cn (F.P.); wangst@buaa.edu.cn (S.W.); mayong@buaa.edu.cn (Y.M.)

2 Department of Aerospace Engineering, University of Illinois, Champaign, IL 61801, USA; qliu38@illinois.edu

* Correspondence: yulichen@buaa.edu.cn

Abstract: The buckling of graphene sheets on substrates can significantly degrade their performance in materials and devices. Therefore, a systematic investigation on the buckling behavior of monolayer graphene sheet/substrate systems is carried out in this paper by both molecular mechanics simulations and theoretical analysis. From 70 simulation cases of simple-supported graphene sheets with different sizes under uniaxial compression, two different buckling modes are investigated and revealed to be dominated by the graphene size. Especially, for graphene sheets with length larger than 3 nm and width larger than 1.1 nm, the buckling mode depends only on the length/width ratio. Besides, it is revealed that the existence of graphene substrate can increase the critical buckling stress and strain to 4.39 N/m and 1.58%, respectively, which are about 10 times those for free-standing graphene sheets. Moreover, for graphene sheets with common size (longer than 20 nm), both theoretical and simulation results show that the critical buckling stress and strain are dominated only by the adhesive interactions with substrate and independent of the graphene size. Results in this work provide valuable insight and guidelines for the design and application of graphene-derived materials and nano-electromechanical systems.

Keywords: graphene; buckling; molecular mechanics; energy method; nano-electromechanical system

---

1. Introduction

Graphene has drawn widespread attention due to their excellent properties in electricity, thermology, optics and mechanics, and thus plays an important role in nano-materials and nano-devices [1–8]. However, as a two-dimensional nano material, graphene sheets are easily wrinkled and buckled under compression, which weakens their mechanical properties significantly. Therefore, in order to make better use of the excellent properties of graphene and improve the properties of graphene-based materials and devices, a comprehensive understanding on the buckling behaviors of graphene sheets is in great demand.

In recent years, experimental and theoretical studies have been carried out to investigate the buckling behavior of graphene sheets [9–22]. Most of them focused on free-standing graphene sheets, and only a few reported the buckling of substrate-supported graphene sheets. In practical applications, graphene sheets are normally used with substrate materials [23–25], and it is therefore of great importance to study the buckling behavior of substrate-supported graphene sheets. Anjomshoa *et al.* [19] and Wang *et al.* [20] investigated the compression behavior of the graphene embedded in polymer matrix. Jiang *et al.* [21] investigated the buckling of monolayer graphene on polyethylene terephthalate (PET) under compression using *in-situ* Raman spectroscopy and atomic force microscopy. They reported a critical compressive strain of around 0.7%. By assuming a simple failure mode, Wilber [22] formulated a continuum model of a graphene sheet supported by a flat rigid substrate and studied the effects of boundary conditions, substrate, and sheet length on the buckling of the sheet based on bifurcation theory. He obtained a theoretical buckling stress much lower than the simulation results, which mainly results from the inaccurate assumption of buckling mode.

However, in previous studies, the buckling modes of substrate-supported graphene sheets are rarely reported. The buckling mode determines the assumption of theoretical model, and it thus affects the critical buckling stress and critical buckling strain significantly. In this paper, the all-atom simulations are carried out to investigate the buckling modes as well as the buckling stress and strain of a monolayer graphene sheet subjected to a uniaxial compressive load on a flat rigid substrate. The effects of sheet size are also studied, and the critical condition for different buckling modes is obtained. Furthermore, the theoretical analysis is carried out based on the energy method to investigate the mechanism of buckling behavior and to reveal the substrate effect on the buckling stress and strain.

2. Atomistic Simulation

In this section, the graphene substrate is taken as an example to simulate the buckling behaviors of graphene sheets. The simulation method and the qualitative conclusion are also applicable to the graphene systems with other substrate materials.

2.1. Model and Method

Figure 1 presents the simulation model of a rectangular monolayer graphene sheet with the length l and the width w on a rigid graphene substrate. The substrate is modeled as a rigid body because it is much thicker than the graphene and thus the deformation is constrained by the lower-layer material. The compressive loads are applied on the zigzag edges of the graphene sheet, as shown in Figure 1. Both zigzag edges, denoted by L and R, are simply supported, which means the out-of-plane displacements (along z-direction) of all the atoms on edges L and R are restrained. Besides, along the armchair-edge direction (x-direction), the displacements of atoms on edge L (or R) are uniform, and their displacements along the zigzag-edge direction (y-direction) are unconfined. When loading, edges L and R move inside along the loading direction, and their displacements are the same in magnitude but opposite in direction.

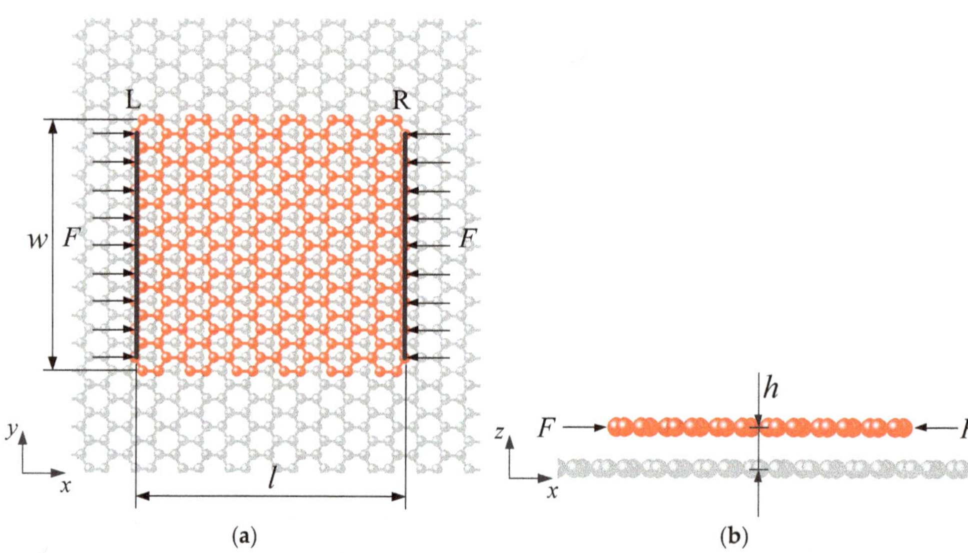

Figure 1. Sketch of a graphene sheet subjected to the compressive force F on an infinite large graphene substrate: (**a**) top view; and (**b**) side view.

The atomistic simulation on the deformation of graphene on rigid substrate under compressive load is carried out by the atomic-scale finite element method (AFEM) [26–28], which is essentially an effective and accurate molecular mechanics method based on the minimum potential energy theorem. The AFEM code has exactly the same formal structure as continuum finite element methods, and therefore can seamlessly be combined with the pre- and post- process codes of finite element software to apply boundary conditions and output atom positions and forces. Hence it has been widely used to solve nanoscale problems, especially for the carbon atom systems [26,27,29–31].

The second-generation Tersoff–Brenner potential [32] is adopted in the simulation to describe the covalent C–C bonds, and the initial length of C–C bond is 0.1425 nm. The van der Waals interactions between graphene sheet and substrate are described by the Lennard–Jones 6–12 potential [33], in which the potential energy between a carbon atom and a substrate atom is:

$$V(d) = 4\varepsilon_{LJ}\left[\left(\frac{\sigma_{LJ}}{d}\right)\right]^{12} - \left[\left(\frac{\sigma_{LJ}}{d}\right)\right]^{6} \quad (1)$$

where d is the distance between the two atoms, ε_{LJ} is the energy at the equilibrium distance $d = \sqrt[6]{2}\sigma_{LJ}$, and σ_{LJ} is the characteristic distance, which has been proved to be the equilibrium distance between an atom and an infinite plane substrate [34]. For graphene substrate, $\varepsilon_{LJ} = 2.39 \times 10^{-3}\ eV$ and $\sigma_{LJ} = 0.34\ nm$ [29,35].

The initial configuration of graphene is generated as an ideal perfect two-dimensional sheet with hexagonal lattices in the plane. The initial distance between the graphene sheet and the substrate is 0.34 nm. Before compression, the graphene is relaxed to reach its minimum energy state, which is the initial equilibrium state. Then, the compressive loadings are applied to edges L and R by controlling the displacements. The deformation and buckling behavior of the substrate-supported graphene sheets are investigated by recording the atom coordinates at each displacement increment, and the total compressive force F is obtained by summing up the loading-direction force components of all the atoms on either of the loading edges.

2.2. Buckling and Post-Buckling Process

Figure 2 presents the relation between compressive force and normalized displacement of the graphene sheet with length l = 5.4 nm and width w = 5.4 nm. In Figure 2, the compressive force F is divided by the sheet width w, which is essentially equivalent to the compressive stress of the graphene sheet. The normalized displacement is defined as the relative displacement Δu_x divided by the total sheet length l, which is essentially the contraction ratio of the graphene sheet and corresponds to the compressive strain of the graphene sheets before buckling. Here, $\Delta u_x = u_x^{\mathrm{L}} - u_x^{\mathrm{R}}$ is the loading displacement of edge L relative to that of edge R, which is a scalar component of displacement vector.

Figure 3 shows the deformed configurations of the graphene sheet at points A–F of the curve in Figure 2. According to the configurations, the curve can be divided into four stages: elastic compression, wrinkling, buckling and folding. Under compressive loading, the graphene first has a linear elastic in-plane deformation until point A. In this stage, both stress and strain of the graphene sheet increases linearly with the contraction, and the slope of the curve in Figure 2 is very high, indicating the extremely high modulus of the graphene. After Point A, the slope of the curve becomes a little lower because the graphene sheet wrinkles slightly on the substrate, and the compressive force keeps increasing until reaching the peak point B. This stage is very short and can only be observed in the graphene sheets with a certain range of length and width. At point B, the three small wrinkles reduces to two larger anti-symmetric waves, as shown in Figure 3c, and the force drops suddenly, which indicates the buckling of the graphene sheet. After that, with the increase of compressive displacement, the two anti-symmetric waves becomes larger and moves outside to the loading edges (Figure 3d,e), and finally converges to one edge suddenly, as shown in Figure 3f. After this sudden change, the sheet is folded and the compressive load drops almost to zero, as shown in Figure 2.

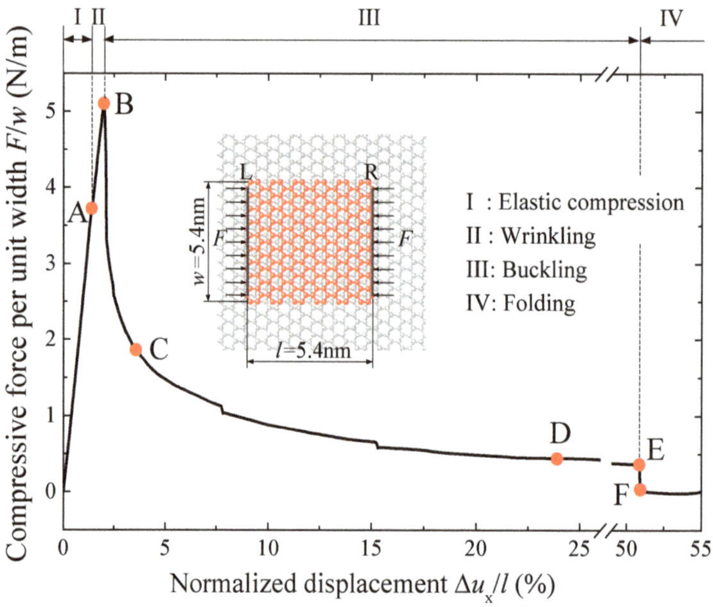

Figure 2. The compressive stress *versus* normalized displacement of the monolayer graphene sheet supported by a graphene substrate.

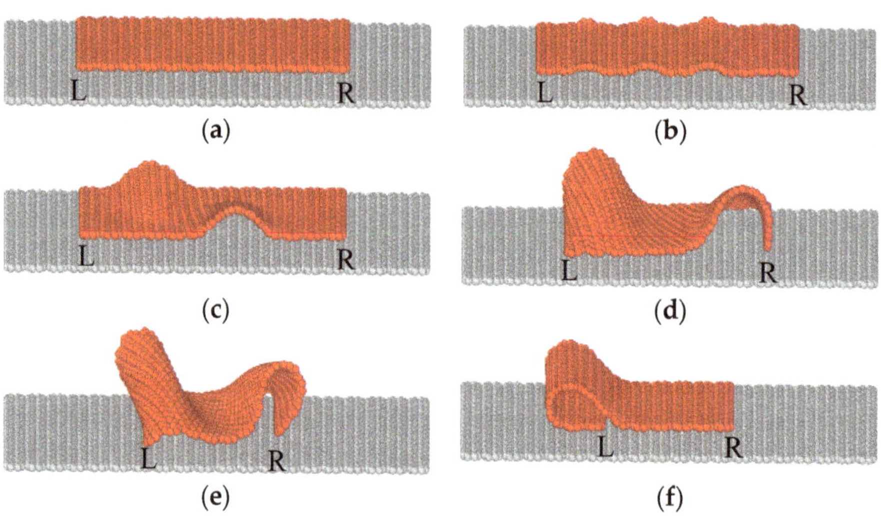

Figure 3. Configurations of the graphene sheet at points A–F of the curve in Figure 2. (**a**) $\Delta u_x/l = 1.37\%$; (**b**) $\Delta u_x/l = 2.00\%$; (**c**) $\Delta u_x/l = 3.71\%$; (**d**) $\Delta u_x/l = 24.09\%$; (**e**) $\Delta u_x/l = 50.89\%$; (**f**) $\Delta u_x/l = 50.93\%$.

It is interesting to find that not all graphene sheets buckle following above mode. Figure 4 presents the deformation and buckling process of a narrow graphene sheet on the substrate. The length l of the narrow graphene sheet is 5.4 nm, same as the case of Figure 2, but the width w is 0.74 nm, much smaller than that of the sheet in Figure 2. Different from Figure 2, the narrow graphene sheet displays only three stages in Figure 4: elastic compression, buckling and folding, and the deformed configurations at points A–E of the curve are presented in Figure 5.

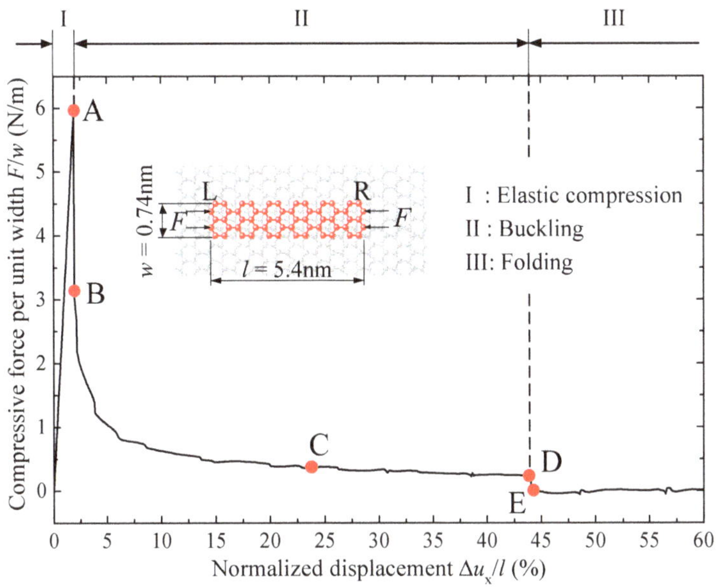

Figure 4. The compressive stress *versus* normalized displacement of the narrow monolayer graphene sheet supported by a graphene substrate.

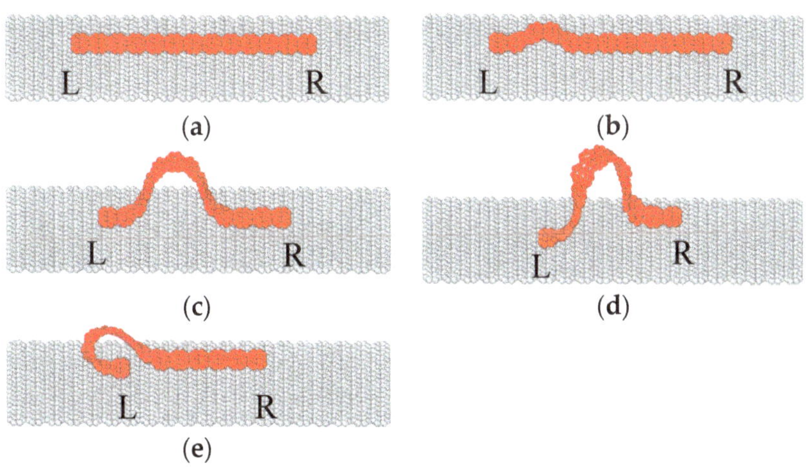

Figure 5. Configurations of the graphene sheet at points A–E of the curve in Figure 4. (**a**) $\Delta u_x/l = 1.85\%$; (**b**) $\Delta u_x/l = 2.59\%$; (**c**) $\Delta u_x/l = 23.72\%$; (**d**) $\Delta u_x/l = 44.00\%$; (**e**) $\Delta u_x/l = 44.11\%$.

In the first stage, both stress and strain of the graphene sheet increases linearly with the compression, and the slope of the curve in Figure 4 is exactly same as that in Figure 2, because the elastic modulus of graphene sheets is a constant and independent of the shape. With the increase of the compressive displacement, the graphene changes directly from flat to buckled configurations, without any wrinkles before buckling. After buckling, only one wave appears. Then, the sheet twists slightly due to the relative transverse movement of the two loading edges, as illustrated in Figure 5d. If the loading edges keep moving inside, the graphene sheet becomes folded suddenly, as shown in Figure 5e.

2.3. Buckling Mode

From Figures 3 and 5 it can be investigated that the size of graphene sheet can affect the buckling mode. In order to have a deep insight into the size effect on the buckling mode of substrate-supported graphene sheets, 70 models of graphene sheets with different sizes are established and simulated, and the buckling modes are presented in Figure 6.

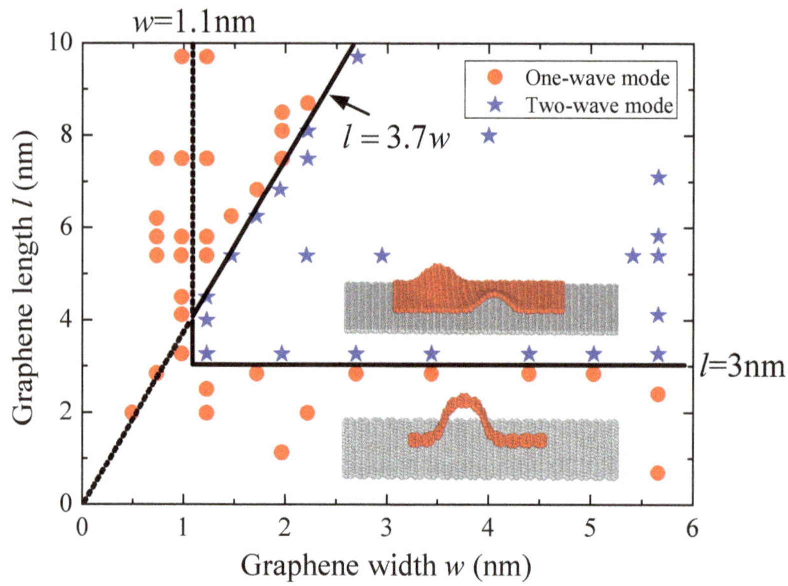

Figure 6. The buckling modes of monolayer graphene sheets with different sizes under uniaxial compression.

For the substrate-supported graphene sheet under uniaxial compression, if the sheet length less than 3 nm or the sheet width less than 1.1 nm, the only possible buckling mode is the one-wave mode similar to that presented in Figure 5. In this case, only one through wave appears when the sheet buckles, because the sheet with width less than 1.1 nm is so narrow that only through waves can be formed, and the sheet with length less than 3 nm is too short to form two or more waves.

Otherwise, if the sheet length is larger than 3 nm and the sheet width is larger than 1.1 nm, the buckling mode depends on the length/width ratio of the graphene sheet. The buckling of graphene sheets with the length/width ratio l/w larger than 3.7 still follows the one-wave mode. However, for the graphene sheets with the length/width ratio l/w less than 3.7, the buckling presents two-wave mode, similar to that illustrated in Figure 3. In this case, wrinkles are likely to appear before buckling and two anti-symmetric waves can be investigated after buckling.

2.4. The Critical Buckling Stress and Strain

The size of graphene sheet can also affect the critical buckling stress and strain, which can be obtained from the peak point of the curves demonstrated in Figures 2 and 4.

From the simulation results, the critical buckling stress, *i.e.*, the critical buckling force per unit width, decreases with increasing sheet length and approaches to 4.39 N/m, as shown in Figure 7. Especially, when the sheet length is larger than 20 nm, the critical buckling stress becomes a constant. This length independence property is consistent with the free-standing monolayer graphene sheet, but the constant buckling stress 4.39 N/m is about 10 times of the free-standing graphene sheets [12]. This difference results from the adhesion energy provided by the substrate. For free-standing graphene sheets, the buckling of graphene sheets is determined only by the strain energy, which is composed of the membrane energy and the bending energy. For the substrate-supported graphene sheets, the existence of adhesion energy suppresses the bending deformation and thus postpones the buckling.

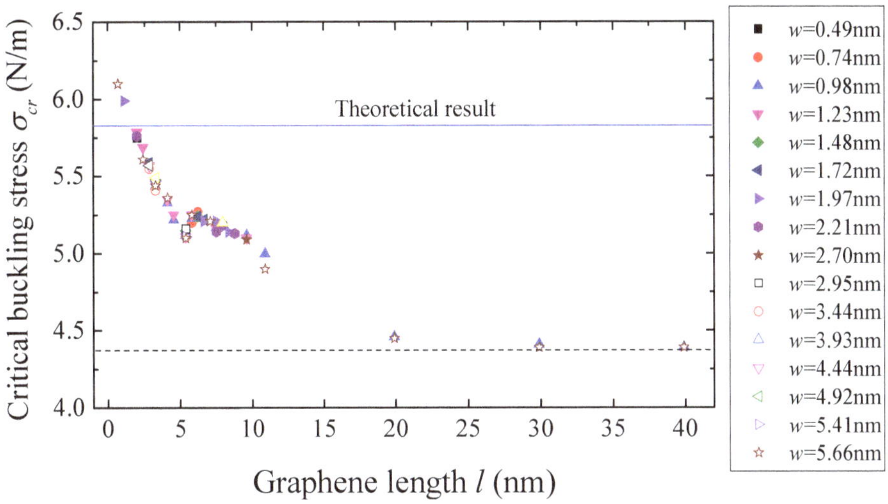

Figure 7. The critical buckling stress *versus* length of the monolayer graphene sheets with different sizes on a graphene substrate.

It is also found from the simulation results that the critical buckling strain is not sensitive to the graphene size, as shown in Figure 8. For the graphene sheet with length larger than 20 nm, the critical buckling strain is almost a constant, which is about 1.58%. This critical buckling strain is higher than the experimental result by Jiang *et al.* [21] because the graphene sheet is perfect in the simulation while it could have defects in experiments.

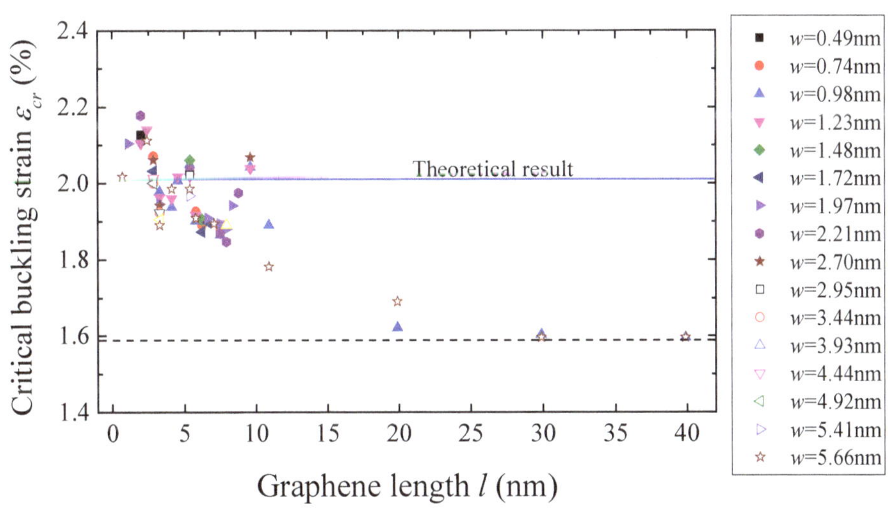

Figure 8. The critical buckling strain *versus* width of the monolayer graphene sheets with different sizes on a graphene substrate.

3. Theoretical Analysis

To obtain a better understanding on the buckling mechanism of the supported graphene sheets, the theoretical analysis is carried out by energy method [36,37].

The deformation and buckling behavior of the graphene sheets results from the competition between two energy terms [25,36–38]: the adhesion energy between graphene and substrate, which tries to keep the graphene sheet flat and attached completely to the substrate, and the strain energy stored in the graphene sheets due to the in-plane compressive deformation and the bending deformation. When the contraction ratio is small, the adhesion energy dominates the total energy of the graphene sheets, and the strain energy is relatively low, so the graphene can keep flat and completely conform to the substrate. With increasing compression, the strain energy becomes larger and larger, and the graphene tends to leave away from the substrate to reduce the membrane energy by bending deformation, and thus the sheet begins to wrinkle and/or buckle.

3.1. Energy Analysis

When the rectangular graphene sheet in Figure 1 is subjected to a uniaxial compression, denote the membrane strain of the sheet by ε_m and the out-of-plane displacement by u_z. Suppose the out-of-plane displacement depends only on x as $u_z = A\cdot\cos(\pi x/\lambda)$, in which A is the corrugation amplitude and λ is half of the wavelength. The membrane strain of graphene sheet can thus be obtained from von Karman nonlinear plate theory [39]:

$$\varepsilon_m = -\frac{\Delta u_x}{l} + \frac{1}{4}\left(\frac{\pi A}{\lambda}\right)^2 \tag{2}$$

The elastic strain energy density of graphene is

$$U_m = \frac{1}{2}Et\varepsilon_m^2 = \frac{1}{2}Et\left(-\frac{\Delta u_x}{l} + \frac{\pi^2 A^2}{4\lambda^2}\right)^2 \tag{3}$$

where $Et = 289$ N/m is the surface Young's modulus, obtained due to the linearity range of compressive force curve (see Figure 2).

The bending energy density is:

$$U_b = \frac{1}{2\lambda}\frac{D}{2}\int_S\left(\frac{d^2u_z}{dxds}\right)^2 ds = \frac{D}{4\lambda}\int_{-\lambda}^{\lambda}\frac{\left(\frac{A\pi^2}{\lambda^2}\cos\left(\frac{\pi}{\lambda}x\right)\right)^2}{\sqrt{1+\left(\frac{A\pi}{\lambda}\sin\left(\frac{\pi}{\lambda}x\right)\right)^2}}dx \tag{4}$$

where s is the curve segment along the free edge of graphene sheet, and D = 0.238 nN· nm [40] is the bending modulus of graphene.

The adhesion energy density between graphene and substrate can be obtained from the integral of Lennard-Jones potential over the area [35]:

$$U_{ad} = \frac{1}{2\lambda}\int_S \rho_c \int_{A_s}\rho_s V(d)\, dA_s ds \tag{5}$$

where ρ_c and ρ_s are the atom area density of graphene and substrate, respectively, and A_s is the infinite substrate area. For graphene substrate, $\rho_s = \rho_c = 4/\left(3\sqrt{3}l_{c\text{-}c}^2\right)$ and $l_{c\text{-}c}^2 = 0.1421$ nm is the equilibrium bond length of graphene. Using $\phi_s = 1.2\pi\rho_c\rho_s\varepsilon_{LJ}\sigma_{LJ}^2$ to denote the adhesion energy density of graphene at the characteristic distance $d = \sigma_{LJ}$, Equation (5) can be rewritten as:

$$U_{ad} = \frac{1}{2\lambda}\int_{-\lambda}^{\lambda}\frac{5\phi_s}{3}\left(\frac{2\sigma_{LJ}^{10}}{5\left(\sigma_{LJ}+A\cos\left(\frac{\pi}{\lambda}x\right)\right)^{10}} - \frac{\sigma_{LJ}^4}{\left(\sigma_{LJ}+A\cos\left(\frac{\pi}{\lambda}x\right)\right)^4}\right)\sqrt{1+\left(\frac{A\pi}{\lambda}\sin\left(\frac{\pi}{\lambda}x\right)\right)^2}dx \tag{6}$$

From Equations (3), (4) and (6), the total energy density of graphene is:

$$U_T = U_m\left(\frac{\Delta u_x}{l}, A, \lambda\right) + U_b(A,\lambda) + U_{ad}(A,\lambda) \tag{7}$$

Obviously, the total energy density U_T is a function of contraction ratio $\Delta u_x/l$, amplitude A and half wavelength λ. For each given contraction ratio $\Delta u_x/l$, the equilibrium state can be obtained by minimizing the total energy:

$$\frac{\partial U_T}{\partial A} = 0 \text{ and } \frac{\partial U_T}{\partial \lambda} = 0 \tag{8}$$

3.2. Numerical Solution

The equation set in Equation (8) can be solved numerically. When the contraction ratio $\Delta u_x/l$ is small, the amplitude $A = 0$ is solved from Equation (8). With increasing $\Delta u_x/l$, at a critical point, the amplitude becomes larger than zero, and the contraction ratio at this point is the critical buckling strain ε_{cr}. For graphene substrate, the critical buckling strain is $\varepsilon_{cr} = 2.01\%$, which is 27% bigger than the result of atomistic simulation in Section 3, as shown by the solid line in Figure 8. Accordingly, the critical buckling stress $\sigma_{cr} = Et\varepsilon_{cr}$ can be obtained as 5.81 N/m, which is presented by the solid line in Figure 7. The difference between the results of atomistic simulation and theoretical analysis attributes to two main factors: the inaccurate buckling mode and the effect of edge-stress. On the one hand, the trigonometric function wave is not accurate enough to describe the buckling mode of graphene sheet, although it has been widely adopted to study the instability problem of plates. The assumptive mode may be stiffer than the real bulking mode. Thus, under the same uniaxial compressive loading, the theoretical model is more difficult to bulk, resulting in higher critical buckling stress and strain than those of simulation result. On the other hand, the effect of edge stress caused by the unsaturation of atomic bonds along the edges tends to accelerate the buckling process of grapheme [41–43], because the reduced bond length along the edges leads to an extra compressive stress [43]. In atomistic simulation, the edge stress is taken into consideration, while it is ignored in the theoretical model. Therefore, in the simulation, the graphene sheet is much easier to buckle, which results in lower critical buckling stress and strain.

Besides, Equations (7) and (8) show that the total energy density U_T and its partial differential $\partial U_T/\partial A$ and $\partial U_T/\partial \lambda$ are all independent of the graphene size, which indicates that the critical buckling stress and strain are both constants depending only on the substrate material.

For different substrate, the adhesion parameters σ_{LJ} and ε_{LJ} and the material density parameter ρ_s are different [44–46]. However, in Equation (6), only two independent parameters affect the adhesion energy, *i.e.*, φ_s and σ_{LJ}, in which φ_s is determined by σ_{LJ}, ε_{LJ} and ρ_s. Therefore, the substrate effect can be presented in Figure 9 by the critical buckling strain *versus* φ_s, for substrates with various σ_{LJ}. The critical buckling strain of some typical substrates, including graphene, Si and SiO_2 are also presented in Figure 9.

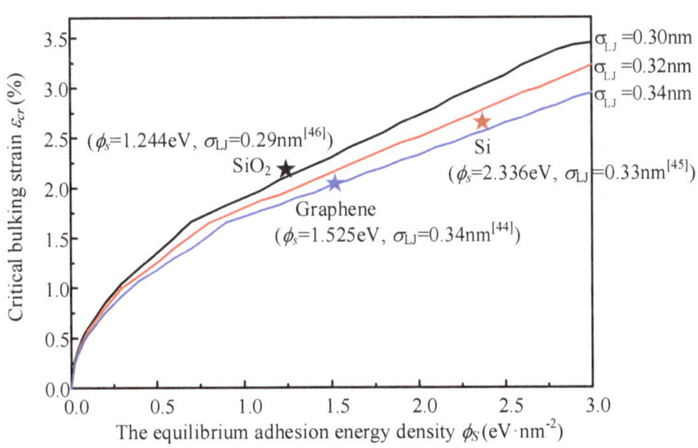

Figure 9. The effect of substrate adhesion on the critical buckling strain. The typical substrate materials are marked by stars.

It can be investigated in Figure 9 that the adhesion energy has significant influence on the critical buckling. It is much more difficult to be buckled if the adhesion energy between graphene and substrate is large.

3.3. Analytical Approximate Solution

The analytical approximate solution is derived to explicitly present the main factors that influence the critical buckling stress and strain as well as their influence tendency.

In order to attain the analytical solution, the bending energy density U_b in Equation (4) is simplified as:

$$U_b \approx \frac{1}{2\lambda}\cdot\frac{D}{2}\int_{-\lambda}^{\lambda}\left(\frac{d^2u_z}{dx^2}\right)^2 dx = \frac{DA^2\pi^4}{4\lambda^4} \tag{9}$$

The adhesion energy density can be simplified by Taylor expansion around $u_z = \sigma_{LJ}$ as:

$$U_{ad} \approx \frac{10\phi_s}{\lambda\sigma_{LJ}^2}\int_{-\lambda}^{\lambda} z^2 dx = \frac{10\phi_s A^2}{\sigma_{LJ}^2} \tag{10}$$

Substituting Equations (9) and (10) into Equations (7) and (8) yields:

$$\lambda^2 = \pi^2\sigma_{LJ}\sqrt{\frac{D}{40\phi_S}} \text{ and } A^2 = -\frac{8D}{Et} + 4\sigma_{LJ}\sqrt{\frac{D}{40\phi_S}}\cdot\frac{\Delta u_x}{l} \tag{11}$$

If A > 0, the graphene sheet buckles, otherwise it keeps flat. Thus, the critical condition can be obtained as:

$$\frac{\Delta u_x}{l} = \frac{4\sqrt{10D\phi_s}}{Et\sigma_{LJ}} \tag{12}$$

Therefore, the critical buckling strain is:

$$\varepsilon_{cr} = \frac{\Delta u_x}{l} = \frac{4\sqrt{10D\phi_s}}{Et\sigma_{LJ}} \tag{13}$$

The critical buckling stress is:

$$\sigma_{cr} = Et\varepsilon_{cr} = \frac{4\sqrt{10D\phi_s}}{\sigma_{LJ}} \tag{14}$$

For the system with graphene substrate, the critical buckling stress and strain obtained from the analytical approximate solution Equations (13) and (14) are $\varepsilon_{cr} = 3.09\%$ and $\sigma_{cr} = 8.90\ N/m$, respectively. Although it deviates from the numerical solution because of the linearization of van der Waals energy, the analytical expression can offer a clear relationship between the critical strain and each influence factor. From Equations (13) and (14), it is easy to know that the critical buckling strain ε_{cr} and critical buckling stress σ_{cr} increase nonlinearly with the increasing equilibrium adhesion energy density φ_s, which agrees with the numerical results shown in Figure 9.

4. Conclusions

In order to study the compressive buckling behavior of the substrate-supported graphene sheets, both molecular mechanics simulations and energy-based theoretical analysis are carried out to investigate the buckling mode and the critical buckling stress and strain of the monolayer graphene sheets. The following conclusions can be drawn.

(1) The buckling mode of the simple-supported graphene sheet is dominated by the graphene size. For graphene sheets with length larger than 3 nm and width larger than 1.1 nm, the buckling mode depends on the length/width ratio only.

(2) The critical buckling stress and strain of the substrate-supported graphene sheet with two simple-supported edges are about 4.39 N/m and 1.58%, which are dominated only by the substrate adhesion and independent of the graphene size.

(3) The substrate effect is revealed by both theoretical analysis and atomistic simulation. The critical buckling stress and strain can be increased by increasing the adhesive interaction of substrate. The existence of commonly-used substrates can make the critical buckling stress and strain increased by more than 10 times relative to the free-standing graphene sheets.

The estimations on the critical buckling stress and strain obtained in this study can provide guidelines for the design of graphene-derived materials and devices. For example, when the graphene sheet is used as conductive coating materials, the allowable stress of the coated devices should be lower than the critical buckling stress so that the graphene sheet has enough contact area with the substrate to display its high conductivity. Studies have shown that the electrical conductivity of Nano-Electromechanical System and the thermal conductivity of heat conduction element both depend on the smoothness of the graphene sheet [47,48] because the buckling behavior can change the electronic structure of the graphene sheets and make the graphene sheet locally separated from the substrate, which hinders the transfer of electronics, heat and load.

On the other hand, the critical buckling behavior can be utilized to design strain/stress sensors and switches [49] because the electrical conductivity changes significantly with the compressive strain at the critical buckling state.

Acknowledgments: Support from the National Natural Science Foundation of China (No. 11202012 and No. 11472027) and the Program for New Century Excellent Talents in University (No. NCET-13-0021) are gratefully acknowledged.

Author Contributions: Kuijian Yang, Shengtao Wang and Qijun Liu conducted all of the simulations and prepared the figures; Fei Pan and Yong Ma developed the theoretical model to explain the results; Yuli Chen supervised the study and wrote the manuscript. All authors reviewed the manuscript.

Conflicts of Interest: The authors declare no conflict of interest.

References

1. Novoselov, K.S.; FalKo, V.I.; Colombo, L.; Gellert, P.R.; Schwab, M.G.; Kim, K.A. A roadmap for graphene. *Nature* **2012**, *490*, 192–200. [CrossRef] [PubMed]
2. Rao, C.N.R.; Sood, A.K.; Subrahmanyam, K.S.; Govindaraj, A. Graphene: The new two-dimensional nanomaterial. *Angew. Chem.* **2009**, *48*, 7752–7777. [CrossRef] [PubMed]
3. Tkachev, S.V.; Buslaeva, E.Y.; Gubin, S.P. Graphene: A novel carbon nanomaterial. *Inorg. Mater.* **2011**, *47*, 1–10. [CrossRef]
4. Feng, X.; Narita, A. Graphene Nanoribbons. In *Encyclopedia of Polymeric Nanomaterials*; Springer-Verlag: Berlin, Germany; Heidelberg, Germany, 2014.
5. Otsuji, T.; Suemitsu, M.; Ryzhii, V. Graphene materials and devices for terahertz science and technology. In Proceedings of the 2013 Asia-Pacific Microwave Conference Proceedings (APMC), Seoul, Korea, 5–8 November 2013; pp. 515–517.
6. Xu, G.Y.; Abbott, J.; Qin, L.; Yeung, K.Y.M.; Song, Y.; Yoon, H. Electrophoretic and field-effect graphene for all-electrical DNA array technology. *Nat. Commun.* **2014**, *5*, 4866. [CrossRef] [PubMed]
7. Echtermeyer, T.J.; Lemme, M.C.; Baus, M.; Szafranek, B.N.; Geim, A.K.; Hurz, H. Nonvolatile switching in graphene field-effect devices. *IEEE Electron Device Lett.* **2008**, *29*, 952–954. [CrossRef]
8. Adán-Más, A.; Wei, D. Photoelectrochemical properties of graphene and its derivatives. *Nanomaterials* **2013**, *3*, 325–356. [CrossRef]
9. Parashar, A.; Mertiny, P. Representative volume element to estimate buckling behavior of graphene/polymer nanocomposite. *Nanoscale Res. Lett.* **2012**, *7*, 1–6. [CrossRef] [PubMed]
10. Ansari, R.; Shahabodini, A.; Alipour, A.; Rouhi, H. Stability of a single-layer graphene sheet with various edge conditions: A non-local plate model including interatomic potentials. *Proc. Inst. Mech. Eng. N J. Nanoeng. Nanosyst.* **2012**, *226*, 51–60. [CrossRef]

11. Ansari, R.; Sahmani, S. Prediction of biaxial buckling behavior of single-layered graphene sheets based on nonlocal plate models and molecular dynamics simulations. *Appl. Math. Modell.* **2013**, *37*, 7338–7351. [CrossRef]
12. Ansari, R.; Shahabodini, A.; Rouhi, H. Prediction of the biaxial buckling and vibration behavior of graphene via a nonlocal atomistic-based plate theory. *Compos. Struct.* **2013**, *95*, 88–94. [CrossRef]
13. Omer, C. Elastic buckling behavior of skew shaped single-layer graphene sheets. *Thin Solid Films* **2014**, *550*, 450–458.
14. Farajpour, A.; Solghar, A.A.; Shahidi, A. Postbuckling analysis of multi-layered graphene sheets under non-uniform biaxial compression. *Physica E* **2013**, *47*, 197–206. [CrossRef]
15. Shi, J.X.; Ni, Q.Q.; Lei, X.W.; Natsuki, T. Nonlocal elasticity theory for the buckling of double-layer graphene nanoribbons based on a continuum model. *Comput. Mater.* **2011**, *50*, 3085–3090. [CrossRef]
16. Pradhan, S.C.; Murmu, T. Small scale effect on the buckling of single-layered graphene sheets under biaxial compression via nonlocal continuum mechanics. *Comput. Mater. Sci.* **2009**, *47*, 268–274. [CrossRef]
17. Natsuki, T.; Shi, J.X.; Ni, Q.Q. Buckling instability of circular double-layered graphene sheets. *J. Phys. Condens. Matter* **2012**, *24*, 135004–135008. [CrossRef] [PubMed]
18. Wang, L.; Zhang, Q. Elastic behavior of bilayer graphene under in-plane loadings. *Curr. Appl. Phys.* **2012**, *12*, 1173–1177. [CrossRef]
19. Anjomshoa, A.; Shahidi, A.R.; Hassani, B.; Jomehzadeh, E. Finite element buckling analysis of multi-layered graphene sheets on elastic substrate based on nonlocal elasticity theory. *Appl. Math. Modell.* **2014**, *38*, 5934–5955. [CrossRef]
20. Wang, G.; Liu, L.; Dai, Z.; Liu, Q.; Miao, H.; Zhang, Z. Biaxial compressive behavior of embedded monolayer graphene inside flexible poly(methyl methacrylate) matrix. *Carbon* **2015**, *86*, 69–77. [CrossRef]
21. Jiang, T.; Huang, R.; Zhu, Y. Interfacial sliding and buckling of monolayer graphene on a stretchable substrate. *Adv. Funct. Mater.* **2014**, *24*, 396–402. [CrossRef]
22. Wilber, J.P. Buckling of graphene layers supported by rigid substrates. *J. Comput. Theor. Nanosci.* **2010**, *7*, 2338–2348. [CrossRef]
23. Pablo, M.; Martin, S.; Pinardi, A.L.; Gonzalo, O.; Martin-gago, J.A. Strain-driven moire' superstructures of epitaxial graphene on transition metal surfaces. *ACS Nano* **2011**, *5*, 5627–5634.
24. N'Diaye, A.T.; Gastel, R.V.; Martinez-Galera, A.J.; Coraux, J.; Hattab, H.; Wall, D. *In situ* observation of stress relaxation in epitaxial graphene. *New J. Phys.* **2009**, *11*. [CrossRef]
25. Zhou, Y.; Chen, Y.; Liu, B.; Wang, S.; Yang, Z.; Hu, M. Mechanics of nanoscale wrinkling of graphene on a non-developable surface. *Carbon* **2015**, *84*, 263–271. [CrossRef]
26. Liu, B.; Huang, Y.; Jiang, H.; Qu, S.; Hwang, K.C. The atomic-scale finite element method. *Comput. Methods Appl. Mech. Engrg.* **2004**, *193*, 1849–1864. [CrossRef]
27. Liu, B.; Jiang, H.; Huang, Y.; Qu, S.; Yu, M.F.; Huang, K.C. Atomic-scale finite element method in multiscale computation with applications to carbon nanotubes. *Phys. Rev. B* **2005**, *72*. [CrossRef]
28. Guo, X.; Leung, A.; He, X.Q.; Jiang, H.; Huang, Y. Bending buckling of single-walled carbon nanotubes by atomic-scale finite element. *Compos. B Eng.* **2008**, *39*, 202–208. [CrossRef]
29. Chen, Y.L.; Liu, B.; Wu, J.; Huang, Y.; Jiang, H.; Hwang, K.C. Mechanics of hydrogen storage in carbon nanotubes. *J. Mech. Phys. Solids* **2008**, *56*, 3224–3241. [CrossRef]
30. Chen, Y.L.; Liu, B.; He, X.Q.; Huang, Y.; Hwang, K.C. Failure analysis and the optimal toughness design of carbon nanotube-reinforced composites. *Compos. Sci. Technol.* **2010**, *70*, 1360–1367. [CrossRef]
31. Chen, Y.L.; Liu, B.; Hwang, K.C.; Huang, Y. A theoretical evaluation of load transfer in multi-walled carbon nanotubes. *Carbon* **2011**, *49*, 193–197. [CrossRef]
32. Brenner, D.W.; Shenderova, O.A.; Harrison, J.A.; Stuart, S.J.; Ni, B.; Sinnott, S.B. A second-generation reactive empirical bond order (REBO) potential energy expression for hydrocarbons. *J. Phys. Condens. Matter* **2002**, *14*, 783–802. [CrossRef]
33. Mandel, F.; Bearman, R.J.; Bearman, M.Y. Numerical solutions of the Percus-Yevick equation for the Lennard-Jones (6–12) and hard-sphere potentials. *J. Chem. Phys.* **1970**, *52*, 3315–3323. [CrossRef]
34. Chong, G.; Gao, G.H.; Yu, Y.X.; Mao, Z.Q. Simulation study of hydrogen storage in single walled carbon nanotubes. *Int. J. Hydrog. Energy* **2001**, *26*, 691–696.
35. Lu, W.B.; Wu, J.; Jiang, L.Y.; Huang, Y.; Hwang, K.C.; Liu, B. A cohesive law for multi-wall carbon nanotubes. *Philos. Mag.* **2007**, *87*, 2221–2232. [CrossRef]

36. Koskinen, P. Graphene cardboard: From ripples to tunable metamaterial. *Appl. Phys. Lett.* **2014**, *104*, 101902–101904. [CrossRef]
37. Lu, Z.; Dunn, M.L. Van der Waals adhesion of graphene membranes. *J. Appl. Phys.* **2010**, *107*, 044301–044305. [CrossRef]
38. Cranford, S.W. Buckling induced delamination of graphene composites through hybrid molecular modeling. *Appl. Phys. Lett.* **2013**, *102*, 031902–031905. [CrossRef]
39. Landau, L.D.; Lifshitz, E.M. *Theory of Elasticity*; Springer: Berlin, Germany; Heidelberg, Germany, 1959; Volume 42, pp. 580–582.
40. Zhang, K.; Arroyo, M. Adhesion and friction control localized folding in supported graphene. *J. Appl. Phys.* **2013**, *113*, 193501–193508. [CrossRef]
41. Ramasubramaniam, A.; Koskinen, P.; Kit, O.O.; Shenoy, V.B. Edge-stress-induced spontaneous twisting of graphene nanoribbons. *J. Appl. Phys.* **2012**, *111*, 054302–054305. [CrossRef]
42. Shenoy, V.B.; Reddy, C.D.; Ramasubramaniam, A.; Zhang, Y.W. Edge-stress-induced warping of graphene sheets and nanoribbons. *Phys. Rev. Lett.* **2008**, *101*, 3382–3390. [CrossRef] [PubMed]
43. Huang, B.; Liu, M.; Su, N.; Wu, J.; Duan, W.; Gu, B.; Liu, F. Quantum manifestations of graphene edge stress and edge instability: A first-principles study. *Phys. Rev. Lett.* **2009**, *102*. [CrossRef] [PubMed]
44. Girifalco, L.A.; Hodak, M.; Lee, R.S. Carbon nanotubes, buckyballs, ropes, and a universal graphitic potential. *Phys. Rev. B* **2000**, *62*, 13104–13110. [CrossRef]
45. Ong, Z.Y.; Pop, E. Molecular dynamics simulation of thermal boundary conductance between carbon nanotubes and SiO_2. *Phys. Rev. B* **2010**, *81*. [CrossRef]
46. Neek-Amal, M.; Peeters, F.M. Strain-engineered graphene through a nanostructured substrate. I. Deformations. *Phys. Rev. B* **2012**, *85*, 2202–2208. [CrossRef]
47. Kerszberg, N.; Suryanarayana, P. *Ab initio* strain engineering of graphene: Opening bandgaps up to 1 eV. *Rsc Adv.* **2015**, *5*, 43810–43814. [CrossRef]
48. Pereira, V.M.; Neto, A.H.C. Strain engineering of graphene's electronic structure. *Phys. Rev. Lett.* **2009**, *103*. [CrossRef] [PubMed]
49. Wang, Y.; Yang, R.; Shi, Z.; Zhang, L.; Shi, D.; Wang, E.; Zhang, G. Super-elastic graphene ripples for flexible strain sensors. *Acs Nano* **2011**, *5*, 3645–3650. [CrossRef] [PubMed]

10

Tuning the Performance of Metallic Auxetic Metamaterials by Using Buckling and Plasticity

Arash Ghaedizadeh [1], Jianhu Shen [1], Xin Ren [1,2] and Yi Min Xie [1,*]

Academic Editor: Geminiano Mancusi

1 Centre for Innovative Structures and Materials, School of Engineering, RMIT University, GPO Box 2476, Melbourne 3001, Australia; s3393723@student.rmit.edu.au (A.G.); jianhu.shen@rmit.edu.au (J.S.); s3460277@student.rmit.edu.au (X.R.)

2 Key Laboratory of Traffic Safety on Track, School of Traffic & Transportation Engineering, Central South University, Changsha 410075, China

* Correspondence: mike.xie@rmit.edu.au

Abstract: Metallic auxetic metamaterials are of great potential to be used in many applications because of their superior mechanical performance to elastomer-based auxetic materials. Due to the limited knowledge on this new type of materials under large plastic deformation, the implementation of such materials in practical applications remains elusive. In contrast to the elastomer-based metamaterials, metallic ones possess new features as a result of the nonlinear deformation of their metallic microstructures under large deformation. The loss of auxetic behavior in metallic metamaterials led us to carry out a numerical and experimental study to investigate the mechanism of the observed phenomenon. A general approach was proposed to tune the performance of auxetic metallic metamaterials undergoing large plastic deformation using buckling behavior and the plasticity of base material. Both experiments and finite element simulations were used to verify the effectiveness of the developed approach. By employing this approach, a 2D auxetic metamaterial was derived from a regular square lattice. Then, by altering the initial geometry of microstructure with the desired buckling pattern, the metallic metamaterials exhibit auxetic behavior with tuneable mechanical properties. A systematic parametric study using the validated finite element models was conducted to reveal the novel features of metallic auxetic metamaterials undergoing large plastic deformation. The results of this study provide a useful guideline for the design of 2D metallic auxetic metamaterials for various applications.

Keywords: mechanical metamaterial; auxetic; buckling; large deformation; plasticity

---

1. Introduction

Four fundamental mechanical properties of materials in isotropic elasticity are Poisson's ratio (ν), Young's modulus (E), shear modulus (G), and bulk modulus (K). Note that these four parameters are interrelated, and the applicability of the theory of elasticity is limited to stress-strain conditions wherein the stress is below the yield point. Young's modulus is the measure of stiffness in the linear elastic range. Poisson's ratio is the least studied among these elastic constants. However, it governs the deformation feature of materials under various loading conditions [1,2]. This property is represented by the negative of the ratio between transverse and longitudinal strains [3]. The majorities of materials have a positive Poisson's ratio that is about 0.5 for rubber and 0.3 for glass and steel [3]. The thermodynamic requirement in the theory of elasticity for a conservative system demonstrates that for homogeneous isotropic materials, the theoretical bound of Poisson's ratio is from −1 to 0.5. Therefore, the existence of materials with negative Poisson' ratio (NPR) has long been accepted and they are known as "auxetic" materials [3,4]. The deformation feature of auxetic materials can be described as an anomalous behavior that indicates that materials expand (contract) in a transverse direction when uniaxially stretched (compressed) [3–7]. Based on their origin, auxetic materials can be classified as naturally occurring or artificial. The exotic mechanical properties of the natural auxetic materials inspired plenty of researchers to search for the underlying mechanisms that cause auxetic behavior. Those findings were applied to the design of artificial materials [1,8].

Many cellular structures were designed to have exceptional properties. When these properties are superior to these found in nature, the artificial cellular material is termed as a metamaterial. Metamaterials gain their uncommon and unique properties from geometrical configurations rather than chemical composition [9,10]. Furthermore, the behavior of metamaterials was normally explained in terms of the intricate interplay between their microstructures and their deformation mechanisms [11,12].

Over the past decades, continuous efforts were put in towards the development of auxetic metamaterials because of their unconventional behavior under uniaxial loading. Their unique properties opened a window towards a wide range of applications such as the design of novel fasteners [13], biomedical applications [14], energy-absorbing devices [15], acoustic dampers [16], membrane filters with variable permeability [17], and the design of composites [18].

Macroscopic auxetic cellular structures with 2D re-entrant honeycombs cells were firstly presented by Gibson and Ashby [19]. Then an artificially designed auxetic foam with a re-entrant cell was reported in the seminal research of Lakes in 1987 [3]. Following that, the interest of researchers was concentrated on periodic 2D auxetic materials through modifying the geometry of microstructures. These consist of composites with star-shaped inclusions [20], a structure formed from lozenge grids [21], a formed structure from square grids [22], square lattice of circular holes [11], and many other 2D topological patterns. Due to the notable challenge in the fabrication of 3D auxetic materials that consist of a microstructure with complicated geometries, only a small number of synthetic 3D auxetic metamaterials have been fabricated [10]. Recently an auxetic cube using "buckliball" as the building cell by Babaee *et al.* [23] and auxetic cube with simple spherical by Shen *et al.* [10] were invented, manufactured, and tested. It is worth noting that, most of these 2D and 3D auxetic metamaterials were constructed with soft elastic base materials. They deformed elastically under uniaxial loading and the deformation was reversible and repeatable. However, the weak stiffness and strength and the low softening temperature of these elastomeric metamaterials prevented them from many applications requiring high stress and high temperature. In contrast to elastomers, metals have higher stiffness, strength, density, and melting point. Copper, gold, platinum, silver, and brass are especially ductile and might be suitable for fabricating novel auxetic metamaterials. There is a great potential for metallic auxetic metamaterials to be used in engineering products.

Only a few researchers studied the auxetic performance of metallic auxetic metamaterials undergoing large plastic deformation. Therefore, knowledge of the deformation features and auxetic performance of metallic metamaterials is very limited. The first work in this area was done by Friis *et al.* [5], who proposed a polymeric and metallic foam with auxetic behavior. Recently, a 2D metallic auxetic periodic structure with low porosity was reported by Taylor *et al.* [24]. Another study in this area was carried out by Dirrenberger *et al.* [25] on the auxetic behavior of the metamaterials undergoing plastic deformation by using an anisotropic compressible plasticity framework as a macroscopic model. Moreover, the behavior of elastoplastic auxetic microstructural arrays was assessed employing a continuum-based micromechanical model by Gilat *et al.* [26].

Inspired by previous works on buckling-induced auxetic elastic and elastoplastic metamaterials [10,11,23,27–30], we expanded the similar design approach to metallic structures. However, the buckling-induced geometrical design lost its auxetic behavior under large plastic deformation when the base material was changed to metals. Therefore, the specific aim of this research is to develop a general approach to designing and tuning the metallic metamaterials with auxetic performance undergoing large plastic deformation. The auxetic performance of metamaterials on any length of scale is dependent on the geometrical features and deformation mechanisms of their microstructures. Hence, the general approach of the metamaterial starts at the microstructure level by employing an existing deformation mechanism that leads to auxetic behavior [10,11]. In this approach, the geometry of microstructure of a regular structure is modified then it is altered by the desired buckling modes. A new category of 2D metallic auxetic metamaterials over a large strain range can be obtained by using different scale factors. The new metallic metamaterial exhibits auxetic behavior with almost constant negative values of Poisson's ratio over a large strain range. This feature is very useful for practical applications in macroscopic metamaterials as well as in nanoscale structures [31,32]. A set of parametric studies using validated finite element (FE) simulation have been done to show the possibility of providing an effective control method to tune the mechanical properties of the 2D metallic auxetic metamaterial during the designing process.

2. A General Approach to Tuning the Performance of 2D Metallic Auxetic Metamaterial

In order to obtain a novel class of 2D metallic auxetic metamaterials with the potential to retain tuneable auxetic behavior under a wide range of applied strains, a general systematic approach was developed based on our previous work [10]. This general approach leads to the possibility of tuning the auxetic behavior and mechanical properties for the targeted applications. The key idea of this methodology is the exploitation of a pattern from the linear buckling analysis to create microstructures of metamaterials with an auxetic response. This methodology can be divided into four steps as described in the following sections. Moreover, the magnitude of alteration of microstructure and plastic properties of the base material were studied as the primary key factors in tuning the performance of 2D metallic auxetic metamaterial. These results are presented in the parametric study section.

2.1. Initial Geometric Design of Microstructures Modified from a Regular Pattern

The auxetic performance of a metamaterial was determined by the geometrical features and deformation mechanisms of its microstructure. According to the geometry of microstructure of metamaterials that exhibited auxetic behavior induced by elastic instability, the ribs of their building cell were slightly thicker in the proximity of their connecting points rather than the middle parts of the ribs [10,11,23,27]. Consequently, rotation occurred at the middle parts of these ribs in clockwise and counter-clockwise directions, to form a deformation mechanism after the buckling of the ribs [33,34]. These observations and mechanisms led us to think that buckling-induced auxetic cellular metamaterials could be produced from a conventional regular cellular lattice through geometrical modification of their microstructures.

In this research the microstructure of a new buckling-induced 2D auxetic metamaterial was produced from an existing conventional regular lattice through the proposed modification. This method consists of moving a small portion of the middle part of their ribs to the proximity of the connecting joints. This geometrical modification on the microstructure was based on the hypothesis that, under deformation, joints behaved as "rigid joints" and rotated relatively to each other after the ribs buckled. The geometric modification of the unit cell of the regular lattice through the proposed method is illustrated in Figure 1b,c. The new buckling-induced 2D auxetic metamaterial was similar to the buckling-induced auxetic metamaterials investigated by Bertoldi *et al.* [11]. According to the buckling pattern, the representative volume element (RVE) contained four unit cells as shown in Figure 1d. Similar to the designing approach of elastic metamaterials [10], the bulk 2D metamaterial was constructed by replicating RVEs along two planar directions, as shown in Figure 1e.

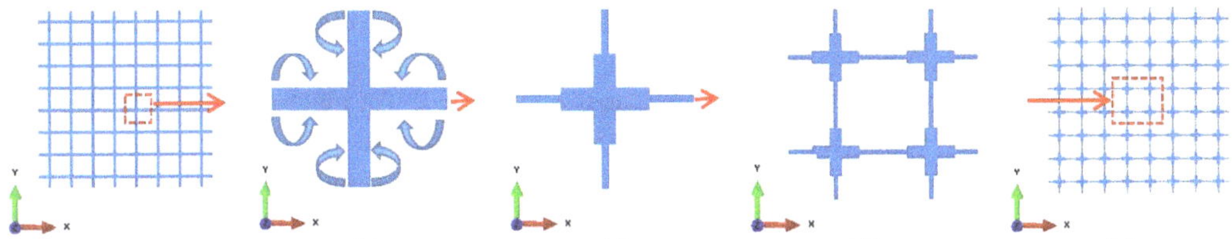

(a) (b) (c) (d) (e)

Figure 1. (**a**) Conventional regular lattice; (**b**) building unit cell of initial regular lattice; (**c**) modified unit building cell of microstructure; (**d**) representative volume element; (**e**) designed bulk metamaterial with cross-shaped microstructures for experiments.

2.2. Investigation of the Instability of Initial 2D Auxetic Metamaterial Using a Linear Perturbation Procedure (Linear Buckling Analysis)

The elastic instability in elastomeric metamaterials triggered a pattern switching phenomenon above a critical load, which led to auxetic behavior under compression [28]. In order to identify the desired buckling mode, finite element analysis (FEA) was performed on the existing cross lattice and the modified periodic cellular metamaterial. The stability of these microstructures was evaluated using commercial finite element software ABAQUS with standard solver (Simulia, Providence, RI, USA) and a linear perturbation procedure was conducted to find the critical loads (eigenvalues) for different buckling modes (Eigen modes). The numerical models were constructed with linear solid elements of the secondary accuracy (element C3D8 with a mesh sweeping seed size of 0.4 mm). It should be noted that the applied boundary conditions influence the buckling modes of the finite sized numerical model. In this study, all degrees of freedom on the top and bottom nodes of the numerical model were constrained except for the nodal movement of the top surface in the direction of the applied load. The analysis was carried out to simulate the standard uniaxial compression. It is worth noting that for purely elastic models the buckling modes were independent of the modulus of elasticity; thus any value of modulus could be used to conduct the buckling analysis. Therefore, the proposed design approach was independent of the influences of the elastic base material. The Young's modulus of elasticity and Poisson's ratio of brass were used as the properties of the base material in numerical simulations.

2.3. Identification of the Desired Buckling Pattern from an Elastic Instability Analysis

As mentioned before, the linear buckling analysis yields a number of eigenvalues and buckling mode eigenvectors corresponding to these eigenvalues. Normally the lowest eigenvalue with the corresponding buckling mode is of interest because the buckling modes with higher eigenvalues are difficult to trigger under quasi-static loading. The desired buckling modes, termed alternating periodic buckling mode, were identified through the deformation features of buckling-induced auxetic metamaterials previously investigated [10,11,23,35]. In most cases, the auxetic behavior of this type of metamaterials was featured by such a mode with alternating mutually orthogonal ellipses. According to this identified principle, the seventh mode shape of initial metamaterial (regular conventional lattice) is similar to the observed buckling pattern that leads to the auxetic behavior. In contrast, the first buckling mode shape of the newly designed metamaterial (with the lowest eigenvalue) can be adopted as the desired buckling mode shape.

2.4. Design New Metamaterials through Superposition of Desired Buckling Mode

Previous research on designing elastic buckling-induced auxetic metamaterials revealed that altering the initial geometry of the microstructure was a way to trigger the deformation to a specific pattern that led to auxetic behavior [10,34]. One set of the output of the buckling analysis was a number of non-dimensional buckling modes. Thus, in order to define the magnitude of alteration of microstructure, a scale factor is required. In this study, the pattern scale factor (PSF) was proposed to add the desired buckling mode shape to the initial geometry of the metamaterial. To quantify the magnitude of the alternation, the PSF was determined using the special geometric feature of an RVE of the microstructure initially designed. The innermost RVE of the bulk metamaterial was selected to quantify the PSF. The final microstructures for metallic auxetic metamaterials were obtained by adding the desired mode shape multiplied by the specified design scale factor (DSF) to the original coordinates. DSF is an independent factor that consists of scaling the maximum displacement amplitude to visualize a deformed shape. When the walls at the center of the RVE touched each other, the corresponding DSF (in this case 0.0092) was defined as PSF = 100%, as shown in Figure 2c. Other magnitudes of PSF were defined accordingly as 0% and 20%, corresponding to DSFs of 0 and 0.00184, respectively. The normalized desired buckling mode and its innermost RVE with different PSFs are presented in Figure 2. Through changing the PSF, a new class of auxetic metamaterials with different auxetic performance was produced [10]. The tuneable auxetic performance was a special feature of this class of auxetic metamaterials and potentially could be implemented in particular applications with a specific negative value of Poisson's ratio. It should be noted that, without this alternation, the deformation of this type of auxetic metamaterial was uniform at the beginning stage under compression, which exhibited a positive Poisson's ratio until buckling occurred [10]. In our previous work [10], the desired buckling mode of the bulk metamaterial was superposed to the original designed microstructures of metamaterial. It resulted in a non-periodic microstructure. The periodicity was achieved by patterning the altered central RVE of the bulk metamaterial in this study. The altered bulk metamaterial was formed by repeating this altered RVE as the unit cell.

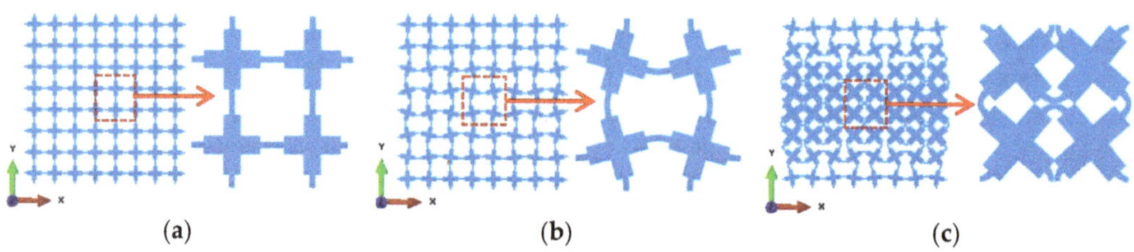

Figure 2. Definition of PSF to alter the initial microstructure of a buckling-induced metamaterial. The normalized desired buckling mode shapes scaled by different corresponding DSFs and their most central RVEs. (**a**) PSF = 0, DSF = 0; (**b**) PSF = 20%, DSF = 0.00184; (**c**) PSF = 100%, DSF = 0.0092.

3. Experiments

3.1. Fabrication of 2D Metamaterials for Experiments

The metallic specimens were fabricated using a high-resolution 3D printing technique. In the first step, the metamaterial was printed in wax then the wax model was put inside liquid plaster. After setting the plaster, it was placed in an industrial oven to melt out the wax. The wax was used as a supporting material to make a plaster mold. Eventually, the molten metal was poured into the plaster mold and was placed at the fixed position to become solid. Raw brass was selected as a representative base material of printed specimens because of its excellent ductility among available materials for 3D printing. The mechanical properties of brass were obtained directly from a standard tensile test on six dog-bone specimens using an MTS (material test system) machine (MTS Company, Eden Prairie, MN, USA), which is shown in Figure 3. The average mechanical properties of brass tests are summarized in Table 1. The influence of the potential anisotropy of base material is not studied here. This potential may be caused by the inevitable inclusion of air bubbles into brass specimens during the manufacturing process. The penetration of tiny bubbles causes very small holes inside the specimens and changes the mechanical properties and density of different printed specimens. However, the excellent agreements between the numerical and experimental results revealed that the influence of potential anisotropy is negligible in this study.

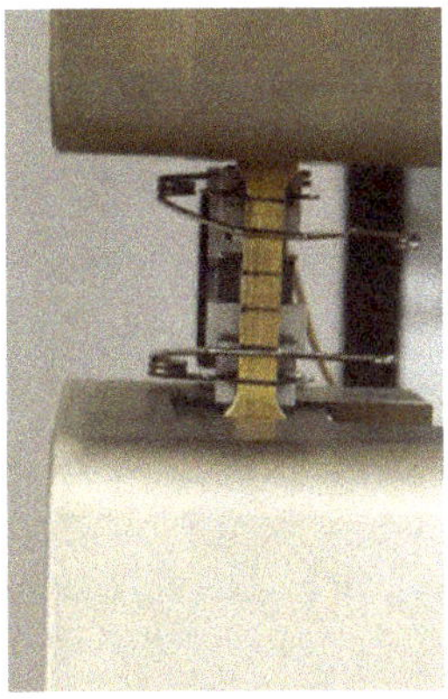

(a)

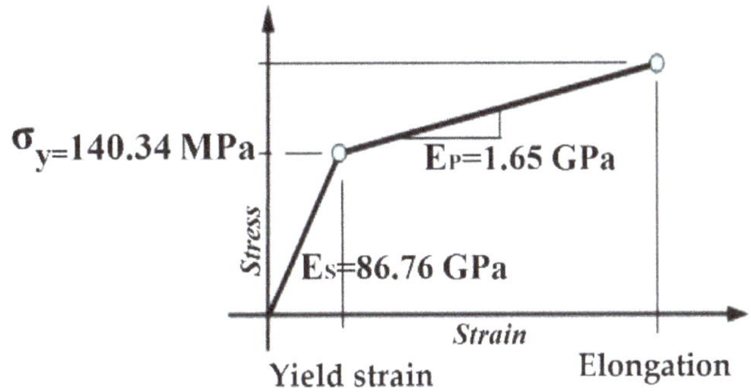

(b)

Figure 3. Tensile test of 3D printed brass dog-bone using MTS machine. (**a**) Front view of brass dog-bone. (**b**) Schematic curve of stress-strain for brass.

Table 1. Material properties of brass.

Modulus of Elasticity (GPa)	Yield Stress (MPa)	Strain Hardening Modulus (GPa)	Density (GPa)
86.76	140.34	1.65	8720

The 2D periodic bulk metamaterials for original buckling-induced design (PSF of 0%) and for an altered one with unit cells with a PSF of 20% are shown in Figure 4.The first 3D printed brass sample is a buckling-induced metamaterial with PSF = 0, as shown in Figure 4a, and the other is the altered one that was superposed by the desired buckling mode with a PSF of 20%, is shown in Figure 4b. The dimensions of the first and second test specimens were height × width × depth = 92.8 mm × 88 mm × 10 mm. Two plates were added to the bulk materials to constrain the degrees of freedom of the top and bottom nodes' surfaces except for a degree of freedom in the direction of load. To check the accuracy of the manufactured samples, the weights of printed models were measured and compared with the weight of our initial design obtained from the numerical mode. The weights of printed samples were slightly heavier than the designed ones. A numerical comparison of models with the different weight obtained by changing the void of RVE confirmed that these small weight differences were negligible. In order to compare the deformation patterns of rubber and metallic metamaterials with initial geometric design (PSF = 0%), a 2D elastomeric specimen was manufactured using silicon-based rubber (Tango plus, In'Tech, Ramsey, MN, USA), as shown in the top-most row of Figure 5. The material properties of silicon-based rubber (Tango plus) were measured in reference [10].

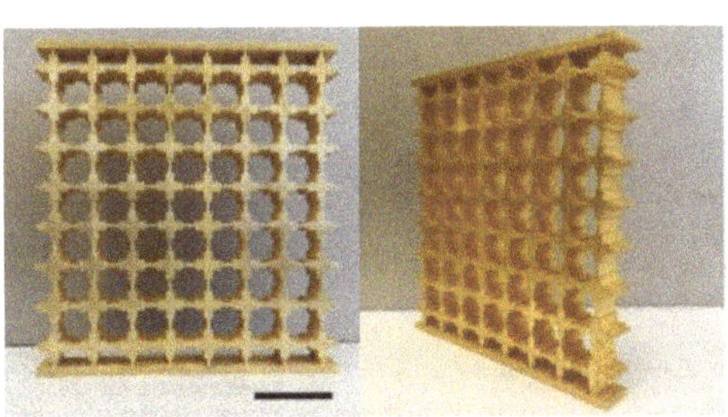

(a)

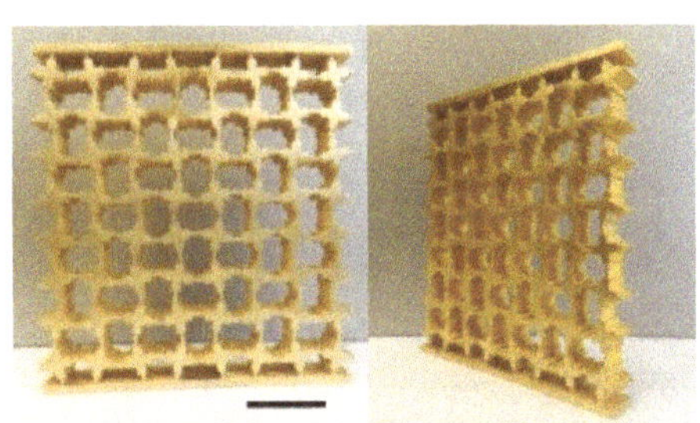

(b)

Figure 4. *Cont.*

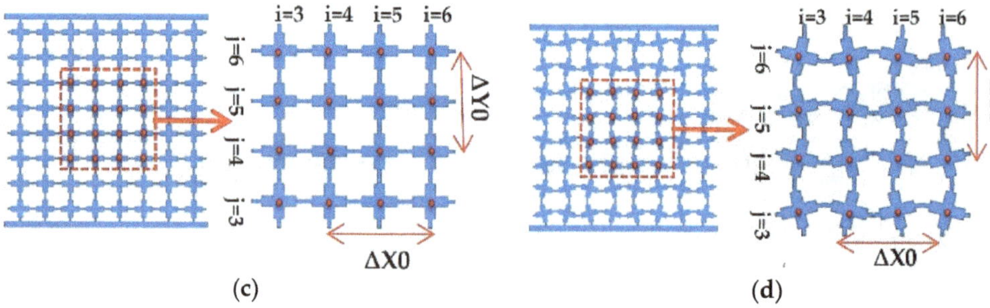

Figure 4. The final design of metamaterial employed for FEA and experimental investigation. (**a**) Front view and perspective view of buckling-induced metamaterial (PSF = 0%; density: 2798.05 kg m^{-3}, overall mass: 228.5 g, mass error: 1.08%, height × width × depth: 92.8 mm × 88 mm × 10 mm, Scale bar: 20 mm); (**b**) front view and perspective view of the metamaterial with altered geometry (PSF = 20%, overall mass: 222.57 g, density: 2725.43 kg m^{-3}, mass error: 2.06%, height × width × depth: 92.8 mm × 88 mm × 10 mm, Scale bar: 20 mm); (**c**) schematic diagram of central region with 16 nodes (PSF = 0%); (**d**) schematic diagram of central region with 16 nodes (PSF = 20%). The image processing was used to measure the horizontal and vertical center-to-center distances of nodes.

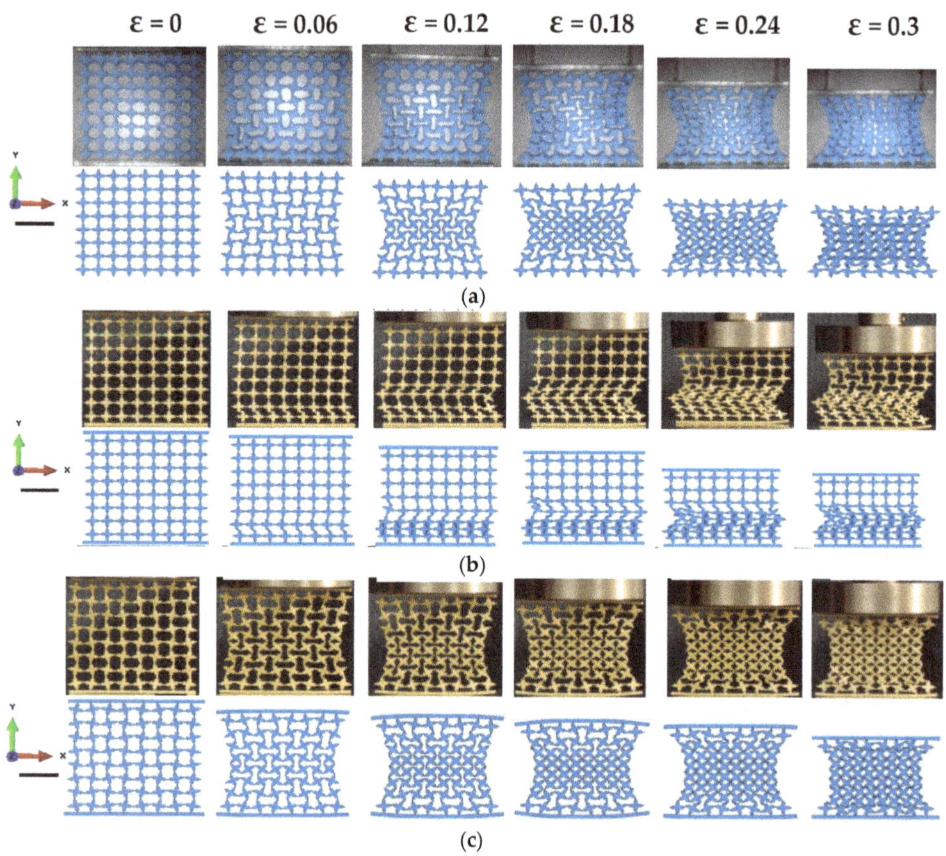

Figure 5. Comparison of deformation patterns of three new designed metamaterials between numerical and experimental results. (**a**) Experimental and numerical results of initial design of auxetic elastic metamaterial (rubber) with PSF = 0%; (**b**) experimental and numerical results of initial design of metallic metamaterial with PSF = 0%; (**c**) experimental and numerical results of the new design of auxetic metallic metamaterial with PSF = 20% (Scale bar: 20 mm, load direction Y, strain rate: 5×10^{-3} S^{-1}).

3.2. Experimental Setup for Uniaxial Compression Testing of 2D Metallic Metamaterials

Quasi-static uniaxial compression tests were carried out to investigate the performance of 2D metallic auxetic metamaterials under compression. The tensile tests were conducted by a Shimadzu machine (Shimadzu Company, Kyoto, Japan) at a fixed strain rate of 5×10^{-3} s^{-1}. The deformation process was captured by two cameras. The first camera was used to capture photos in the lateral direction of our 2D test specimens every 30 s. The second camera was employed to record a performance video of the test process. The nonlinear response of metamaterials was considered to be in the range of effective auxetic strain. In other words, the focus of this research was on the behavior of metamaterials, whereas the value of nominal applied strain was below the densification strain. Densification strain was defined as the maximum value of applied strain that satisfies the condition of maximum energy efficiency, as used by Shen *et al.* [36]. The experimental value of Poisson's ratio was calculated through image processing from the nodes in the central parts of ligaments, as shown in Figure 4c,d. Indeed, at a specific value of nominal strain, the $(x_{i,j},\ y_{i,j})$ coordinates of the centroid point of each unit cell were determined. The row and column indices are represented by $1 \ll i \ll \mathbf{8}$ and $1 \ll j \ll \mathbf{8}$, respectively. The central area under consideration consists of 16 nodes; these are designated with a red dashed line indicating $3 \ll i \ll 6$ and $3 \ll j \ll 6$, as shown in Figure 4c,d. The horizontal and vertical centroid-to-centroid distances were calculated by using coordinates $(x_{i,j}, y_{i,j})$, $\Delta x_{i,j} = x_{i+1,j} - x_{i-1,j}$ and $\Delta y_{i,j} = y_{i,j+1} - y_{i,j-1}$. Also, the center to center distances between the points of undeformed unit cells before compression were defined by $\Delta x\,(0) = \Delta y\,(0) = 11\ mm$. Equation (1) was employed to calculate the local values of the engineering strain Poisson's ratio.

$$\upsilon_{i,j} = -\frac{\dfrac{\Delta X_{i,j}}{\Delta x\,(0)}}{\dfrac{\Delta Y_{i,j}}{\Delta y\,(0)}} \tag{1}$$

Eventually, four values of Poisson's ratio were calculated from the central nodes under consideration, and the average of them was computed at each specific value of nominal strain. The deformation pattern of elastic buckling-induced auxetic metamaterial with a rubber base material is shown in Figure 5a, which is similar to the pattern observed by Bertoldi *et al.* [11].

3.3. Experimental Results

As expected, the deformation pattern of a brass metamaterial undergoing plastic flow was different from that of a rubber sample with only elastic deformation. When the base material was changed to metal, the auxetic behavior disappeared, and a new non-auxetic pattern similar to the global buckling pattern for elastic bulk materials [10] was observed, as shown in Figure 5b. This pattern usually yielded a zero or positive Poisson's ratio. In contrast, the test specimen with altered microstructures (PSF = 20%) exhibited obvious auxetic behavior, as shown in Figure 5c. It developed from a localized buckling pattern and was similar to mutual ellipses with long orthogonal axes. The results for all samples are presented in Figure 6 as a function of nominal strain. These experimental results revealed the significant influence of base materials with nonlinear properties or metal plasticity on the auxetic behavior of the buckling-induced auxetic materials. They also proved the effectiveness of designing the metallic auxetic metamaterials using a buckling mode to alter the initial geometry of buckling-induced metamaterials. By applying different values of PSF, a family of metallic metamaterials were generated, which exhibited obvious auxetic behavior nearly in the full range of its deformation process, as illustrated in Figure 6. A detailed analysis on the deformation process of the metallic specimen was conducted to reveal the reason behind the loss of auxetic behavior in Figure 5b. At the beginning of the compression process ($\varepsilon < 0.06$), the initial localization of plastic deformation occurred within the bottom layer of cells, and they were crushed firstly. As compression progressed, more localization took place within the central layers until all layers of the specimen were crushed and a non-auxetic deformation pattern was formed, as shown in Figure 5b. The observed crushing process also showed that the localized plastic deformation did not propagate layer by layer. This deformation localization triggered a different pattern without auxetic behavior. Thus, the key feature of the observed pattern in the brass specimen with PSF = 0% was the localization of plastic deformation. Inherently, deformation localization in one layer was caused by the sudden decrease of loading modulus of the metallic base material. The starting layer for this localization was determined by the manufacture imperfection in the brass specimen.

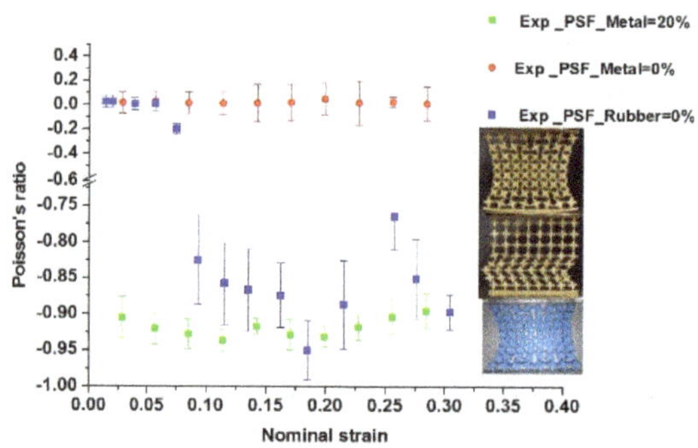

Figure 6. Comparison of auxetic and non-auxetic behavior of new design of auxetic metallic metamaterial with PSF = 20%, initial design of metallic metamaterial with PSF = 0 and initial design of auxetic elastic metamaterial (rubber) with PSF = 0% (Scale bar: 20 mm, load direction Y, strain rate: 5×10^{-3} S^{-1}).

4. FE Simulations

4.1. Post-Buckling Analysis and Validation of Numerical Models

FE simulations were conducted to further clarify the loss of auxetic behavior of metallic buckling-induced metamaterials and to reveal the features observed in experiments on the new metallic auxetic metamaterials. A nonlinear post-buckling analysis on all of those metamaterials under uniaxial compression was carried out using ABAQUS/Explicit (Simulia, Providence, RI, USA). It should be noted that the results of post-buckling analyses were affected by the effects of complex self-contact and large deformations. Remarkably, the complex self-contact was an unavoidable factor for large deformation analysis [37]. To overcome this obstacle, the explicit simulation was carried out with a prescribed velocity profile. The mechanical properties of brass were set as a bilinear material model in FEA. In numerical models, the classical metal plasticity model was used to define the yield, hardening rule, and inelastic flow of the metamaterials at relatively low temperatures. The plasticity behavior of this model was simplified as a bilinear isotropic hardening behavior. The von Mises yield surface is used to measure isotropic yielding by defining the exact value of yield stress as a function of uniaxial equivalent plastic strain. The inertia effect was minimized by defining amplitude to apply the velocity on top of the FE model as it was used in reference [37]. The velocity was applied gradually and the acceleration at the beginning and ending of compression process was zero [10].

At the first stage of post-buckling analysis, the FE simulations were carried out utilizing 3D solid elements, the same as the previous buckling analysis. The deformation process of the newly designed metallic auxetic metamaterial with PSF = 20% is shown in Figure 5c. A comparison of the deformation process between the experiments and the numerical model revealed an excellent agreement between them, as shown in Figure 5.

Three-dimensional shell elements were used to reduce the computational costs. Four nodes—standard, linear square element with second order accuracy (S4R)—were used with at least five integration points at the minimum links in the FE model. The FE simulation results with shell elements were validated by comparing with experimental results and also with the result of previous post-buckling analyses with 3D solid elements, as shown in Figure 7a. They agreed with each other in general trend and features. However, the level of stress corresponding to FE results was higher than the stress level from the experimental results. These differences were attributed to the imperfection of the printed specimen in our experiments, which was supported by the large variation of properties, listed in our previous work [29].

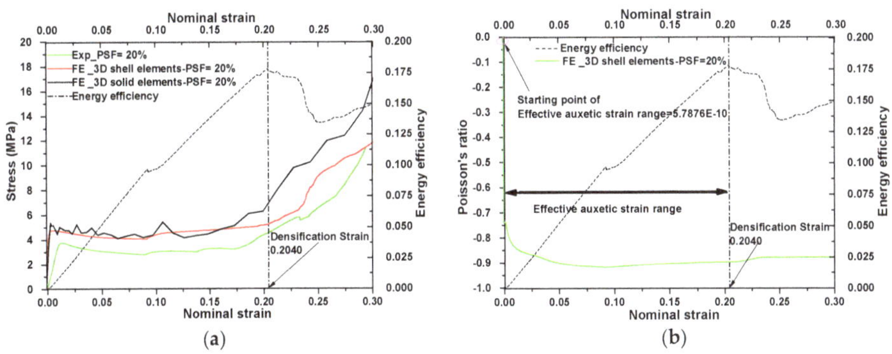

Figure 7. (**a**) Comparison of nominal stress-strain curves of the auxetic metamaterial between experiment and FE results (FE models with 3D shell elements and 3D solid elements) and using energy efficiency method to find densification strain; (**b**) defining the lower bound and upper bound of effective auxetic strain range.

4.2. Defining a Strain Range for Auxetic Metamaterials Using the Energy Efficiency Method

In this study, the effective auxetic strain range was defined as the strain between the starting point for negative value of Poisson's ratio shown in Figure 7b and the densification point from stress-strain shown in Figure 7a,b. The starting point of effective auxetic strain range was defined as the first point that Poisson's ratio began to decline as shown in Figure 7b.

The auxetic behavior of the metallic metamaterial started after application of a very small strain in the compression test, and this value was close to zero. The endpoint of the effective auxetic range is densification strain; this physically corresponds to the start point, from which a sharply rise of stress is observed in the stress-strain relationship as shown in Figure 7a. It is shown that from the densification strain onwards, the stress level in stress-strain curve rises sharply. The term "effective" is used to indicate two important features of the metallic auxetic metamaterials within the described strain range for their potential applications. One is that the nominal Poisson's ratio is negative and similar, and the other is that the stress level over the entire deformation range is similar and below the densification strain. A relative constant plateau stress and similar Poisson's ratio over the entire deformation range from the FE results was calculated using an energy efficiency method, as was used in references [36,38]. In this method, firstly the absorbed energy is defined as the total area under the stress-strain curve. Then, the energy efficiency parameter is found by dividing the absorbed energy by the stress itself, as shown below:

$$E(\varepsilon) = \frac{\int_0^{\varepsilon} \sigma(\varepsilon)\, d_{\varepsilon}}{\sigma} \tag{2}$$

When the $E(\varepsilon)$ reaches the maximum point, the corresponding strain is the densification strain. The densification strain is marked by a dashed line in Figure 7a,b. As a special feature, the fact that the densification strains from the experimental data are similar to those from the FE results also proved the accuracy of the FE models.

4.3. Parametric Studies

After the numerical models were validated, parametric studies were carried out to confirm the cause of loss of auxetic behavior for buckling-induced metallic specimens identified in the experiments. As for the pattern-altered specimen, the dominant factors to control the performance of metallic auxetic metamaterials were identified and qualified using the validated FE models. Based on experimental observations, the reason for the loss of auxetic behavior for metallic specimens was related to the plastic deformation of the buckling-induced auxetic metamaterials. The localization of the plastic deformation caused by buckling at an RVE length scale resulted in a global deformation mode rather than the desired buckling pattern in all RVEs. It was difficult to verify this assumption directly. However, two methods to prevent the localization of plastic strain could be applied to buckling-induced design. The first method is eliminating buckling by introducing a large imperfection; the second method is the increase of strain hardening modulus of its base material close to the elastic modulus. If those methods should be effective, the assumption would be proved indirectly. As mentioned before, the buckling mode alternation of the microstructure could be used as an innovative approach to design new metallic auxetic metamaterials. To enhance and facilitate its potential applications, an individual tuning method should be provided to assist with the design of a metallic metamaterial with a prescribed performance. Based on the feature of metamaterials, the change of Poisson's ratio can be achieved by adjusting the geometrical configuration of microstructures, while other mechanical properties can be achieved by changing the elastoplastic properties of its base material. It is difficult to individually change those features by experimental methods, so the validated FE model was used to do the parametric study.

In this study, the magnitude of alteration was represented by PSF. As mentioned before, the PSF was a key parameter to alter the initial topology of RVE with desired buckling mode. The range of variation of PSF was between 0% and 100%. A series of systematic simulations were conducted to study the influence of PSF on auxetic performance of the new metallic auxetic metamaterials. The elastoplastic properties of brass were defined identically to the material properties in the experiment. The results are presented in Figure 8, in which the effective auxetic strain range is marked with round dots.

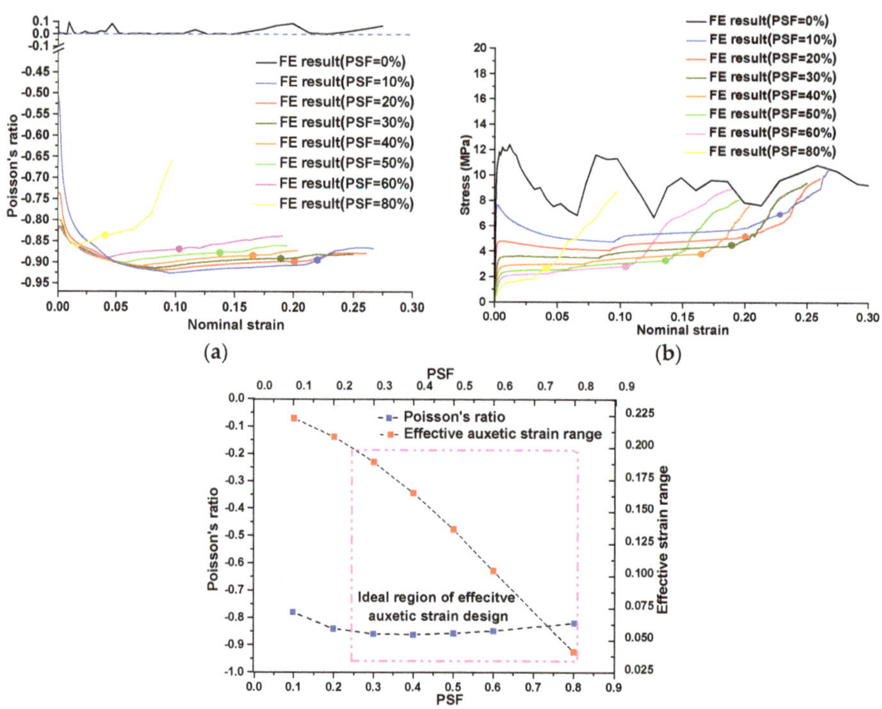

Figure 8. Parametric study on influence of PSF on auxetic behavior. (**a**) Evaluations of Poisson's ratio as function of applied strain corresponding to different PSF; (**b**) stress-strain curves corresponding to different PSF; (**c**) average values of Poisson's ratio and effective auxetic strain ranges corresponding to different PSF.

According to these results, it was determined that the increase of PSF resulted in the decrease of the densification strain and thus the effective auxetic strain range for auxetic behavior. This phenomenon was obvious, as shown in Figure 8c. By increasing the PSF, the mutual ellipses in RVE became smaller, thus the internal walls of the elliptical parts touched each other at a smaller strain. Therefore, the densification strains decreased accordingly.

The nonlinear response of Poisson's ratio with respect to compressive strain is shown in Figure 8a with different PSF values. The negative values of Poisson's ratio did not change much with respect to PSF, especially when the PSF are in the range of 10% to 60%. This finding revealed that the value of Poisson's ratio is dominated by the deformation patterns initiated by the buckling mode applied. In this PSF range, the effective auxetic strain range can be individually designed with a prescribed negative value of Poisson's ratio required by the aimed application, as illustrated in Figure 8c. An ideal region for individually altering the effective auxetic strain is defined and illustrated with a dashed rectangle in Figure 8c. Within this region, Poisson's ratio maintained a constant negative value and the effective auxetic strain can be individually changed by the value of PSF.

As mentioned previously, the strain hardening of the base material could be used to prevent the localization of the plastic deformation caused by buckling at RVEs. Plastic strain hardening is a common phenomenon for metallic materials, which can be simplified as a linear relationship with respect to plastic strain. It is represented by the slope of this line, termed the strain hardening modulus (Ep), in the stress-strain curves shown in Figure 9c. In most cases, the stress-strain curve of a metallic material is determined by the elastic region (Es) as well as the plastic region (Ep). In this study, a simple ratio was defined as Ep/Es to characterize the effect of plastic hardening on the auxetic performance of our designed metamaterials. This ratio was termed the "plastic hardening ratio." The variation range of this ratio is between 0 and 1.

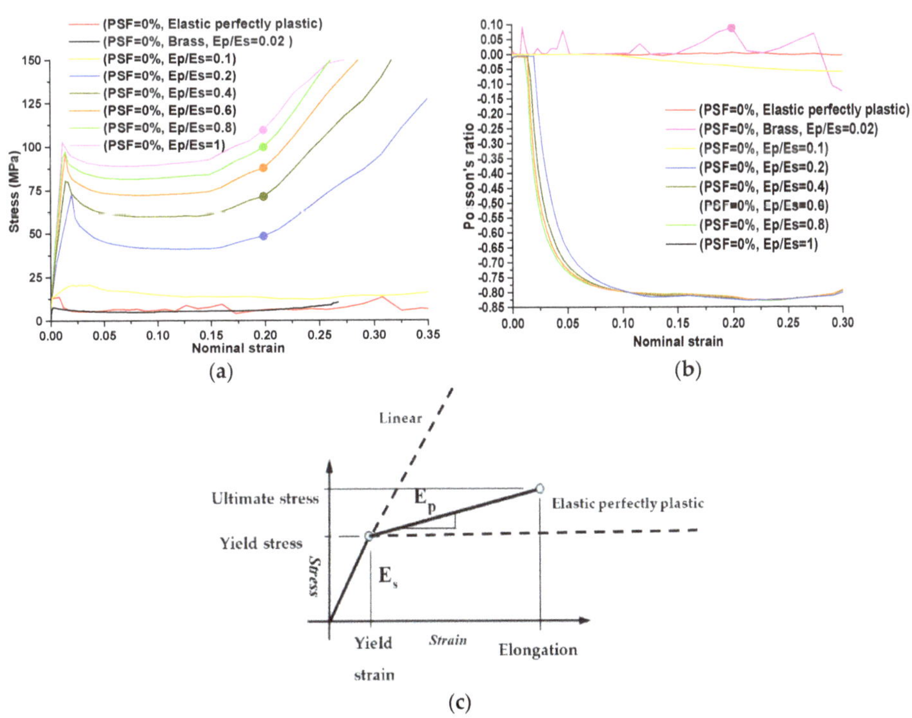

Figure 9. The results of a parametric study on the influence of plastic hardening on the recovery of auxetic behavior. (**a**) Evaluations of Poisson's ratio as a function of applied strain corresponding to different hardening ratios for metamaterials with PSF = 0%.; (**b**) stress-strain curves corresponding to different hardening ratios for metamaterials with PSF = 0%; (**c**) schematic curve of stress-strain for metallic materials (strain hardening ratio).

A ratio of 0 represents perfectly plastic material and a ratio of 1 represents purely elastic material. The result for a plastic hardening ratio of 1 with buckling-induced metamaterials confirmed that increasing the plastic hardening ratio will restore the auxetic behavior of buckling-induced metamaterials. As shown in Figure 5a, the experimental rubber specimen with PSF = 0% exhibited auxetic behavior under compression as a purely elastic material or when Ep/Es = 1. This finding was further validated by a numerical investigation on the initial geometry of buckling-induced metamaterial with PSF = 0%. The FE results for different values of Ep/Es at PSF = 0 are presented in Figure 9a,b. It is shown that the response of a metamaterial changed from non-auxetic to auxetic by increasing the plastic hardening ratio. Based on this evidence, the reason behind the loss of auxetic behavior was attributed to the localization of the plastic deformation caused by buckling at RVEs. To evaluate the effect of plasticity on the auxetic behavior of our newly designed metallic auxetic metamaterials, the hardening ratio was varied in the range of 0 to 1. The results of this numerical parametric study for the newly designed buckling-induced metamaterial with PSF = 20% are presented in Figure 10 and for other values of PSF are shown in Figure 11. Also, it should be mentioned that when Ep/Es tends to zero, the response is closer to the rotation of rigid squares and so the Poisson's ratio tends to be −1 [39]. According to the numerical results shown in Figures 10 and 11 it can be concluded that the difference of the calculated negative value of Poisson's ratio for models with different strain hardening ratios were negligible. Also, the effective auxetic strain ranges for models with different hardening ratios were similar and the difference among these ranges can be neglected. However, the stress level increased considerably when the hardening ratios increased from 0 to 1. Based on this observation, metallic auxetic metamaterials undergoing large deformation were dominated by the topology of the microstructures; this was independent of the properties of the base material.

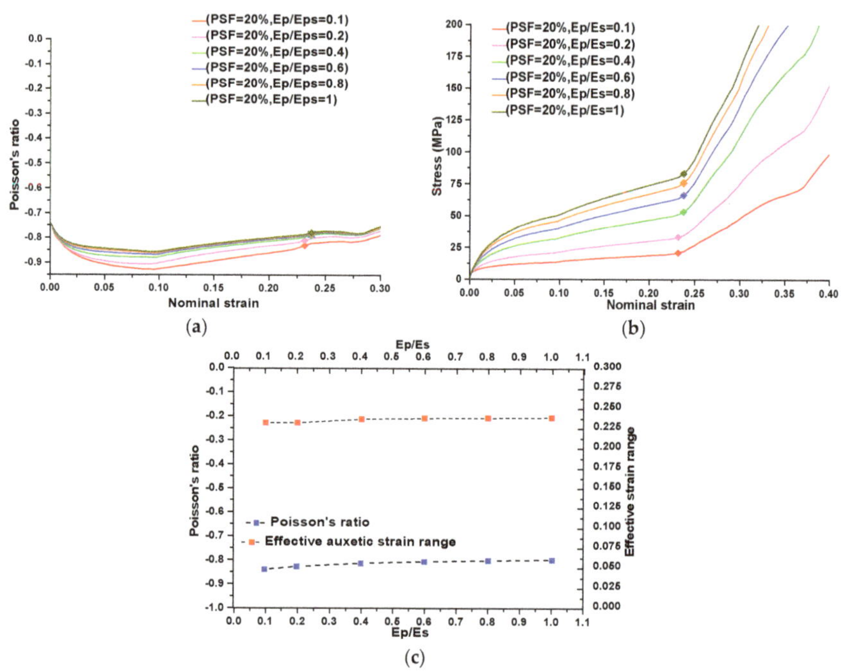

Figure 10. The results of a parametric study on the influence of plastic hardening on the auxetic behavior of a metamaterial with PSF = 20%. (**a**) Evaluations of Poisson's ratio as a function of applied strain corresponding to different hardening ratios; (**b**) stress-strain curves corresponding to different hardening ratios; (**c**) average values of Poisson's ratio and effective auxetic strain ranges corresponding to different strain hardening ratios for metamaterials with PSF = 20%.

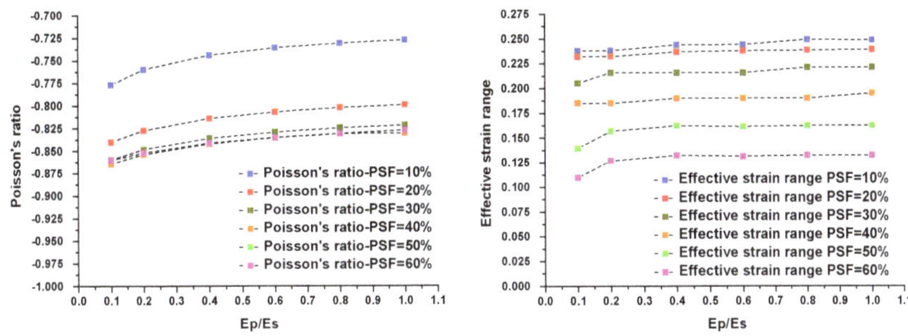

(a) (b)

Figure 11. The results of a parametric study on the influence of pattern scale factors on auxetic behavior and effective auxetic strain range. (**a**) Evaluations of Poisson's ratio as the function of hardening ratio corresponding to pattern scale factors; (**b**) evaluations of effective auxetic strain range as the function of hardening ratio corresponding to different pattern scale factors.

5. Discussion and Conclusions

The findings of this study demonstrated a fundamental way to design a new class of 2D metallic auxetic metamaterials. The new design approach consisted of two important improvements compared with the previous approach for elastomer [10,11]. The first improvement was the inclusion of a geometric modification of the conventional structure to generate a new buckling-induced metamaterial. The second improvement was altering the initial geometry of a non-metallic auxetic metamaterial at the RVE level. The FE models were validated by experiments. A systematic parametric study was carried out using validated FE models on the effects of the plastic properties of the base material and the pattern scale factor.

It is interesting to note that a similar approach to that of the PSF has been also used by other researchers to direct the deformation path of soft metamaterials [40]. The effect of boundary conditions on the auxetic performance of our 2D auxetic metamaterial was also investigated. From the results of those investigations, the following conclusions can be drawn:

(1) The deformation process of the metallic auxetic metamaterial is influenced by boundary conditions; however, the auxetic behavior and its performance corresponding to different deformation processes are insensitive to boundary conditions.
(2) The auxetic behavior of our designed metallic metamaterial remains within the effective auxetic strain range.
(3) The effective auxetic strain can be controlled individually by PSF while maintaining a similarly negative value of Poisson's ratio, especially when the PSF is in the range of 10% to 60%.
(4) The stiffness and strength of the metallic auxetic metamaterials can be individually controlled through adjustment of the properties of the base materials, while the negative value of Poisson's ratio remains relatively constant.
(5) The loss of auxetic behavior in the metallic buckling-induced metamaterial is attributed to the localization of plastic collapse of RVEs.
(6) The results from FE simulation confirmed that the buckling-induced metamaterial would have auxetic behavior after the plastic hardening modulus is large than a certain value determined by the microstructures of metamaterials. These results confirm that the effectiveness of increasing plastic hardening ratio will restore the auxetic behavior of buckling-induced metamaterials.

Except for those solid conclusions based on our numerical and experimental results, there are some concerns relating to the 2D microstructures of our metallic auxetic metamaterials. The mechanism of the resultant metallic auxetic metamaterials is the rotating of joints/squares. The analytical model of rotating interconnected squares was proposed independently and simultaneously by Ishibashi & Iwata [41] and Grima & Evans [39]. Thus, most of the auxetic behavior of the metallic auxetic metamaterial can be predicted by their model. To some extent, the PSF used to define the topology of the RVE can be equivalent to the rotation angle of the square units in the analytical model. The insensitive negative value of Poisson's ratio of −0.9 with respect to PSF can be approximately explained by the negative value of Poisson's ratio of −1 in a model of the rotating of joints/squares. However, there is a noticeable advantage to using PSF to design the microstructures of metallic auxetic metamaterials, besides tuneability. The volume fraction does not vary significantly through changing the PSF, as shown in Figure 12a. Thus, it can be revealed that a set of 2D auxetic metallic metamaterials that are obtained with different PSFs have similar volume fractions.

It should be noted that the Poisson's ratio was measured in the center region of the specimens both in experiments and FE simulations. The nominal strain of the overall specimen was using to plot all strain-related curves. This method was used in all recent experimental-related research on auxetic structures and materials [10,11,23,29]. This measurement resulted in the insensitivity of the measured Poisson's ratio with respect to strain localization. It will enlarge the negative strain range for auxetic behavior, evidenced by the constant negative Poisson's ratio of −0.9 after its effective strain of 0.2 in Figures 6 and 7. Based on the observed deformation of a metallic auxetic metamaterial that is presented in Figure 5c ($\varepsilon = 0.2040$), the walls of central rows of unit building cells are touching. However, the unit building cells of the top and bottom rows are not compacted completely. As mentioned before, the base martial of the specimen was made of ductile metal. Beyond the auxetic strain range, the metallic metamaterial can still be compressed further by plastic deformation of those not fully compacted layers. While the local strains in the central region remain nearly constant or are slightly compressed, the Poisson's ratio will remain negative. Thus, the measurement of the Poisson's ratio in the center region may not be representative of the whole specimen when the strain localization occurs, as shown in Figure 5b. We present those data just for comparison purposes; the accuracy of this measurement should be checked further. For a similar reason, we used the densification strain to define the end point for an effective strain range.

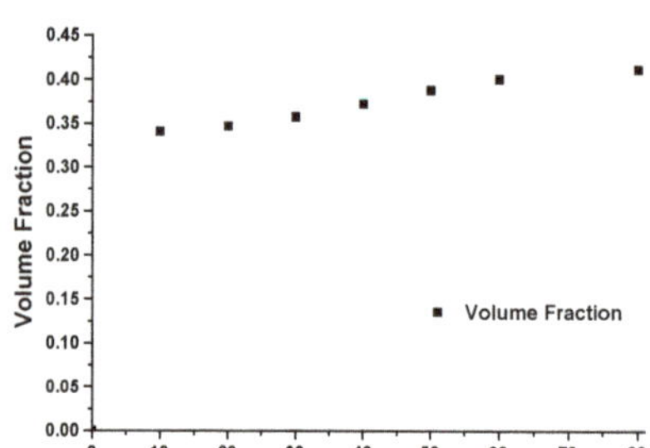

(a)

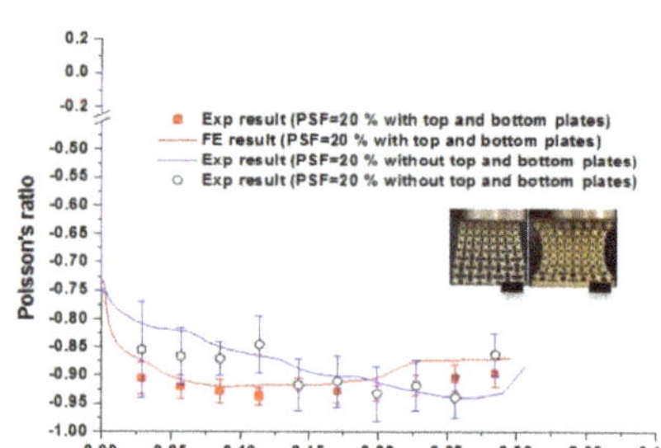

(b)

Figure 12. (**a**) Relation between PSF and corresponding volume fraction; (**b**) comparison of auxetic behaviors between buckling-induced design with PSF = 0%, buckling pattern-altered design without plates (Scale bar: 20 mm, load direction Y, strain rate: 5×10^{-3} S^{-1}).

In order to study the influence of the boundary on the auxetic performance of a designed specimen, the top and bottom plates were removed so that the top and bottom surfaces of a new 2D designed metamaterial with PSF = 20% were not constrained. An experiment was conducted to reveal its effect on the deformation process. At the beginning of the test, a localized deformation occurred near the loading end. With further compression, a localized lateral inwards deformation developed from the top moving surface toward the bottom fixed surface. Interestingly, this deformation process led to obvious auxetic behavior. The auxetic *versus* strain curves of two specimens were put together and the results were presented in Figure 12b. Interestingly, the overall trend and the average values of this set of Poisson's ratio curves were similar. These observations illustrated that although the deformation process of our metallic auxetic metamaterial under compression was significantly influenced by boundary conditions, the auxetic performance was insensitive to the boundary conditions.

It was revealed that the enhancement of the plastic-hardening ratio leads to restoring the auxetic behavior of buckling-induced auxetic metamaterials. According to Figure 9b, this effect is obvious when the hardening modulus is in the range of Ep/Es = 0 and Ep/Es = 0.3. Out of this range, the auxetic performance of models with different hardening ratios was similar and their differences could be neglected. This result is in accordance with the finding that the auxetic effect persists and becomes even stronger when the hardening modulus is in the range of h = 100 MPa and h = 1000 MPa by Dirrenberger *et al.* [25]. However, regarding the metallic auxetic metamaterials with altered microstructures, the influence of an enhancement of the plastic-hardening ratio on auxetic performance is negligible.

Acknowledgments: This work was supported by the Australian Research Council (DP140100213).

Author Contributions: Yi Min Xie and Jianhu Shen initiated and supervised this study. Arash Ghaedizadeh and Jianhu Shen carried out the analysis and experiments. Xin Ren provided technical assistance for FE Simulations. Arash Ghaedizadeh wrote the draft manuscript. Yi Min Xie, Jianhu Shen, and Xin Ren edited and revised the manuscript.

Conflicts of Interest: The authors declare no conflict of interest.

References

1. Evans, K.E.; Alderson, A. Auxetic materials: Functional materials and structures from lateral thinking! *Adv. Mater.* **2000**, *12*, 617–628. [CrossRef]
2. Fozdar, D.Y.; Soman, P.; Lee, J.W.; Han, L.-H.; Chen, S. Three dimensional polymer constructs exhibiting a tunable negative poisson's ratio. *Adv. Mater.* **2011**, *21*, 2712–2720. [CrossRef] [PubMed]
3. Lakes, R. Foam structures with a negative Poisson's ratio. *Science* **1987**, *235*, 1038–1040. [CrossRef] [PubMed]
4. Evans, K.E.; Nkansah, M.A.; Hutchinson, I.J.; Rogers, S.C. Molecular network design. *Nature* **1991**, *353*, 124. [CrossRef]
5. Friis, E.A.; Lakes, R.S.; Park, J.B. Negative Poisson's ratio polymeric and metallic foams. *J. Mater. Sci.* **1988**, *23*, 4406–4414. [CrossRef]
6. Evans, K.E.; Caddock, B.D. Microporous materials with negative Poisson's ratios. I. Microstructure and mechanical properties. *J. Phys. D Appl. Phys.* **1989**, *22*, 1877–1882. [CrossRef]
7. Evans, K.E.; Caddock, B.D. Microporous materials with negative Poisson's ratios. II. Mechanisms and interpretation. *J. Phys. D Appl. Phys.* **1989**, *22*, 1883–1887. [CrossRef]
8. Yang, W.; Li, Z.M.; Shi, W.; Xie, B.H.; Yang, M.B. Review on auxetic materials. *J. Mater. Sci.* **2004**, *39*, 3269–3279. [CrossRef]
9. Critchley, R.; Corni, I.; Wharton, J.A.; Walsh, F.C.; Wood, R.J.K.; Stokes, K.R. A review of the manufacture, mechanical properties and potential applications of auxetic foams. *Phys. Status Solidi B* **2013**, *250*, 1963–1982. [CrossRef]
10. Shen, J.; Zhou, S.; Huang, X.; Xie, Y.M. Simple cubic three-dimensional auxetic metamaterials. *Phys. Status Solidi B* **2014**, *251*, 1515–1522. [CrossRef]
11. Bertoldi, K.; Reis, P.M.; Willshaw, S.; Mullin, T. Negative Poisson's ratio behaviour induced by an elastic instability. *Adv. Mater.* **2010**, *22*, 361–366. [CrossRef] [PubMed]
12. Mancusi, G.; Feo, L. A refined finite element formulation for the microstructure-dependent analysis of two-dimensional (2D) lattice materials. *Materials* **2013**, *6*, 1–17. [CrossRef]
13. Choi, J.B.; Lakes, R.S. Design of a fastener based on negative Poisson's ratio foam. *Cell. Polym.* **1991**, *10*, 205–212.
14. Scarpa, F. Auxetic materials for bioprostheses. *IEEE Sig. Proc. Mag.* **2008**, *25*, 128–126. [CrossRef]
15. Chen, C.P.; Lakes, R.S. Micromechanical analysis of dynamic behavior of conventional and negative Poisson's ratio foams. *J. Eng. Mater. Technol.* **1996**, *118*, 285–288. [CrossRef]
16. Scarpa, F.; Ciffo, L.; Yates, J. Dynamic properties of high structural integrity auxetic open cell foam. *Smart Mater. Struct.* **2004**, *13*. [CrossRef]
17. Alderson, A.; Rasburn, J.; Ameer-Beg, S.; Mullarkey, P.G.; Perrie, W.; Evans, K.E. An auxetic filter: A tuneable filter displaying enhanced size selectivity or defouling properties. *Ind. Eng. Chem. Res.* **2000**, *39*, 654–665. [CrossRef]

18. Sun, Y.; Pugno, N. Hierarchical fibers with a negative Poisson's Ratio for tougher composites. *Materials* **2013**, *6*, 699–712. [CrossRef]
19. Gibson, L.J.; Ashby, M.F.; Schajer, G.S.; Robertson, C.I. The mechanics of two-dimensional cellular materials. *Proc. R. Soc. Lond. A* **1982**, *382*, 25–42. [CrossRef]
20. Theocaris, P.S.; Stavroulakis, G.E.; Panagiotopoulos, P.D. Negative Poisson's ratios in composites with star-shaped inclusions: A numerical homogenization approach. *Arch. Appl. Mech.* **1997**, *67*, 274–286. [CrossRef]
21. Smith, C.W.; Grima, J.N.; Evans, K.E. A novel mechanism for generating auxetic behaviour in reticulated foams: Missing rib foam model. *Acta Mater.* **2000**, *48*, 4349–4356. [CrossRef]
22. Gaspar, N.; Ren, X.J.; Smith, C.W.; Grima, J.N.; Evans, K.E. Novel honeycombs with auxetic behaviour. *Acta Mater.* **2005**, *53*, 2439–2445. [CrossRef]
23. Babaee, S.; Shim, J.; Weaver, J.C.; Chen, E.R.; Pate, N.; Bertoldi, K. Soft metamaterials with negative Poisson's Ratio. *Adv. Mater.* **2013**, *25*, 5044–5049. [CrossRef] [PubMed]
24. Taylor, M.; Francesconi, L.; Gerendás, M.; Shanian, A.; Carson, C.; Bertoldi, K. Low porosity metallic periodic structures with negative Poisson's ratio. *Adv. Mater.* **2014**, *26*, 2365–2370. [CrossRef] [PubMed]
25. Dirrenberger, J.; Forest, S.; Jeulin, D. Elastoplasticity of auxetic materials. *Comp. Mater. Sci.* **2012**, *64*, 57–61. [CrossRef]
26. Gilat, R.; Aboudi, J. Behavior of elastoplastic auxetic microstructural arrays. *Materials* **2013**, *6*, 726–737. [CrossRef]
27. Bertoldi, K.; Boyce, M.C.; Deschanel, S.; Prange, S.M.; Mullin, T. Mechanics of deformation-triggered pattern transformations and superelastic behavior in periodic elastomeric structures. *J. Mech. Phys. Solids* **2008**, *56*, 2642–2668. [CrossRef]
28. Mullin, T.; Willshaw, S.; Box, F. Pattern switching in soft cellular solids under compression. *Soft Matter* **2013**, *9*, 4951–4955. [CrossRef]
29. Ren, X.; Shen, J.; Ghaedizadeh, A.; Tian, H.; Xie, Y.M. Experiments and parametric studies on 3D metallic auxetic metamaterials with tuneable mechanical properties. *Smart Mater. Struct.* **2015**, *24*. [CrossRef]
30. Singamaneni, S.; Tsukruk, V.V. Buckling instabilities in periodic composite polymeric materials. *Soft Matter* **2010**, *6*, 5681–5692. [CrossRef]
31. Ho, D.T.; Park, S.-D.; Kwon, S.-Y.; Park, K.; Kim, S.Y. Negative Poisson's ratios in metal nanoplates. *Nat. Commun.* **2014**, *5*. [CrossRef] [PubMed]
32. Jiang, J.W.; Park, H.S. Negative Poisson's ratio in single-layer black phosphorus. *Nat. Commun.* **2014**, *5*. [CrossRef] [PubMed]
33. Mullin, T.; Deschanel, S.; Bertoldi, K.; Boyce, M.C. Pattern transformation triggered by deformation. *Phys. Rev. Lett.* **2007**, *9*. [CrossRef] [PubMed]
34. Wu, G.; Xia, Y.; Yang, S. Buckling, symmetry breaking, and cavitation in periodically micro-structured hydrogel membranes. *Soft Matter* **2014**, *10*, 1392–1399. [CrossRef] [PubMed]
35. Körner, C.; Liebold-Ribeiro, Y. A systematic approach to identify cellular auxetic materials. *Smart Mater. Struct.* **2015**, *24*. [CrossRef]
36. Shen, J.; Xie, Y.M.; Huang, X.D.; Zhou, S.W.; Ruan, D. Mechanical properties of luffa sponge. *J. Mech. Behav. Biomed. Mater.* **2012**, *15*, 141–152. [CrossRef] [PubMed]
37. Hanssen, A.G.; Hopperstad, O.S.; Langseth, M.; Ilstad, H. Validation of constitutive models applicable to aluminium foams. *Int. J. Mech. Sci.* **2002**, *44*, 359–406. [CrossRef]
38. Avalle, M.; Belingardi, G.; Montanini, R. Characterization of polymeric structural foams under compressive impact loading by means of energy-absorption diagram. *Int. J. Impact. Eng.* **2001**, *25*, 455–472. [CrossRef]
39. Grima, J.N.; Alderson, A.; Evans, K.E. Auxetic behaviour from rotating rigid units. *Phys. Status Solidi B* **2005**, *242*, 561–575. [CrossRef]
40. Wu, G.; Cho, Y.; Choi, I.-S.; Ge, D.; Li, J.; Han, H.N.; Lubensky, T.; Yang, S. Directing the deformation paths of soft metamaterials with prescribed a symmetric units. *Adv. Mater.* **2015**, *27*, 2747–2752. [CrossRef] [PubMed]
41. Ishibashi, Y.; Iwata, M.A. Microscopic model of a negative poisson's ratio in some crystals. *J. Phys. Soc. Jpn.* **2000**, *69*, 2702–2703. [CrossRef]

11

First-Principles Study of Mo Segregation in MoNi(111): Effects of Chemisorbed Atomic Oxygen

Yanlin Yu [1], Wei Xiao [1,2], Jianwei Wang [1] and Ligen Wang [1,3,*]

Academic Editor: Federico Bella

1 General Research Institute for Nonferrous Metals, Beijing 100088, China; yuyanlin_121@163.com (Y.Y.); wxiao@ustb.edu.cn (W.X.); jswjw@sina.com (J.W.)
2 School of Materials Science and Engineering, University of Science and Technology Beijing, Beijing 100083, China
3 Power Environmental Energy Research Institute, Covina, CA 91722, USA
* Correspondence: lg_wang1@yahoo.com

Abstract: Segregation at metal alloy surfaces is an important issue because many electrochemical and catalytic properties are directly correlated to the surface composition. We have performed density functional theory calculations for Mo segregation in MoNi(111) in the presence of chemisorbed atomic oxygen. In particular, the coverage dependence and possible adsorption-induced segregation phenomena are addressed by investigating segregation energies of the Mo atom in MoNi(111). The theoretical calculated results show that the Mo atom prefers to be embedded in the bulk for the clean MoNi(111), while it segregates to the top-most layer when the oxygen coverage is thicker than 1/9 monolayer (ML). Furthermore, we analyze the densities of states for the clean and oxygen-chemisorbed MoNi(111), and see a strong covalent bonding between Mo d-band states and O p-states. The present study provides valuable insight for exploring practical applications of Ni-based alloys as hydrogen evolution electrodes.

Keywords: density-functional theory calculation; surface segregation; hydrogen evolution electrode; oxygen chemisorptions; water electrolysis

---

1. Introduction

Adding a second metal into a pure metal catalyst can provide a great opportunity to tailor the properties of the catalyst [1–4]. Therefore, bimetallic systems have attracted considerable attention due to their importance in basic science and industry [5–8]. Bimetallic systems are much more complicated than pure metal catalysts since one element may segregate to the surface and lead to surface composition enrichment/depletion compared to the bulk. The adsorbate-induced surface segregation of metallic alloys under the reaction conditions and, thus, the changes in local atomic composition and surface structure have been predicted and demonstrated to occur for a number of bimetallic systems [9,10]. So for a given bimetallic configuration which exhibits a desired property, it is crucial to know whether the particular configuration is stable under the operating environment for a specific application.

Many theoretical and experimental studies have been performed to investigate the surface segregation for various bimetallic systems [11–29]. The recent experimental and theoretical works focusing on the description of bimetallic and ternary, extended and nanosized, alloy surfaces under reactive gas phase environments were reviewed by Zafeiratos, *et al.* [19] and by Guesmi [20]. On the theory side, Guesmi and co-workers had performed extensive first-principles calculations to investigate the transition metal segregation behaviors for the gold-based bimetallic systems in the presence of various gas adsorptions [11,12,21–23]. Specifically, the authors found that segregation of Pd on

PdAu(111) is oxygen coverage–dependent and that Pd atoms tend to be in the bulk for the clean surface while they segregate to the surface in the presence of more than 1/3 ML of oxygen [12]. The surface phase diagrams were investigated within the first-principles atomistic thermodynamics framework by addressing the effect of the bulk alloy and the gas phase reservoir [13,24–26]. The first-principles-based cluster expansion technique [27] had also been employed to model the surface ordering and segregation of alloys in a reactive environment [28,29]. This approach allows us to explore enormous different atomic configurations. Although the latter two methodologies are very useful for studying the alloy surface ordering and segregation in the presence of adsorbates, the direct first-principles calculations [11,12,21–23] are certainly able to predict the segregation tendency of alloy systems. The advantage of the direct first-principles calculation method is that it is accurate and computationally more affordable, while its drawback is that it cannot predict what equilibrium phases and structures will form under given conditions.

Among the bimetallic systems, nickel-based alloys are especially interesting because of their applications in a number of catalytic reactions including the hydrodeoxygenation of esters [30], methanation reaction [31], and propane reforming [32]. In the context of hydrogen evolution reaction (HER), recent experiments showed that adding some Mo element into nickel-based electrodes could improve the electro-catalytic activity [33]. Previously, the calculations without any gas adsorption and the experiments in a ultra high vacuum (UHV) chamber showed that it is difficult for Mo atoms to segregate to the surfaces in NiMo alloys [34–36]. However, several previous experiments found that Mo atoms segregate to the NiMo alloy surfaces and induce an enrichment of the surface Mo composition under the reaction conditions [17,34]. The composition and structure of Ni-2 at % Mo(100), Ni-2 at % Mo(110) and Ni-6 at % Mo(110) surfaces were studied by Auger electron spectroscopy (AES) and low-energy electron diffraction (LEED) [34]. The authors obtained the molybdenum pre-enriched surfaces by annealing the samples in purified hydrogen. We also experimentally investigated the Mo surface segregation by energy-dispersive spectroscopy (EDS) in the NiMoCo electrolysis electrode materials [17]. In our experiments, the Ni at 20.8 wt % and Mo at 1.6 wt % Co foam electrodes were prepared by the electro-deposition method and decarbonized at atmospheric conditions to remove the residual carbon due to the polyurethane foams, followed by a reduction process in hydrogen atmosphere for removing the surface oxides. This Mo surface enrichment (40 wt % Mo for the surface *vs.* the nominal 20.8 wt % Mo in the bulk) could be because of the presence of chemisorbed atomic oxygen during the decarbonation of the MoNi electrodes [17]. Obviously, detailed theoretical calculations are desirable and helpful to understand the experimental observation.

In this paper, we investigate the effects of chemisorbed atomic oxygen on the segregation behavior of the Mo element in MoNi(111) by performing first-principles calculations. The (111) surface is likely the dominant facet for the NiMo electrolysis electrodes since, among various Ni surfaces, Ni(111) is the most stable one. Based on the previous studies [22,35] we believe that, if we consider other orientations, the segregation behaviors will not change. This is because the surface segregation has a weak dependence on the orientation [22]. Mo segregation on NiMo(111) and NiMo(100) in vacuum has been calculated and the results show that the Mo atom has the same segregation tendency for the two surfaces [35,36]. In the work by Sansa, *et al.* [22], the authors found that the stronger adsorption energies of M impurities on the (100) alloy surfaces compared to the (111) surfaces do not induce a better segregation toward the (100) facet. The authors argued that the adsorption anisotropy is mostly generated by the matrix metal Au and slightly depends on the chemical nature of M. Inconsistent with previous studies [35,36], our calculations show that the surface segregation of the Mo element does not occur in the vacuum, namely for the clean MoNi(111) it is energetically unfavorable for Mo atoms to segregate onto the surface. In the presence of chemisorbed atomic oxygen, the Mo element is found to segregate to the alloy surface when the coverage is thicker than 1/9 ML. By carefully analyzing the densities of states, we see that the preference of Mo atoms to be embedded in the Ni bulk is mainly governed by surface energy considerations and atomic size effects. We also find a strong covalent bonding between Mo d-states and O p-states. The remainder of the paper is organized as follows.

In Section 2, the theoretical methods and computational details are described. Section 3 presents the calculated results of segregation energies, and electronic structure analyses. A brief summary is given in Section 4.

2. Computational Methods

The first-principles calculations were performed within a density functional theory using the Vienna Ab-initio Simulation Package (VASP) [37–39]. The electron-ion interaction was described using the projector augmented wave method (PAW) [40,41] and the exchange correlation potential using the Perdew–Burke–Ernzerhof (PBE) functional method [42]. The energy cutoff for the plane wave basis set was 450 eV for all the calculated systems. Spin polarization was taken into account in the calculations and the Methfessel and Paxton [43] was employed to determine electron occupancies with a smearing parameter σ of 0.14 eV. The convergence criteria for the electronic self-consistent iteration and the ionic relaxation loop were set to 10^{-5} eV and 0.02 eV/A, respectively.

To simulate metallic surfaces, a slab supercell was employed. All the calculations presented in this work were based on slabs of 54 atoms, containing six atomic layers representing a 3×3 supercell, separated by 15 A of vacuum space. MoNi(111) alloy systems corresponding to the substitution of one Ni atom by one Mo atom in the first, second, third or fourth nickel layer are presented in Figure 1, respectively. The Mo atom plays a role as a prober to determine whether it prefers to stay on the surface or in the bulk. As pointed out above, the drawback of this method is that it cannot predict the equilibrium phases and structures. However, investigation of the equilibrium phases and structures under certain given conditions is beyond the scope of the present study. Chemisorbed atomic oxygen was located on only one side of the slab. The influence of the resulting electric dipole on the computed energy values was estimated to be very small according to standard methods [44], and had thus been neglected. The atoms in the top four layers and the chemisorbed atomic oxygen were allowed to relax, while the atoms in the bottom two layers were fixed at the bulk geometry positions. Brillouin zone integrations were performed using Monkhorst–Pack grids [45] of $4 \times 4 \times 1$ for slab calculations. The gas-phase oxygen molecule was simulated through a large supercell with dimensions of $12 \times 12 \times 12$ A^3.

In order to characterize the segregation behavior of the Mo atom in nickel, the segregation energy (E_{segr}) was defined as the energy difference between the states with the Mo atom located at the upper surface layer and in the bulk. According to this definition, the segregation energies were calculated according to the following equation:

$$E_{segr} = E_{MoNi(Mo,x-layer)} - E_{MoNi(Mo,4th-layer)} \quad (1)$$

where $E_{MoNi(Mo,x-layer)}$ represents the total energy of the MoNi alloy system with the Mo atom located in the upper x nickel layers (x = 1, 2 or 3), and $E_{MoNi(Mo,4th\text{-}layer)}$ represents the total energy of the MoNi alloy system with the Mo atom located in the fourth nickel layer, which corresponds to the presence of the Mo atom in the "bulk" nickel matrix. For the oxygen adsorption cases, $E_{MoNi(Mo,x\text{-}layer)}$ and $E_{MoNi(Mo,4th\text{-}layer)}$ are the total energies of the slabs with O atoms adsorbed on the top layer and the Mo atom located at the corresponding atomic layer. According to our calculations, a chemisorbed atomic oxygen prefers to occupy a three-fold fcc hollow site on MoNi(111) and Ni(111). For higher oxygen coverage cases (such as 2/9, 3/9, and 4/9 ML oxygen coverages corresponding to the adsorption of two, three and four oxygen atoms, respectively), we considered all possible oxygen adsorption configurations and calculated the segregation behavior by using the most stable adsorption configurations.

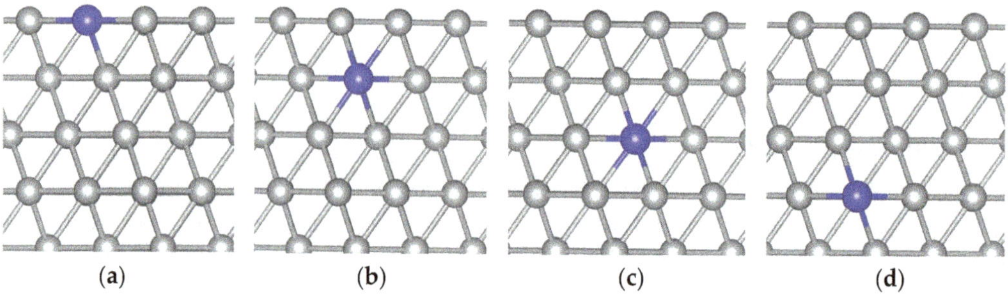

Figure 1. MoNi(111) alloy systems showing one Mo monomer substituting one Ni atom in the (**a**) first; (**b**) second; (**c**) third; and (**d**) fourth nickel layer. Only the four top layers are shown. Gray and blue balls represent Ni and Mo atoms, respectively.

3. Results and Discussion

We have done an exhaustive and extensive search on the lowest-energy O adsorption patterns (or arrangements) for each oxygen coverage with the Mo atom located at various atomic layers. The most stable O adsorption configurations for Mo located in the top-most atomic layer are shown in Figure 2. For the 1/9 ML oxygen coverage case, the O atom prefers to occupy a fcc site near the surface Mo atom. The two O atoms occupy a fcc site and an hcp site next to the Mo atom, respectively, for the oxygen coverage of 2/9 ML It is less stable by 0.16 eV for the two O atoms to occupy the two fcc sites. For the higher oxygen coverage cases (such as 3/9 ML and 4/9 ML) O atoms first occupy the fcc sites next to the surface Mo atom and then those fcc sites far away from the Mo atom. We find that the subsurface Mo atom prefers to stay away from the adsorbed O atoms; for instance, when the Mo atom is located at the second atomic layer, O atoms occupying the fcc sites far away from the top position of the Mo atom are energetically more favorable. When the Mo atom is located at the third or lower atomic layer, it imposes a negligible effect on the O adsorption, as we can see from Figure 3, that the segregation energies do not rely on the oxygen coverage.

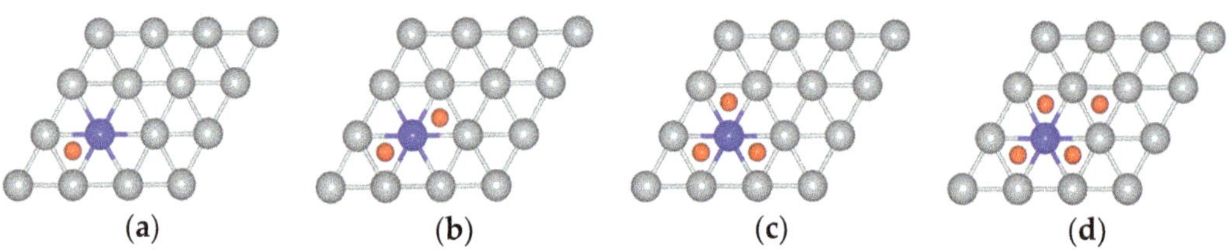

Figure 2. Top views of the most stable adsorption configurations for the oxygen coverage of (**a**) 1/9; (**b**) 2/9; (**c**) 3/9 and (**d**) 4/9 with Mo located at the top-most atomic layer of the slab. Only the top-most metal atomic layer and adsorbed O atoms are shown. Gray, blue and red balls represent Ni, Mo and O, respectively.

The calculated segregation energies for clean and oxygen- adsorbed surfaces are given in Table 1 and also plotted in Figure 3. In absence of adsorbed gas, *i.e.*, under vacuum conditions, the segregation energy for Mo located at the top-most layer is positive and has a value of 0.79 eV. This indicates that Mo does not segregate to the top-most surface layer. We can attribute this behavior to the smaller surface energy of nickel compared to that of molybdenum (experimental values are 2.45 eV/atom for Ni and 3.00 eV/atom for Mo [46]). Interestingly, the Mo atom prefers to occupy a site below the top-most surface layer. This oscillatory phenomenon is quite common in alloying systems [47–51]. That the Mo atoms located at the second layer are energetically favorable can be understood by two facts: (1) since it is not on the top-most layer it can avoid the higher surface energy of Mo; and (2) by locating at the second layer, the system can largely release the elastic energy due to the atomic size mismatch.

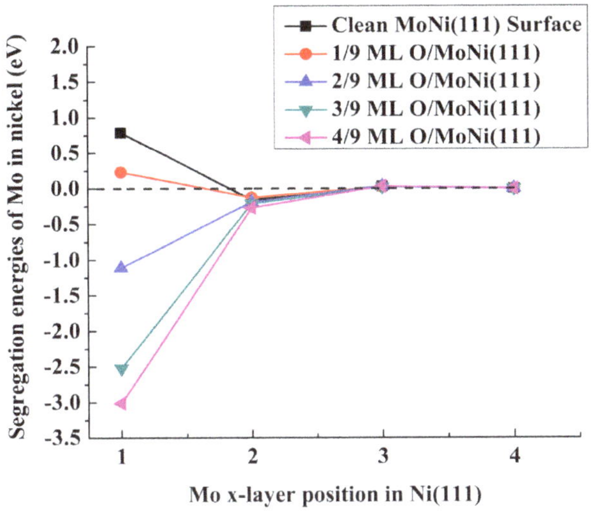

Figure 3. Evolution of the segregation energies (eV) of Mo atom from the nickel "bulk" (fourth layer) to upper layers toward the surface in the presence of different atomic oxygen coverage.

At the oxygen coverage of 1/9 ML, although the segregation energy drops to 0.23 eV from the value of 0.79 eV for the clean surface, the Mo atom still prefers to locate inside of the bulk. With increasing the oxygen coverage to 2/9 ML or thicker, the Mo atom becomes more stable in the top-most layer than in the lower layer and in the bulk. This result indicates that the presence of chemisorbed oxygen on the surface of a MoNi alloy electrode, such as during the decarbonation of the electrode, may cause the surface to be Mo-rich, *i.e.*, having a concentration higher than the nominal concentration in the alloy. This explains our recent experimental observation that the electrode surface is Mo-rich [17]. We can understand the O adsorption-driven segregation of Mo as follows. There are two factors that favor Mo to segregate onto the top-most layer. The first one is that segregation of the Mo atom onto the top-most layer helps to release the elastic energy caused by the size mismatch. Another factor is that the Mo-O bond is stronger than the Ni-O bond (from the chemical rubber company (CRC) handbook [52], the bond strengths for diatomic Mo-O and Ni-O molecules are 145.1 kcal/mol and 93.6 kcal/mol at 298 K, respectively). There exists a factor that is against the segregation of Mo onto the first surface layer. This unfavorable factor is that the Mo surfaces have a larger surface energy than the corresponding Ni surfaces. The final segregation behavior is determined by the competition between the two favorable factors and the unfavorable factor.

Table 1. Calculated segregation energies for MoNi alloy configurations with different adsorbed atomic oxygen coverage. The reported segregation energies of molybdenum in O-MoNi(111) systems are calculated by considering the most energetically stable adsorbed oxygen configurations.

Position of the Mo Atom	Atomic Oxygen Sub-Monolayer on MoNi(111) Alloy				
	0	1/9 ML	2/9 ML	3/9 ML	4/9 ML
	E_{tot} (eV)				
First layer	0.79	0.23	−1.11	−2.52	−3.01
Second layer	−0.16	−0.13	−0.19	−0.21	−0.27
Third layer	0.03	0.02	0.03	0.02	0.03
Fourth layer	0	0	0	0	0

The d-band densities of states (DOS) for Mo and its nearest neighboring nickel atoms in the top-most surface layer with and without oxygen adsorption were calculated. Figure 4 shows the d-band DOS for both molybdenum and nickel atoms in the MoNi alloy compared to their corresponding DOS in pure metals. The d-band DOS for Ni(111) and Mo(110) are presented because they are the most stable surfaces for the metals. Compared to the pure metal surface atom, the nickel d-band DOS center is slightly shifted up towards the Fermi level in the alloy case (Figure 4a), while the molybdenum d-band DOS center is shifted towards the lower energy region, away from the Fermi level (Figure 4b). According to the d-band center model developed by Hammer, *et al.* [53], these d-band center shifts closely correlate to oxygen adsorption–induced Mo surface segregation as we have observed above. There exists an additional peak at the higher edge of the d-band DOS for the alloy case (Figure 4a). This peak indicates the formation of a covalent Ni-Mo bond [12]. From Figure 4b we can see that the Mo d-band not only shifts to the lower energy region, it also becomes narrower than in the pure metal, and the empty states are fewer. This means that the Mo atom gains some electrons from its neighboring nickel atoms, which is consistent with the d-band DOS of Ni shifting up and having more empty states in the alloy case (Figure 4a).

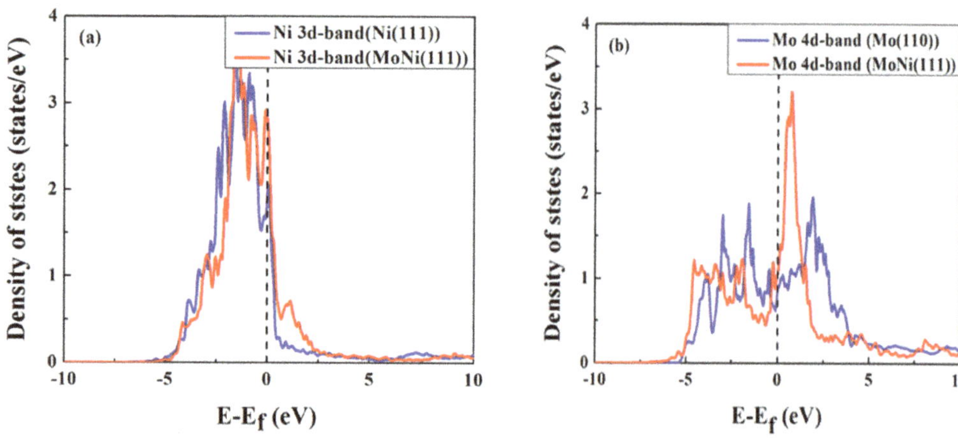

Figure 4. Calculated density of states (DOS) of the Mo and Ni atoms in the alloy and pure metal surfaces. (**a**) d-band DOS of Ni atom in the top-most atomic layer for the alloy and pure Ni surfaces; (**b**) d-band DOS of Mo atom in the top-most atomic layer for the alloy and pure Mo surfaces.

The d-band DOS for the Mo atom located at the top-most surface layer, the second layer and the third layer without oxygen adsorption are presented in Figure 5a. From Figure 5a, we see the shape and position of the Mo d-band DOS do not significantly change no matter which layer the Mo atom is located at. This indicates that the preference of Mo atoms to locate in the Ni bulk is mainly governed by surface energy considerations and atomic size effects (2.39 A for Mo and 2.22 A for Ni [54]) [12]. Figure 5b shows that the d-band DOS of Mo in MoNi(111) interacting with oxygen are drastically modified compared to those for the clean surface (*i.e.*, without oxygen adsorption). The d-band DOS for the oxygen adsorption case are largely widened. The narrower DOS for the surface atoms are easy to understand, given they have a smaller coordination number relative to the bulk atoms. For the adsorption cases, it can be considered that as the adsorbate increases the coordination number for the surface atoms, it therefore broadens the DOS [55]. From Figure 5b, we can see that the broadening of the Mo d-band DOS causes some d-states to move to the lower energy region upon O adsorption. This effect lowers the system energy and leads to better stability of the system. Furthermore, we also see that the O-Mo anti-bonding states, located above the Fermi level, are largely unoccupied and that the O p-states and Mo d-states have a large overlap, thus allowing a strong covalent bonding and hybridization interaction between molybdenum and the oxygen atom.

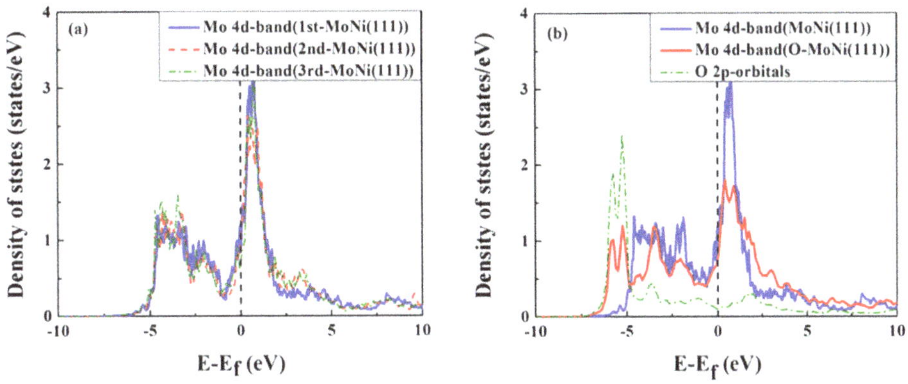

Figure 5. (**a**) d-band DOS for the Mo atom located in the first, second and third MoNi(111) layers in absence of oxygen; (**b**) d-band DOS for the Mo atom located in the top-most atomic layer with and without oxygen adsorption and oxygen p-band DOS.

4. Conclusions

We have performed density functional theory calculations to investigate the effects of chemisorbed atomic oxygen on the segregation behavior of the Mo element in MoNi(111). In particular, the coverage dependence and possible adsorption-induced segregation phenomena are addressed by calculating the segregation energies of the Mo atom in the upper layers of MoNi(111). The theoretical calculated results show that the Mo atom prefers to be embedded in the bulk in the clean MoNi(111), while it segregates to the top-most layer when the oxygen coverage is thicker than 1/9 ML For the clean MoNi(111) we see that the d-band center of Ni atoms surrounding the Mo atom shifts up to the Fermi energy and the Mo d-band becomes narrower with its center shifted down away from the Fermi energy. The shape and position of the Mo d-band DOS do not significantly change, no matter which layer the Mo atom is located at. This indicates that the preference of Mo atoms to be embedded in the Ni bulk is mainly governed by surface energy considerations and atomic size effects. The Mo d-band DOS for the oxygen adsorption cases are largely widened and lead to a strong covalent bonding between the molybdenum atom and the oxygen atom. The present study provides valuable insight for exploring practical applications of Ni-based alloys as hydrogen evolution electrodes.

Acknowledgments: This work was supported by Special foundation for institute of technology research and development under 2013EG115003 and 2014EG115002.

Author Contributions: Ligen Wang and Yanlin Yu conceived and designed the experiments; Yanlin Yu performed the experiments; Wei Xiao and Jianwei Wang analyzed the data; Yanlin Yu and Ligen Wang wrote the paper.

Conflicts of Interest: The authors declare no conflict of interest.

References

1. Lendzion-Bielun, Z.; Narkiewicz, U.; Arabczyk, W. Cobalt-based catalysts for ammonia decomposition. *Materials* **2013**, *6*, 2400–2409. [CrossRef]
2. Liu, J.; Xu, C.; Liu, C.; Feng, W.; Liu, H.; Jing, J.; Li, Z. Impact of Cu-Pt nanotubes with a high degree of alloying on electro-catalytic activity toward oxygen reduction reaction. *Electrochim. Acta* **2015**, *152*, 425–432. [CrossRef]
3. Lloyd, A.C.; Noël, J.J.; Mcintyre, S.; Shoesmith, D.W. Cr, Mo and W alloying additions in Ni and their effect on passivity. *Electrochim. Acta* **2004**, *49*, 3015–3027. [CrossRef]
4. Yarulin, A.; Berguerand, C.; Alonso, A.O.; Yuranov, I.; Kiwi-Minsker, L. Increasing Pt selectivity to vinylaniline by alloying with Zn via reactive metal-support interaction. *Catal. Today* **2015**, *256*, 241–249. [CrossRef]

5. Bowker, R.H.; Ilic, B.; Bo, A.C.; Reynolds, M.A.; Murray, B.D.; Bussell, M.E. Carbazole hydrodenitrogenation over nickel phosphide and Ni-rich bimetallic phosphide catalysts. *Appl. Catal. A* **2014**, *482*, 221–230. [CrossRef]
6. Iizuka, Y.; Inoue, R.; Miura, T.; Morita, N.; Toshima, N.; Honma, T.; Oji, H. Chemical environment of Ag atoms contained in AuAg bimetallic catalysts and the generation of the catalytic activity for CO oxidation. *Appl. Catal. A* **2014**, *483*, 63–75. [CrossRef]
7. Kang, J.; Wang, R.; Wang, H.; Liao, S.; Key, J.; Linkov, V.; Ji, S. Effect of Ni Core structure on the electrocatalytic activity of Pt-Ni/C in methanol oxidation. *Materials* **2013**, *6*, 2689–2700. [CrossRef]
8. Zhu, H.; Anjum, D.H.; Wang, Q.; Abou-Hamad, E.; Emsley, L.; Dong, H.; Laveille, P.; Li, L.; Samal, A.K.; Basset, J.M. Sn surface-enriched Pt-Sn bimetallic nanoparticles as a selective and stable catalyst for propane dehydrogenation. *J. Catal.* **2014**, *320*, 52–62. [CrossRef]
9. Andersson, K.J.; Federico, C.V.; Jan, R.; Ib, C. Adsorption-driven surface segregation of the less reactive alloy component. *J. Am. Chem. Soc.* **2009**, *131*, 2404–2407. [CrossRef] [PubMed]
10. Menning, C.A.; Chen, J.G. General trend for adsorbate-induced segregation of subsurface metal atoms in bimetallic surfaces. *J. Chem. Phys.* **2009**, *130*, 363–366. [CrossRef]
11. Dhouib, A.; Guesmi, H. DFT study of the M segregation on MAu alloys (M = Ni, Pd, Pt) in presence of adsorbed oxygen O and O_2. *Chem. Phys. Lett.* **2012**, *521*, 98–103. [CrossRef]
12. Guesmi, H.; Louis, C.; Delannoy, L. Chemisorbed atomic oxygen inducing Pd segregation in PdAu(111) alloy: Energetic and electronic DFT analysis. *Chem. Phys. Lett.* **2011**, *503*, 97–100. [CrossRef]
13. Kitchin, J.R.; Reuter, K.; Scheffler, M. Alloy surface segregation in reactive environments: First-principles atomistic thermodynamics study of $Ag_3Pd(111)$ in oxygen atmospheres. *Phys. Rev. B Condens. Matter Mater. Phys.* **2008**, *77*, 439–446. [CrossRef]
14. Mainardi, D.S.; Balbuena, P.B. Monte Carlo simulation of Cu-Ni nanoclusters: Surface segregation studies. *Langmuir* **2001**, *17*, 2047–2050. [CrossRef]
15. Oetelaar, D.; Nooij, W.; Oerlemans, S.; Gon, D. Surface segregation in supported Pd-Pt nanoclusters and alloys. *J. Phys. Chem. B* **1998**, *102*, 3445–3455. [CrossRef]
16. Mainardi, D.S.; Balbuena, P.B. Surface segregation in bimetallic nanoclusters: Geometric and thermodynamic effects. *Int. J. Quantum Chem* **2001**, *85*, 580–591. [CrossRef]
17. Wang, J.W.; Wang, Y.F.; Wang, L.G.; Yu, Y.L.; Zhang, J.G.; Wang, L.S. Optimization of the electrocatalytic properties of the NiMoCo foam electrode for water electrolysis by post-treatment processing. *Rare Met.* **2015**, *34*, 802–807. [CrossRef]
18. Zhang, X.M.; Liu, J.C.; Tang, J.G.; Li, L.; Chen, M.A.; Liu, S.D.; Zhu, B. Element segregation on the surfaces of pure aluminum foils. *Appl. Surf. Sci.* **2010**, *256*, 7300–7304. [CrossRef]
19. Zafeiratos, S.; Piccinin, S.; Teschner, D. Alloys in catalysis: Phase separation and surface segregation phenomena in response to the reactive environment. *Catal. Sci. Technol.* **2012**, *2*, 1787–1801. [CrossRef]
20. Guesmi, H. Theoretical insights on the effect of reactive gas on the chemical ordering of gold-based alloys. *Gold Bull.* **2013**, *46*, 213–219. [CrossRef]
21. Juárez, M.F.; Soldano, G.; Guesmi, H.; Tielens, F.; Santos, E. Catalytic properties of Au electrodes modified by an underlayer of Pd. *Surf. Sci.* **2015**, *631*, 235–247. [CrossRef]
22. Sansa, M.; Dhouib, A.; Guesmi, H. Density functional theory study of CO-induced segregation in gold-based alloys. *J. Chem. Phys.* **2014**, *141*. [CrossRef] [PubMed]
23. Zhu, B.; Thrimurthulu, G.; Delannoy, L.; Louis, C.; Mottet, C.; Creuze, J.; Legrand, B.; Guesmi, H. Evidence of Pd segregation and stabilization at edges of AuPd nano-clusters in the presence of CO: A combined DFT and DRIFTS study. *J. Catal.* **2013**, *308*, 272–281. [CrossRef]
24. Sun, Q.; Reuter, K.; Scheffler, M. Effect of a humid environment on the surface structure of $RuO_2(110)$. *Phys. Rev. B Condens. Matter Mater. Phys.* **2003**, *67*, 920–925. [CrossRef]
25. Silva, A.M.; Achete, C.A.; Capaz, R.B. First-principles study of oxygen-induced copper segregation in $Cu_3Au(111)$. *Chem. Phys.* **2013**, *410*, 99–102. [CrossRef]
26. Wang, L.; Kuklja, M.M. First-principles study of small aluminum clusters: Oxygen adsorptions, oxidation and phase stability. *J. Phys. Chem. Solids* **2010**, *71*, 140–144. [CrossRef]
27. Zunger, A.; Wang, L.G.; Hart, G.L.W.; Sanati, M. Obtaining Ising-like expansions for binary alloys from first principles. *Model. Simul. Mater. Sci. Eng.* **2002**, *10*, 685–706. [CrossRef]

28. Chen, W.; Schmidt, D.; Schneider, W.F.; Wolverton, C. Ordering and oxygen adsorption in Au–Pt/Pt(111) surface alloys. *J. Phys. Chem. C* **2011**, *115*, 17915–17924. [CrossRef]
29. Han, B.C.; Ven, A.V.D.; Ceder, G.; Hwang, B.J. Surface segregation and ordering of alloy surfaces in the presence of adsorbates. *Phys. Rev. B Condens. Matter Mater. Phys.* **2005**, *72*. [CrossRef]
30. Kukushkin, R.G.; Bulavchenko, O.A.; Kaichev, V.V.; Yakovlev, V.A. Influence of Mo on catalytic activity of Ni-based catalysts in hydrodeoxygenation of esters. *Appl. Catal. B* **2015**, *163*, 531–538. [CrossRef]
31. Yuan, C.; Nan, Y.; Wang, X.; Wang, J.; Lv, D.; Li, X. The SiO_2 supported bimetallic Ni-Ru particles: A good sulfur-tolerant catalyst for methanation reaction. *Chem. Eng. J.* **2015**, *260*, 1–10. [CrossRef]
32. Malaibari, Z.O.; Croiset, E.; Amin, A.; Epling, W. Effect of interactions between Ni and Mo on catalytic properties of a bimetallic Ni-Mo/Al_2O_3 propane reforming catalyst. *Appl. Catal. A* **2015**, *490*, 80–92. [CrossRef]
33. Xun, T.; Li, X.; Yang, C.; Lu, J.; Lin, Z. Noble fabrication of Ni-Mo cathode for alkaline water electrolysis and alkaline polymer electrolyte water electrolysis. *Int. J. Hydrog. Energy* **2014**, *39*, 3055–3060.
34. Marcus, P.; Chandler, C.; Chadli, H.; Wynblatt, P. Surface composition and structure of nickel-molybdenum single crystal alloys. *Appl. Surf. Sci.* **1989**, *37*, 33–43. [CrossRef]
35. Nilekar, A.U.; Ruban, A.V.; Mavrikakis, M. Surface segregation energies in low-index open surfaces of bimetallic transition metal alloys. *Surf. Sci.* **2009**, *603*, 91–96. [CrossRef]
36. Ruban, A.; Skriver, H.; Nørskov, J.K. Surface segregation energies in transition-metal alloys. *Phys. Rev. B Condens. Matter Mater. Phys.* **1999**, *59*, 15990–16000. [CrossRef]
37. Kresse, G.; Furthmüller, J. Efficiency of ab-initio total energy calculations for metals and semiconductors using a plane-wave basis set. *Comput. Mater. Sci.* **1996**, *6*, 15–50. [CrossRef]
38. Kresse, G.; Furthmüller, J. Efficient iterative schemes for *ab initio* total-energy calculations using a plane-wave basis set. *Phys. Rev. B Condens. Matter Mater. Phys.* **1996**, *54*, 11169–11186. [CrossRef]
39. Kresse, G.; Hafner, J. *Ab initio* molecular-dynamics simulation of the liquid-metal-amorphous-semiconductor transition in germanium. *Phys. Rev. B Condens. Matter Mater. Phys.* **1993**, *49*, 14251–14269. [CrossRef]
40. Blöchl, P.E. Projector augmented-wave method. *Phys. Rev. B Condens. Matter Mater. Phys.* **1994**, *50*, 2665–2668. [CrossRef]
41. Kresse, G.; Joubert, D. From ultrasoft pseudopotentials to the projector augmented-wave method. *Phys. Rev. B Condens. Matter Mater. Phys.* **1999**, *59*, 1758–1775. [CrossRef]
42. Perdew, J.P.; Burke, K.; Ernzerhof, M. Generalized gradient approximation made simple. *Phys. Rev. Lett.* **1996**, *77*, 3865–3868. [CrossRef] [PubMed]
43. Methfessel, M.; Paxton, A.T. High-precision sampling for Brillouin-zone integration in metals. *Phys. Rev. B Condens. Matter Mater. Phys.* **1989**, *40*, 3616–3621. [CrossRef]
44. Neugebauer, J.; Scheffler, M. Adsorbate-substrate and adsorbate-adsorbate interactions of Na and K adlayers on Al(111). *Phys. Rev. B Condens. Matter Mater. Phys.* **1992**, *46*, 16067–16080. [CrossRef]
45. Monkhorst, H.J.; Hendrik, J.; James, D. Special points for Brillouin-zone integrations. *Phys. Rev. B Condens. Matter Mater. Phys.* **1976**, *13*, 5188–5192. [CrossRef]
46. DeBoer, F.R.; Boom, R.; Miedema, A.R. *Cohesion in Metals*, 2nd ed.; North-Holland Physics Publishing: Amsterdam, The Netherlands, 1989; pp. 657–660.
47. Florencio, J.; Ren, D.M.; Tsong, T.T. Absolute composition depth-profiles in surface segregation of Pt-Rh alloys. *Surf. Sci.* **1996**, *345*, 29–33. [CrossRef]
48. Helfensteyn, S.; Luyten, J.; Feyaerts, L.; Creemers, C. Modelling surface phenomena in Pd-Ni alloys. *Appl. Surf. Sci.* **2003**, *75*, 844–849. [CrossRef]
49. Kuntze, J.; Speller, S.; Heiland, W.; Deurinck, P.; Creemers, C.; Atrei, A.; Bardi, U. Surface structure and segregation profile of the alloy Au_3Pd(110): Experiment and theory. *Phys. Rev. B Condens. Matter Mater. Phys.* **1999**, *60*, 9010–9018. [CrossRef]
50. Ren, D.M.; Qin, J.H.; Wang, J.B.; Tsong, T.T. Oscillatory compositional depth profiles in surface segregation of a Pt-Rh alloy. *Phys. Rev. B Condens. Matter Mater. Phys.* **1993**, *47*, 3944–3946. [CrossRef]
51. Tsong, T.T.; Ren, D.M.; Ahmad, M. Atomic-layer by atomic-layer compositional depth profiling: Surface segregation and impurity cosegregation of Pt-Rh and Pt-Ru alloys. *Phys. Rev. B Condens. Matter Mater. Phys.* **1988**, *38*, 7428–7435. [CrossRef]
52. Kerr, J.A.; Trotman-Dickenson, A.F. Strengths of chemical bonds. In *CRC Handbook of Chemistry and Physics*, 62nd ed.; Weast, R.C., Astle, M.J., Eds.; CRC Press, Inc.: Boca Raton, FL, USA, 1982; pp. F180–F190.

53. Hammer, B.; Morikawa, Y.; Nøskov, J.K. CO chemisorption at metal surfaces and overlayers. *Phys. Rev. Lett.* **1996**, *76*, 2141–2144. [CrossRef] [PubMed]
54. Allinger, N.L.; Zhou, X.; Bergsma, J. Molecular mechanics parameters. *J. Mol. Struct. THEOCHEM* **1994**, *312*, 69–83. [CrossRef]
55. Wang, L.G.; Tsymbal, E.Y.; Jaswal, S.S. Structural and magnetic properties of clean and methylthiolate-adsorbed Co(0001) surfaces: A first-principles study. *J. Magn. Magn. Mater.* **2005**, *286*, 119–123. [CrossRef]

12

Aminopropyl-Silica Hybrid Particles as Supports for Humic Acids Immobilization

Mónika Sándor [1], Cristina Lavinia Nistor [2,*], Gábor Szalontai [3], Rusandica Stoica [2], Cristian Andi Nicolae [2], Elvira Alexandrescu [2], József Fazakas [1], Florin Oancea [2] and Dan Donescu [2]

Academic Editor: Silvia Gross

1 Faculty of Chemistry and Chemical Engineering, Babes-Bolyai University, Arany János Street, No.11, Cluj-Napoca 400028, Romania; sandormonika16@gmail.com (M.S.); fazakaseniko7@gmail.com (J.F.)
2 National Research & Development Institute for Chemistry and Petrochemistry—ICECHIM, Splaiul Independentei No.202, 6th district, PO Box 35-174, Bucharest 060021, Romania; rusandica.stoica@gmail.com (R.S.); cristian.nicolae@icechim-pd.ro (C.A.N.); elvira.alexandrescu@icechim-pd.ro (E.A.); florino@ping.ro (F.O.); ddonescu@chimfiz.icf.ro (D.D.)
3 NMR Laboratory, Institute of Materials Engineering, University of Pannonia, Egyetem Street, No. 10, Veszprém 8200, Hungary; szalontai.gabor@solidnmr.hu
* Correspondence: cristina.nistor@icechim-pd.ro

Abstract: A series of aminopropyl-functionalized silica nanoparticles were prepared through a basic two step sol-gel process in water. Prior to being aminopropyl-functionalized, silica particles with an average diameter of 549 nm were prepared from tetraethyl orthosilicate (TEOS), using a Stöber method. In a second step, aminopropyl-silica particles were prepared by silanization with 3-aminopropyltriethoxysilane (APTES), added drop by drop to the sol-gel mixture. The synthesized amino-functionalized silica particles are intended to be used as supports for immobilization of humic acids (HA), through electrostatic bonds. Furthermore, by inserting beside APTES, unhydrolysable mono-, di- or trifunctional alkylsilanes (methyltriethoxy silane (MeTES), trimethylethoxysilane (Me_3ES), diethoxydimethylsilane (Me_2DES) and 1,2-bis(triethoxysilyl)ethane (BETES)) onto silica particles surface, the spacing of the free amino groups was intended in order to facilitate their interaction with HA large molecules. Two sorts of HA were used for evaluating the immobilization capacity of the novel aminosilane supports. The results proved the efficient functionalization of silica nanoparticles with amino groups and showed that the immobilization of the two tested types of humic acid substances was well achieved for all the TEOS/APTES = 20/1 (molar ratio) silica hybrids having or not having the amino functions spaced by alkyl groups. It was shown that the density of aminopropyl functions is low enough at this low APTES fraction and do not require a further spacing by alkyl groups. Moreover, all the hybrids having negative zeta potential values exhibited low interaction with HA molecules.

Keywords: aminosilane supports; humic acids immobilization; silica hybrid particles; sol-gel process

---

1. Introduction

Hybrid organic/inorganic materials derived through sol-gel processing have been extensively studied and reviewed [1,2]. These hybrid materials combine the properties of inorganic (e.g., heat resistance, retention of mechanical properties at high temperature, and low thermal expansion) and of organic (e.g., flexibility, low dielectric constant and ductility) compounds in one material [2]. The reactive organosilane compounds have been used for modifying and tailoring the surface properties of inorganic materials, thereby the material will contain the desired functional groups on its surface.

Mono-, di- and trifunctional organosilanes were used as reactive compounds for functionalization, having one, two or three hydrolysable groups [3]. Depending on the reaction conditions, chemistry of the organosilane, and surface history, a number of different structures can be produced on the surface [3–5].

The attachment of humic acids (HA) to the silica hybrid particles can lead to the formation of materials with very interesting properties. It is well known that, in soil, humic acids are important chelators, combining minerals into organic compounds that are more available to plants. They also tie up toxins, making them less available to plants. Thus, remediation capacity of HA immobilized on silica surfaces was already investigated in previous studies [5–14]. It was shown that these HA-SiO_2 hybrids are effective for remediation of water contaminated with aromatic hydrocarbons [7,11,14], metal ions [6,8,12,13] or radioactive iodine [6,8]. These can also be used to reduce the amount of chemical oxygen demand in wastewater [14]. The aminopropyl functionalized and humic substances activated silica particles have also been used for preparation of supports to immobilize enzymes, which can be widely used in analytical chemistry, and the food, pharmaceutical and chemical industries [15].

Despite the wide variety of silane derivatives, available for surface modification, the majority of studies concerning immobilization of HA on silica supports have employed aminosilanes, in particular 3-aminopropyltriethoxysilane (APTES), because of its intrinsic chemical properties [9,16]. After grafting the amino functional groups in organic solvents, it was found that not all Si-OH groups were derivatized with amino-alkoxysilanes [6,8]. For this reason, there have been attempts to block the -OH residual groups with reactive compounds having simple substituents such as trimethylchlorosilane [6,8].

A similar phenomenon was observed in the case of the immobilization of humic acid on aminopropyl-silica, where only a fraction of the amino groups reacted with the humic acid molecules, due to the steric hindrance [9,10]. The remaining amino groups were blocked with small substituents such as acetic anhydride [9] and sodium acetate [10].

Recent studies related to increasing of interaction efficiency to organically derivatized silica, have shown that this can be achieved by incorporating non-reactive silanes [17]. The capacity of functionalization with trialkoxysilane [5–10] or monoalkoxysilane derivatives was previously studied [10]. But, in all the reported situations, derivatization process was performed in organic solvents after a careful dehydration.

The present work is focused on preparation and characterization of a series of aminosilica hybrid particles, prepared by sol-gel technique in aqueous medium, which can be used as supports for humic acids (HA) immobilization. Furthermore, we investigated the effect of grafting, beside 3-aminopropyltrimethoxysilane (APTES), of mono-, di- and trifunctional alkylsilanes, in order to ensure a better access of the large HA molecules to the free amino groups. Thus, the resulted aminosilane supports contain, on one hand, aminopropyl chains with free NH_2 groups, acting like coupling agents for the humic acids molecules, and, on the other hand, alkylsilane derivatives, as spacers for the -NH_2 groups. Two sorts of humic acids (a commercial product from Aldrich (St. Louis, MO, USA) (noted here AHA) and a lab-made humic acid, extracted from peat (noted here PHA)) were used for evaluation of the efficiency of coupling on the newly synthesized aminosilane supports.

2. Results and Discussions

2.1. CHN Elemental Analysis (Determination of the Mass Fraction of Carbon, Hydrogen and Nitrogen)

The carbon percentage contents are listed in Table 1 and vary according to the precursor's molar ratio used in the sol-gel process. Obviously, the highest carbon content was measured for silica particles synthesized only from APTES (sample 1), while the lowest, for the particles containing only TEOS in the sol-gel mixture (sample 2), Grafting of APTES on the preformed SiO_2 particles (sample 3) increases the carbon weight share in the final hybrid.

For samples in which, besides TEOS and APTES, alkyl-derivatives were also added to the sol-gel reaction (samples 4–8), the carbon content is slightly lower. Between them, the highest carbon content was found in the Me_2-DES modified sample5. Grafting of the different alkyl derivatives (methyl, dimethyl, trimethyl) on the surface of silica particles lowers the weight share of aminopropyl groups in the whole reaction system, leading to lower carbon content. These results were also further confirmed by the thermal analyses, the weight loss decreasing for the TEOS/APTES/alkyl silane samples.

Table 1. Composition of the different silica systems and the corresponding carbon content and weight loss.

Samples	Silica Systems	Molar Ratio	Carbon (%)	Weight Loss 25–250 °C (%)	Weight Loss 250–450 °C (%)	Weight Loss 450–700 °C (%)	Inorganic Residue at 700 °C (%)
1	APTES	-	27.63	23.1	7.5	22.1	47.3
2	TEOS	-	0.94	9.6	2.3	2.3	85.8
3	TEOS/APTES	20/1	5.98	11.0	4.4	4.1	80.5
4	TEOS/APTES/MeTES	20/1/1	3.27	9.6	3.7	2.9	83.8
5	TEOS/APTES/Me_2DES	20/1/1	4.74	9.4	3.5	3.0	84.1
6	TEOS/APTES/Me_3ES	20/1/1	3.83	11.0	3.6	3.2	82.2
7	TEOS/APTES/BETES	20/1/0.5	4.42	8.5	3.5	2.7	85.3
8	TEOS/APTES/Me_3ES	10/1/1	5.01	13.6	4.6	5.1	76.7
9	TEOS/BETES	10/0.5	2.96	10.6	4.1	2.2	83.1

2.2. *Thermogravimetric Analysis (TGA)*

The TGA results were consistent with previously presented data; the weight loss being in general proportional to the carbon content, the observed small deviations being caused by the removal of the adsorbed water in the first temperature range. Sample 1 (only APTES particles), with the highest carbon content, showed the highest value of weight loss, while for sample 2 (only TEOS particles), having the lowest carbon content, was showed the lowest value of the weight loss.

Figure 1 shows the TG profiles of different aminopropyl functionalized silica systems. The most important differences were observed in the first temperature range (25–250 °C), which corresponds to the release of water and other solvents. Having a hydrophilic character and a less cross-linked network, sample 1 shows the highest weight loss due to water removal.

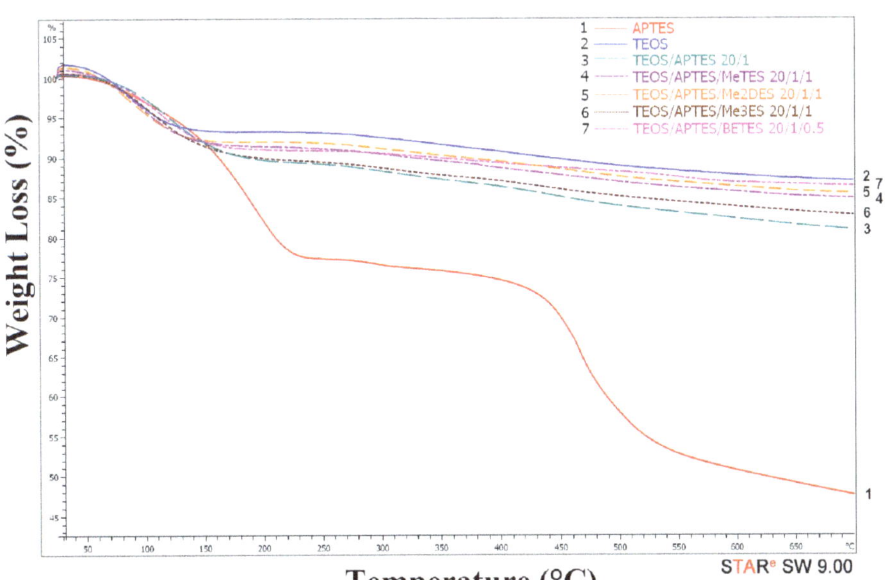

Figure 1. Thermogravimetric (TG) curves of aminopropyl-silica particles.

The second temperature range (250–450 °C) was due to the degradation of aminopropyl and of alkyl chains (when alkyl silanes were added to the sol-gel system). The third temperature range (450–700 °C) corresponds to dehydroxylation of Si-OH residual functions, which led to the formation of Si-O-Si groups. Sample 1 (only APTES), which has a bulky aspect after drying, exhibits a larger weight loss due to the higher content of uncondensed Si-OH functions.

From thermogravimetric analysis it was shown that the thermal behavior of silica hybrids was affected by the nature and concentration of the used organosilanes. Thus, the value of the inorganic residue measured at 700 °C decreased accordingly to the increase in the content of organic compounds (Table 1).

2.3. Particle Size Measurements (Dynamic Light Scattering (DLS) Technique)

In Table 2 are presented the values of the average hydrodynamic diameters recorded for samples diluted in distilled water. For sample 1, the measured average particle diameter is 580 nm. The measurement precision for this sample was lower because the sol dispersion was very complex, containing aggregates of various sizes. After the drying of this sol-gel system a bulky material was obtained (see later the Environmental Scanning Electronic Microscopy (ESEM) results). Sample 2 has an average diameter of 549 nm and a narrow size distribution, indicating the successful synthesis of uniform Stöber silica particles (see later the ESEM images).

Table 2. Particle size and zeta potential values for the studied supports.

Samples	Silica Systems	Molar Ratio	Average Diameter (nm)	Zeta Potential (mV)
1	APTES	-	580	−19
2	TEOS	-	549	−45
3	TEOS/APTES	20/1	691	15
4	TEOS/APTES/MeTES	20/1/1	1471	52
5	TEOS/APTES/Me_2DES	20/1/1	858	27
6	TEOS/APTES/Me_3ES	20/1/1	653	13
7	TEOS/APTES/BETES	20/1/0.5	716	53
8	TEOS/APTES/Me_3ES	10/1/1	558	−5
9	TEOS/BETES	10/0.5	590	−48

According to the DLS measurements, by grafting the aminopropyl functions on the surface of preformed silica particles, the hydrodynamic diameter of the resulted particles increases from 549 (sample 2) to 691 nm (sample 3). When alkyl derivatives have been used, the particles size depends on the addition of alkyl-silanes beside APTES in the reaction system.

When BETES is used as spacer (sample 7), the recorded average diameter (716 nm) is just slightly higher than the one recorded for sample 3 (691 nm). Though, a narrower size distribution is shown for sample 7 when compared with sample 3, being already well known that BETES acts as a cross-linking agent in the sol-gel reaction, leading to a more compact silica network (Figure 2). It also should be noted that, by introduction of the methyl or dimethyl spacers, the size increases significantly and the distribution becomes much broader. This is probably due to the increased ability of particles to form aggregates due to the hydrophobic interaction induced by the methyl groups (Figure 2). Therefore, the highest values of the average hydrodynamic diameter (Table 2) were registered for samples 4 (1471 nm) and 5 (858 nm), respectively. These two samples showed a very hydrophobic behavior, floating on the top of the aqueous dispersion and requiring an increased effort to be homogenized for the DLS analyze.

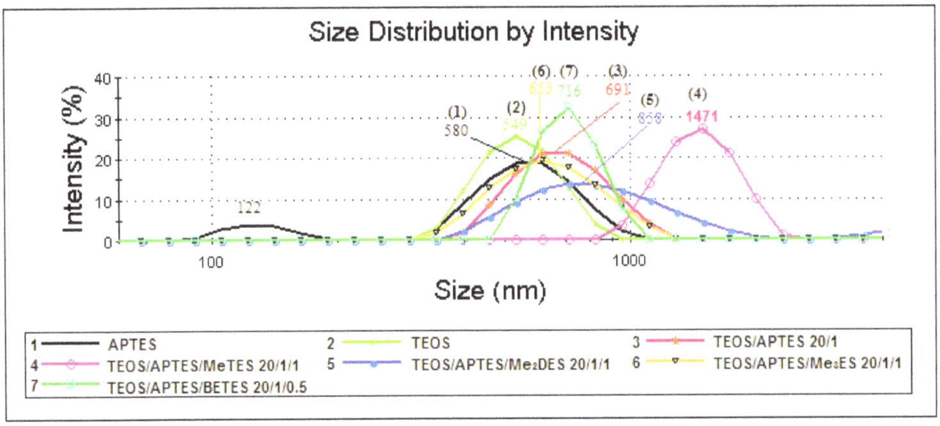

Figure 2. Size distribution by intensity of aminopropyl silica particles.

A surprising result was obtained for sample 6 (where Me_3ES has been used as spacer) for which both the average diameter (653 nm) and size distribution are almost similar to that obtained for sample 3 (691 nm), indicating, in agreement with the C-13 ssNMR (solid state Nuclear Magnetic Resonance) data (see later), the absence of integration of thetrimethyl derivative in the formed silica network.

2.4. Zeta Potential Measurements (Laser Doppler Velocimetry (LDV) Technique)

In Figure 3, zeta potential values (the potential that exists at the particle's hydrodynamic share or slipping plane boundary) are illustrated for samples dispersed in distilled water. As we expected, for samples 2 and 9, the value of zeta potential is highly negative (−45 mV and −48 mV, respectively), due to the high density of residual Si-OH present on the surface of silica particles.

By grafting the aminopropyl functions onto the surface of the preformed TEOS silica particle, the surface charge of particle was modified. As a result, the value of zeta potential shifts towards positive values (samples 3–7, Table 2). In addition, the alkyl-aminopropyl hybrids (samples 4–7) show an increase of zeta potential values, with the increase of the probability of interaction of the Si-OH from the particles surface with the available reactive ethoxy groups from the alkyl-silanes. In these conditions, in distilled water, the $-NH_2$ functions from APTES can be protonated in an increasingly higher extent. Thus, co-grafting (beside APTES) of di- or tri-functional alkyl-silanes (MeTES—sample 4 and BETES—sample 7) led to the highest values of zeta potential (+52 and +53 mV, respectively). The hybrid with the bi-functional co-precursor (Me_2DES—sample 5) exhibited a lower zeta potential (+27 mV), while the use of the mono-functional co-precursor (Me_3ES—sample 6) led to a value of zeta potential (+13 mV) very similar with the one recorded for the hybrid prepared with APTES and TEOS only (sample 3, +15 mV).

As mentioned above (Section 2.3), for samples 4 and 5 a hydrophobic behavior was noticed (the silica powder was floating). This change of the hybrid properties proves the structure modification of the surface of the reference APTES/TEOS hybrid (sample 3). When Me_3ES was used as co-precursor beside TEOS and APTES (sample 6), the expected hydrophobic behavior, thought to be given by the insertion of the trimethyl functions in the final hybrid network, was not noticed anymore. All these observations come to strengthen the idea that Me_3ES does not efficiently connect to the silica network.

For samples 1 and 8, by significantly increasing the aminopropyl density, this hydrophobic chains tend to self-associate and to orientate towards the interior of silica particles. A phase-inversion mechanism occurs, the $-NH_2$ functions being mostly hidden inside the silica hybrid frame, while towards water remain predominantly exposed Si-O charges. Thus, for the both samples, the value of zeta potential remains negative (−19 mV and −5 mV, respectively), but less negative than the one recorded for samples without APTES, since the amino functions are still partially able to interact with the surrounding medium.

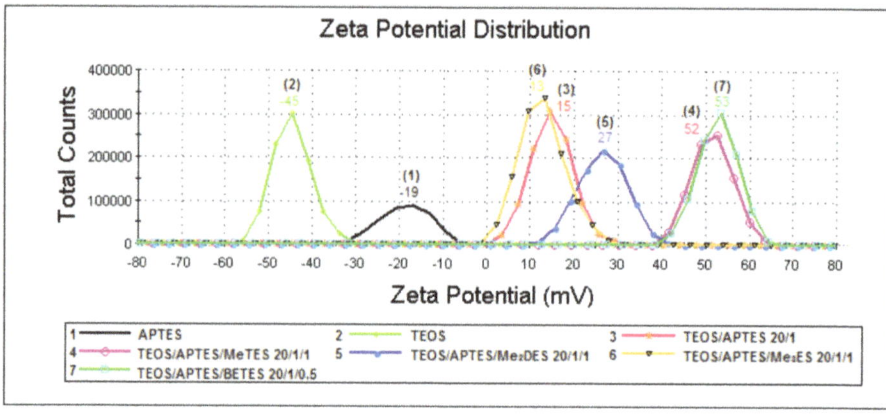

Figure 3. Zeta potential distribution of aminopropyl silica particles.

2.5. *Environmental Scanning Electronic Microscopy (ESEM)*

ESEM images of some of the studied aminosilica hybrids are shown in Figure 4. Sample 1, obtained from the sol-gel reaction of APTES only, is a film forming hybrid, of which cross-section can be seen in Figure 4a. In Figure 4b the uniform and spherically shaped SiO_2 particles resulted from the TEOS reaction (sample 2) are clearly shown. By grafting a small amount of APTES on these preformed SiO_2 particles (1 mole of APTES at 20 moles of TEOS), no significant change in the particles shape or size could be seen (Figure 4c).

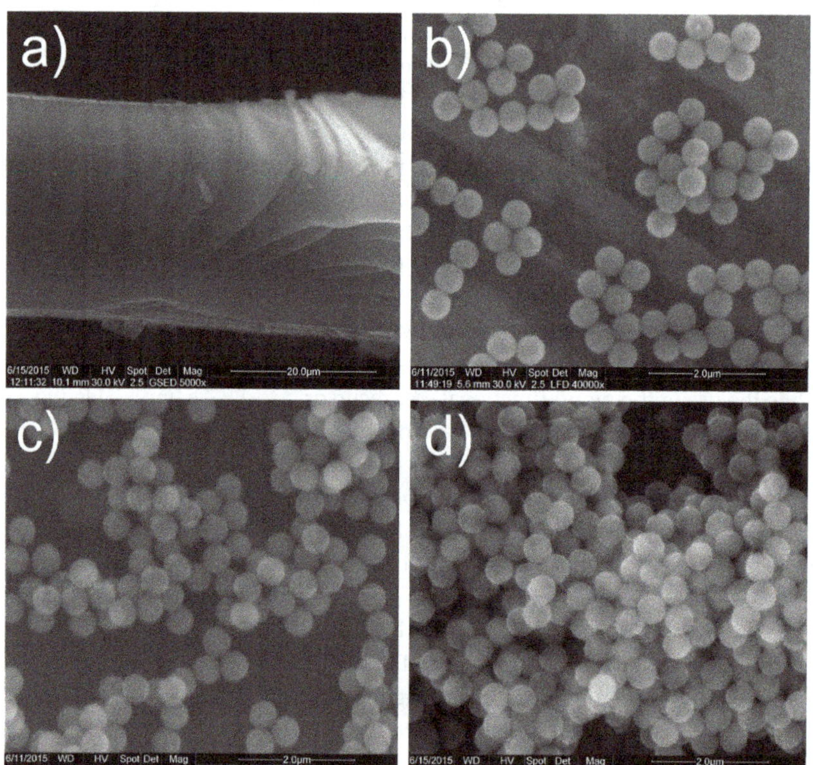

Figure 4. ESEM images of: (**a**) APTES film (Sample 1); (**b**) pristine TEOS silica particles (Sample 2); (**c**) APTES-grafted silica particles (Sample 3); and (**d**) MeTES and APTES—grafted silica particles (Sample 4).

Unlike the hydrodynamic diameter (measured by DLS technique—see Section 3.3), which was measured in an aqueous dispersion, where the organic chains from the particle's surface were found in an extended state and where aggregates were allowed to form and to affect the measurement's result, the particles size estimated by ESEM analyze seems to not change after grafting the aminopropyl or alkyl chains onto particle's surface. Thus, for sample 4 (Figure 4d), a closer packing of the methyl- and aminopropyl-functionalized particles is noticed due to the hydrophobic interactions. However, the size of the individual particles seems unchanged after drying.

2.6. Fourier Transformed Infrared Spectroscopy (FTIR) Spectra

Figure 5 shows the FTIR spectra in the region 4000–400 cm^{-1} of the pristine TEOS silica and of aminopropyl-grafted silica hybrids. Because APTES used for functionalization of the silica particles have a very small weight share compared with TEOS used to obtain the initial particles (TEOS/APTES = 20/1 molar ratio), its specific vibrations are overlapped by the stronger vibrations of the inorganic silica network.

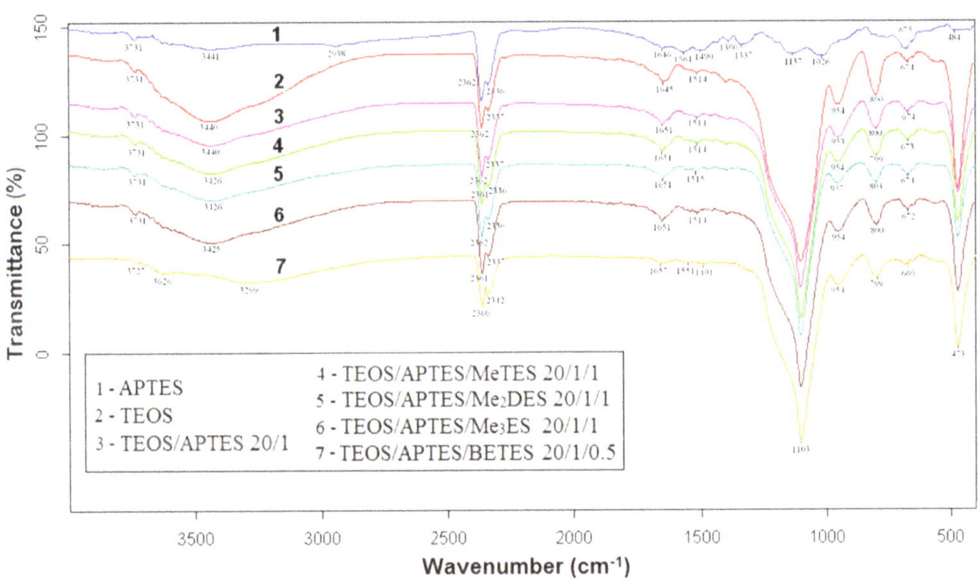

Figure 5. FTIR spectra of aminopropyl functionalized silica particles.

Thus, the broad band covering the range from 3000 to 3800 cm^{-1} can be assigned to hydrogen bonded and partially hydrated silanols (Si-OH). This band was caused by overtones or combinations of vibrations of Si-OH or H_2O and is generally composed of stretching modes.

For samples grafted with aminopropyl, the silanol groups that are shown in the spectrum of sample 2 (the broad band at ~3440 cm^{-1}) were partially consumed by reaction with APTES.

The region of 3000–2800 cm^{-1} corresponds to symmetrical and asymmetrical stretching vibrations of CH_2 from aminopropyl segment. In 1680–1620 cm^{-1} region, the band at ~1630 cm^{-1} characteristic to NH_2 deformation in primary amines (due to the formation of NH_3^+ as the carbon dioxide is adsorbed onto the silica hybrid), is overlapped by the band at ~1650 cm^{-1}, assigned to C=O stretching in carbonyl compounds. Peak 1491 cm^{-1} was attributed to N-H vibration (bending mode) in the primary amine group (R-NH_2). The band at 1390 cm^{-1} have been assigned to the bicarbonate species (C=O bending). In the presence of water, one mole of amine can chemisorb one mole of CO_2, leading in the end to the formation of a bicarbonate group [18–20].

The main bands in the region 1300–400 cm^{-1} are associated with the combination of vibrations of silica network: the strong band that emerged at ~1103 cm^{-1} (SiO_4 ring asymmetric stretching),

~954 cm^{-1} (Si-O-Si symmetric stretching within the SiO_4 tetrahedron), ~800 cm^{-1} (O-Si-O symmetrical stretching), and ~473 cm^{-1} (Si-O-Si bending in SiO_4 tetrahedron). As it was expected, the intensities of these absorption bands decrease with reducing the weight percentage of TEOS in the final hybrid.

2.7. Solid State Nuclear Magnetic Resonance (ssNMR)

Valuable structural information about the obtained amino-silica supports can be obtained by means of ^{29}Si [21] and ^{13}C [22] ssNMR. It also proved extremely useful in characterization of sol-gel-derived materials [23]. This technique has been used under the condition of cross polarization (CP) and magic angle spinning (MAS) [24] for structural characterization of silica particles used. In general, in the CP/MAS spectra, the Q^4groups are heavily underrepresented [4].

The ^{29}Si NMR spectra of un-grafted (pristine) SiO_2 particles (sample 2) and hybrid silica particles (samples 3–6) are shown in Figure 6a. The assignments of the signals to the D, T and Q sites [21] are also shown. The spectrum of sample 2 contains three signals at chemical shifts of −92.3, −100.9, and −110.6 ppm, which have been, respectively, assigned to (i) silicon atoms with two attached hydroxyl groups, designated as Q^2; (ii) silicon atoms with one attached hydroxyl groups, designated as Q^3; and (iii) silicon atoms with no hydroxyls, designated as Q^4.

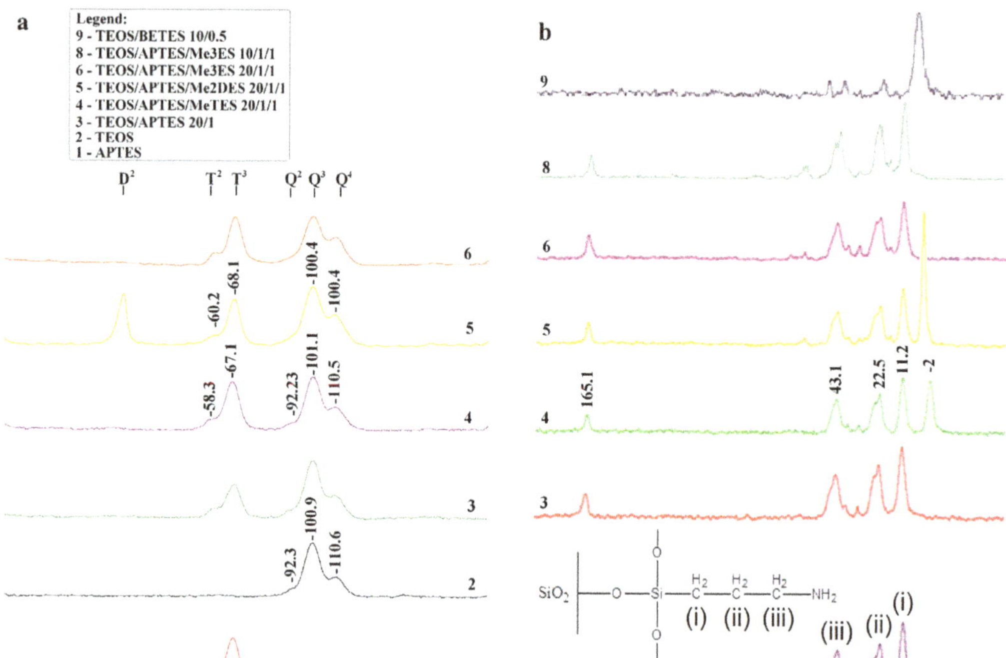

Figure 6. (**a**) ^{29}Si CP/MAS–NMR; and (**b**) ^{13}C CP/MAS–NMR solid state spectra of pristine and modified silica particles.

The attachment of di- and trifunctional co-precursors leads to the appearance of extra signals. Thus, signal at −20 ppm is assigned to -O_2SiR_2 structures, designated as D^2, the signal around −60 ppm is assigned to -$O_2Si(OH)R$ structures, designated as T^2, and signal around −68 ppm is assigned to -O_3SiR structures, designated as T^3. The signal at about −20 ppm, shown in the sample 5 spectrum, is due to the covalently attached Me_2DES molecules (a D^2 group), while the T^2 and T^3 signals indicate the attachment of the hydrolyzed APTES (and MeTES for sample 4) to the preformed TEOS-based silica network. Addition of monofunctional ligands (e.g., Me_3ES) could not be confirmed (see the spectrum of sample 6, which is identical to that of sample 3).

The spectrum of sample 1 (only APTES-based homogenous bulky material), apart from the spinning side bands, contains the same T signals as the APTES-grafted SiO_2 particles (samples 3–6). As expected, the spectrum exhibits only trifunctional Si atoms, thus here we are facing with an irregular (random) or one of the possible cage structures formed in a self-condensation process.

Figure 6b shows the ^{13}C CP/MAS spectra obtained for APTES or TEOS/APTES hybrids, (spectra 1 and 3–8). In the ^{13}C NMR spectrum of sample 1 three signals are seen at 43 ppm (shoulder at 45.6), 22.5 ppm (shoulder at 25.2) and 11 ppm in accordance with the expectations. Note, however, that the two high-frequency signals consist of two overlapping peaks. This suggests that carbons (ii) and (iii) (see Figure 6b) exist in two different chemical environments. In this respect, the presence of the 165.1 ppm signal in spectrum of all APTES derivatives gives some clue. Leal *et al.* [18] showed that the adsorption of CO_2 on the aminopropyl-modified silica material is due to the formation of carbamate species, where two amino groups from the silica surface are used up per carbon dioxide adsorbed molecule. If so, the 165.1 ppm signal can be assigned to a carbamate carbon atom [25].

For samples 4 (TEOS/APTES/MeTES) and 5 (TEOS/APTES/Me_2DES), the appearance of a new signal at −2.3 and 1.7 ppm, assigned to the methyl groups, indicates the integration of the hydrolyzed mono- and di-alkyl silica derivatives in the TEOS-based silica network, respectively. However, regardless of the ratio used, the more bulky monofunctional silanes (Me_3ES, samples 6 and 8) did not bind to the silica surface (Figure 6b). In a control sample, prepared only from TEOS and BETES (sample 9), the presence of Si-CH_2- groups was detected (see the strong signal at 1.7 ppm), while the 165.1 ppm carbonyl signal was no longer observed. This can be taken as further confirmation that only the amino-functionalized hybrids (all APTES containing samples) have the ability to capture the atmospheric CO_2, unlike the ligands without the terminal amine group, such as the TEOS/BETES hybrid.

2.8. Thermogravimetric Analysis Coupled with Mass Spectroscopy (TG-MS)

The contamination/reaction with atmospheric CO_2 of amino-functionalized hybrids was evaluated by TG-MS analysis. The weight loss for the examined samples during controlled heating (following a specific heating profile) is represented in Figure 7. The largest weight loss in the interval 0–200 °C (Figure 1) was observed for sample 1, obtained through sol-gel reactions of APTES only and which contains about twenty times more NH_2 groups than the other samples. In the same temperature interval, CO_2 is eliminated from all samples except sample 2.

When compared to sample 1, sample 3 shows a lower CO_2 weight loss as it has a much lower density of amino groups. A lower CO_2 weight loss was also measured for sample 10 (a control sample synthesized from TEOS/bis(aminopropyl)trimethoxysilane (BAPTMS) 2/1).

Containing secondary amines the BATMS/TEOS hybrid exhibits a lower reactivity towards CO_2. In 0–200 °C interval, the weight loss recorded for samples 3 and 10 takes place in two steps, at temperatures below and above 100 °C. One possible explanation is that a part of CO_2 is physically adsorbed on the surface (is removed in the first step at about 85 °C) and another part is chemically bounded (is lost in the second step at about 165 °C). The CO_2 weight loss between 400 and 600 °C can be due to the thermo-oxidative decomposition of the organic groups found in the final hybrids.

2.9. Porosimetry

Figure 8 compares the nitrogen adsorption-desorption isotherms of two selected aminosilica supports (samples 3 and 4). The resulted isotherms can be assigned as type II, according to the IUPAC classification. The observed hystereses are due to the presence of textural pores (inter-particles pores). As expected, the BET (Brunauer–Emmett–Teller) specific surface areas recorded for the two samples are very small (6 m^2/g for the both studied materials) and so is the total pore volume (0.01642 cm^3/g for sample 3 and 0.01389 cm^3/g for sample 4). Thus, our aminopropyl-grafted SiO_2 particles are not mesoporous and it is unlikely that the large and complex HA molecules can be adsorbed in the particle's cavities.

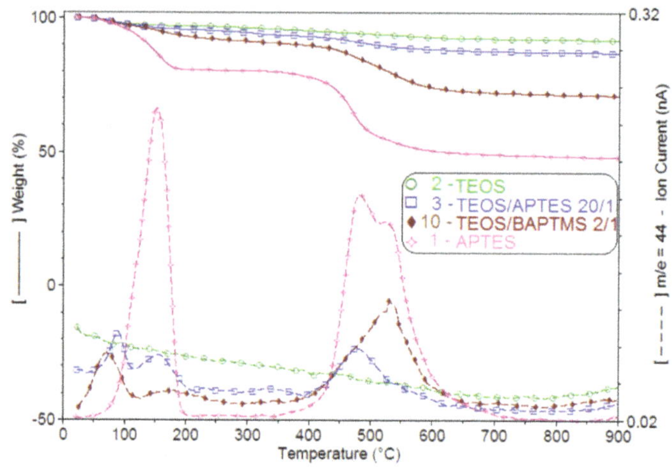

Figure 7. CO_2 weight loss (TG-MS diagrams) recorded for representative aminosilica supports.

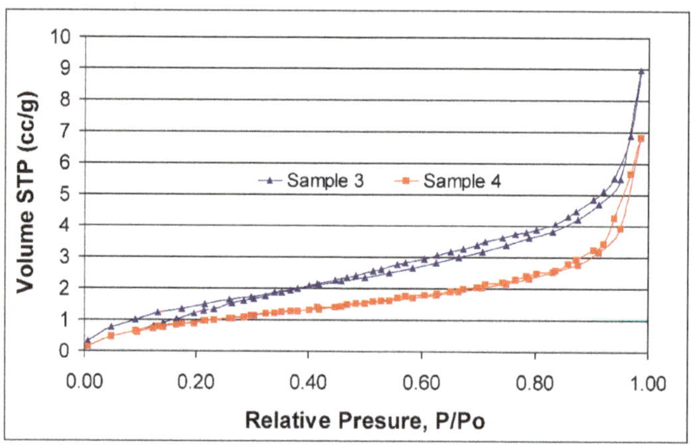

Figure 8. Nitrogen adsorption-desorption isotherms for two selectedaminopropyl supports.

2.10. The Humic Acid (HA) Immobilization Tests of the Novel Aminosilica Supports

Two sorts of humic acids—AHA and PHA—were selected for testing the functionalization efficiency of the synthesized aminosilica hybrids. The relative concentrations of HA immobilized on various silica supports are represented in Figure 9. Obviously, the lowest concentration of up-taken HA was determined for sample 2, having only Si-OH groups on the surface. Sample 1, synthesized only from APTES, shows an almost twice higher concentration of chemically immobilized HA if compared with sample 2, due to its amino functions. In previous studies [9,10] it was shown that the large molecules of humic acids are sterically hindered to attach to the amino functions of the aminosilica supports. More than that, through zeta potential measurements, we showed above (see Section 2.4.) that a significant amount of the amino functions are hidden inside the silica frame resulted from self-condensation of APTES, while towards the exterior remains mostly Si-O ions, less reactive with HA molecules. Therefore, it is not surprising that for the hybrid (sample 3) (TEOS/APTES 20/1), which has a much lower density of the aminopropyl groups on the surface than sample1 and has a positive value of zeta potential (most of the amino functions are located toward the exterior of the silica particles), the HA-binding ability is greatly improved.

Despite our expectations, for samples where methyl-, dimethyl- or ethyl-groups were used as spacers for the amino-functions (samples 4, 5 and 7), the improvements of the immobilization capacity are not impressive enough (~2% for AHA and ~6% for PHA immobilization) to justify the usefulness of aminopropyl spacing by alkyl groups. So, the density of aminopropyl functions seem to be small enough at this low APTES molar ratio (TEOS/APTES = 20/1) not to sterically hinder the interaction with the HA molecules. Further spacing by alkyl groups is not therefore an absolute need (see also Figure 10). In addition, the inefficient grafting of Me_3ES co-precursor on the silica hybrid network is also confirmed by the coupling tests, in which similar (sample 6) or lower concentration (sample 8) of immobilized HA was determined compared with the reference support obtained from only TEOS and APTES (sample 3).

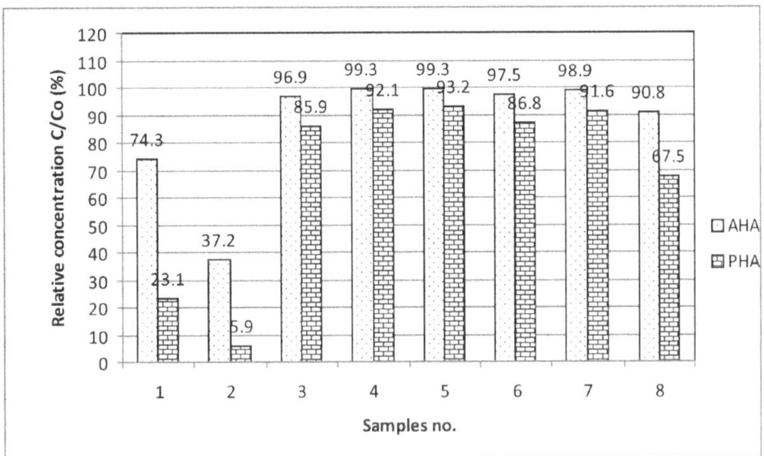

Figure 9. Relative Concentration of immobilized humic acids (AHA and PHA, respectively) as a function of silica support type.

A very important observation is that all the samples which showed a negative value of zeta potential exhibit a lower capacity of immobilizing HA by ionic bonds [5,26]. This is mainly because the HA molecules are electrostatically bounded on the aminosilica supports [9].

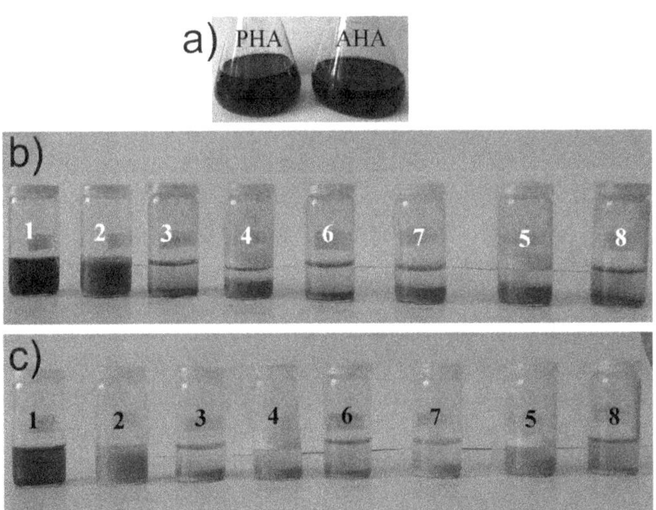

Figure 10. (**a**) Pictures of the two humic acid stock solutions and immobilization tests of (**b**) AHA; and (**c**) PHA, respectively, on various silica supports.

3. Experimental Section

3.1. Materials

Tetraethyl orthosilicate (TEOS), 3-aminopropyltriethoxysilane (APTES), methyltriethoxy silane (MeTES), trimethylethoxysilane (Me_3ES) and bis(3-trimethoxisilylpropyl)amine (BAPTMS), diethoxydimethylsilane (Me_2DES), 1,2-bis(triethoxysilyl)ethane (BETES) from Sigma-Aldrich, St. Louis, MO, USA were used as received. Ammonium hydroxide (NH_4OH 25% solution) (Chimreactiv, Bucharest, Romania) and ethanol (Chimopar, Bucharest, Romania) were used without further purification. The commercial sort of humic acid (AHA), used for testing the coupling efficiency of the amino-silica supports, was purchased from Aldrich. The other sort of humic acid (PHA) was extracted in the laboratory from a local peat deposit as in reference [27].

3.2. Characterization Techniques

The structure and morphology of the synthesized amino-silica particles were characterized by CP/MAS, NMR, FTIR, elemental analysis (CHN), DLS, LDV, TGA, TG coupled with MS, ESEM and N_2 adsorption-desorption technique. The coupling efficiency of humic acids on the aminosilica supports was evaluated by UV-VIS technique.

Elemental analysis (CHN)—Elemental composition was determined using a Thermo FlashEA® 1112 analyzer (PerkinElmer Instruments LLC, Shelton, CT, USA).

Fourier transformed infrared spectroscopy (FTIR): Samples were analyzed as pellets (0.001 g of silica sample diluted in 1 g KBr), by transmission FTIR, using a Tensor 37 (Bruker, Munich, Germany). The measurements were recorded in the 4000–400 cm^{-1} range.

Solid State Nuclear Magnetic Resonance (ssNMR): All the NMR spectra were recorded at room temperatures (~303 K) on a Bruker Avance II 400 spectrometer (Bruker, Munich, Germany), with proton frequency of 400.13 MHz (9.38 T) equipped with a 4 mm CP/MAS probes. All solid-state experiments included ramped-amplitude cross polarization and magic-angle spinning. The contact time varied between 2 and 3 ms. The relaxation delays were 2–4 s, and the number of scans varied between 2 and 4 k. Between 100 and 120 mg of sample was used to fill up the zirconia rotors. The ^{13}C spectra were recorded under high-power proton decoupling (SPINAL-64 spin decoupling scheme). The rotation speed varied between 3000 and 6000 Hz for the ^{29}Si and ^{13}C nuclides, respectively. For the measurements we have used standard NMR methods (Software: Bruker Biospin Topspin 2.1.3., Bruker, Munich, Germany): C-13 and Si-29 variable amplitude CP/MAS. The chemical shifts were externally referenced to glycine and tris(trimethylsilyl)silane (TTMSS).

The particle hydrodynamic medium diameter and size distribution were measured by DLS (Dynamic Light Scattering) technique in aqueous dispersion (0.02 g of dry sample was re-dispersed in 25 mL distilled water, using a ultrasonic bath for 10 min.), employing Zetasizer Nano ZS ZEN 3600 instrument, manufactured by Malvern Instruments Ltd. (Malvern, UK). The device has a "red" He-Ne laser (633 nm) and can perform measurements of particle size in the range 0.6 to 6000 nm. Disposable polystyrene cells DTS0012 (Malvern Instruments Ltd., Malvern, UK) were used.

The same instrument, Zetasizer Nano ZS ZEN 3600, was used to perform zeta potential measurements, based on electrophoretic mobility measurement technique, LDV (Laser Doppler Velocimetry). Disposable cell DTS 1060 (Malvern Instruments, Malvern, UK) were used. The same dispersions prepared for DLS measurements were also used for zeta potential analysis.

ESEM images of silica supports were obtained with a scanning electron microscope FEI Quanta 200 (Philips, Eindhoven, The Netherlands). The samples were dispersed in distilled water (0.02 g of milled sample in 25 mL of distilled water), deposited on the corresponding sample carrier by dropping with a pipette and let them dry in air.

The nitrogen adsorption-desorption isotherms were recorded at −196 °C, using a NOVA 2200e Automated Gas Sorption instrument (Quantachrome Instruments, Hartley Wintney, UK). Prior to measurements, samples were degassed under vacuum at 100 °C for 5 h. The Brunauer–Emmett–Teller specific surface areas of samples were calculated from adsorption data at a relative pressure range of 0.08–0.3. The total pore volumes were evaluated from the adsorbed amount of nitrogen at a relative pressure of 0.986.

Thermogravimetric Analyzes were performed in air (60 mL/min), using a SDTA 851^{e}Mettler Toledo instrument (Mettler-Toledo GmbH, Schwerzenbach, Switzerland). The TG diagrams were measured in the 25–700 °C range with an increment of 10 °C/min.

In order to check contamination of the resulted aminosilica supports with atmospheric carbon dioxide, TG-MS measurements were also performed. A thermoanalyzer (TA instruments, New Castle, DE, USA) combined with a quadrupole mass spectrometer (Pfeiffer Vacuum, Aßlar, Germany) by a capillary inlet system allowed simultaneous DTA, TG and evolved gas analysis of samples. Thus, thermogravimetric analyses were performed in helium with a SDT Q600 instrument (TA Instruments, New Castle, DE, USA), module DSC-TGA (standard sample size: 9.5–10.5 mg; pan: alumina; purge gas: helium 5.0, 100 mL/min; method: ramp 10 °C/min to 900 °C), while for the evolved gas analysis (EGA) a Quadrupole Mass Spectrometer ThermoStar TM GSD 301 T instrument (Pfeiffer Vacuum, Asslar, Deutschland) was used (quartz glass capillary, heated up to 200 °C).

The capacity of aminosilica supports to capture the HA molecules was evaluated by UV-VIS spectroscopy. The absorbance measurements for the filtered dispersions were recorded with a Nicolet Evolution 500 instrument (Thermo Electron Corp., Waltham, MA, USA). A buffer solution having pH = 7.5 was prepared from 700 mL KH_2PO_4 0.1M and 575 mL of NaOH 0.1M. Thereafter, 100 mL of stock solutions of humic acids in buffer were prepared at a concentration of 1 g/L. The resulted stock solutions were further used for the acid immobilizing tests. The results shown in Figure 9 are the average of 3 measurements for each sample. All the materials subjected to the humic acid (HA) coupling tests were previously milled. Accurately weighed quantities (0.5 g) of the milled silicas were added to 5 mL of HA stock solution and were left in contact for 20 h. Then, by means of syringes equipped with 0.2 μm filters, 1 mL of solution was collected from the supernatant of each sedimented sample and the absorbency was recorded by UV-VIS technique at a fixed wavelength (254 nm) [9]. Concentrations of the HA remained in the filtered solution were calculated from the measured absorbencies and by means of the equations resulted from the corresponding calibration curves. Subtracting the concentration of the free HA from the initial concentration of HA present in the stock solution resulted in the concentration of HA chemically attached onto the silica supports.

3.3. *Synthesis of the Silica Hybrid Supports*

A method adapted from reference [1] was used for the synthesis of the amino-functionalized silica particles, through a two-step sol-gel method.

Step 1. Preparation of Pristine Silica Particles (Preformed SiO_2 Particles)

Forty milliliters of ammonia and 200 mL of ethanol were introduced in a flask equipped with a mechanical stirrer, condenser and addition funnel. 10 g (0.048 moles) of TEOS diluted in 40 mL of ethanol were rapidly added to this mixture and kept under stirring (~200 rot/min) for another two hours at room temperature.

Step 2. Preparation of Amino-Functionalized Silica Particles (NH_2-SiO_2 Particles)

Over the dispersion prepared in Step 1, 0.532 g of APTES (0.0024 moles) diluted in 4 mL ethanol were continuously added for two hours. This amount represents an APTES/TEOS molar ratio of 1/20. The co-precursors/TEOS molar ratio was chosen according to the method described by Brambilla *et al.* [28]. The surface modification of the preformed SiO_2 particles was previously performed by other groups [29,30], but our preparation procedure is different from what was previously reported. Thus, Jung *et al.* [29] obtained amino-functionalized silica particles only in ethanol medium and after the re-dispersion of the pristine SiO_2 particles. Also, Meera *et al.* [30] prepared castor oil based polyurethane (PU)–silica nanocomposite films by adding amino functionalized silica particles to the polyurethane mixture. Their amino functionalized silica particles were prepared by a method similar to that used by us, but they did not study the effect of the alkyl spacers over the particles ability to immobilize humic acids.

In some cases, the alkyl derivatives (spacers for the amino functions)—methyltriethoxy silane (MeTES), trimethylethoxysilane (Me_3ES), diethoxydimethylsilane (Me_2DES) and 1,2-bis(triethoxysilyl) ethane (BETES)—were added together with APTES to the reaction system. The final mixture was further stirred for two more hours and then allowed to rest overnight. Subsequently, one part from the resulted dispersion was kept in a closed bottle and another part has been transferred to a Petri dish and allowed to dry at room temperature for 3–4 d. Finally it was dried under vacuum at 40 °C.

For preparation of sample 1 (only APTES), 10.6 g APTES diluted in 40 mL EtOH were introduced over a mixture of 40 mL NH_4OH 25% and 200 mL EtOH. The reaction was kept under stirring for 4 h and left over night. The next day the whole mixture was a homogenous and transparent dispersion. After the removal of the volatile compounds, the resulted solid was transparent and difficult to be milled. This reference sample was prepared in order to demonstrate the effect of sterical hindrance over the ability to immobilize the HA, as a result of the high density of the amino groups.

4. Conclusions

Amino-silica supports for humic acid immobilization were developed by grafting of the amino functions (alone or spaced by alkyl functions) onto preformed mono-dispersed SiO_2 particles (549 nm). Zeta potential measurements showed that, by increasing the APTES weight share in the reaction system, negative values of zeta potential were obtained. Due to their hydrophobic character, the aminopropyl chains tend to self-associate and to orientate towards the interior of the silica particle (a water repellent behavior). By this phase-inversion mechanism, the -NH_2 functions are partially hidden inside the silica frame, while towards the water are predominantly exposed Si-O^- charges, lowering the ability of the resulted material to interact with HA molecules. Meanwhile, all the APTES/TEOS = 1/20 hybrid materials, having a low density of the free aminopropyl functions on the particles surfaces, showed positive values of zeta potential and were able to efficiently interact with external large HA molecules, as confirmed by the immobilization tests.

The presence of the aminopropyl chains and methyl, dimethyl and ethylene moieties on the silica surface was evidenced using ^{29}Si and ^{13}C ssNMR spectroscopic methods. The type of bonded silyl species (-O_4Si, -O_3SiR, -O_2SiR_2), resulted by the organosilanization reaction, was confirmed by ^{29}Si CP/MAS technique. The ^{13}C CP/MAS spectra proved that the organic groups are indeed present in the relevant samples. Integration of the bi- and tri-functional alkyl silanes (Me_2DES and MeTES) in the silica frame was also confirmed by the properties change of the final particles (both sample exhibited a strong water repelling behavior—the resulted silica powders were floating on the water). It was also evidenced that, perhaps for sterical reasons and for its lower reactivity (only one Si-OEt reactive group), the monofunctional derivative (Me_3ES) did not connect to the silica network.

The HA immobilization tests revealed that the density of aminopropyl functions is low enough at this low APTES fraction (APTES/TEOS = 1/20) not to satirically hinder the HA complex. In this condition, the spacing of the amino groups by alkyls was successfully achieved, but no significant differences in the HA immobilization capacity were observed.

The structure and properties of synthesized aminosilica materials showed the potential of these novel supports to be successfully applied for the immobilization of different bio-molecules (as was demonstrated here for humic acids).

Acknowledgments: This study was supported by the Romanian Ministry of Education and Research (CNDI-UEFISCDI), in the frame of the Programme Parternship PN2 P4 PCCA 2013, by project "PN-II-PT-PCCA-2013-4-0995/160/2014 (MAIA)" and by project "PN-II-PT-PCCA-2013-4-0846/159/2014 CERES".

Author Contributions: Dan Donescu, Cristina Lavinia Nistor and Mónika Sándor were involved in designing the aim of this manuscript, performed the experiments and prepared the manuscript. Gábor Szalontai performed the Solid State NMR measurements and analyzed the data. Rusandica Stoica, Cristian Andi Nicolae,

Elvira Alexandrescu and József Fazakas helped with several analyses. Florin Oancea contributed as manuscript advisor. Each contributor was essential to the production of this work.

Conflicts of Interest: The authors declare no conflict of interest.

References

1. Brinker, C.J.; Scherer, G.W. *Sol-Gel Science: The Physics and Chemistry of Sol-Gel Processing*; Academic Press: San Diego, CA, USA, 1990.
2. Brinker, C.J. Hydrolysis and condensation of silicates: Effects on structure. *J. NonCryst. Solids* **1988**, *100*, 31–50. [CrossRef]
3. Fadeev, A.Y.; McCarthy, T.J. Self-assembly is not the only reaction possible between alkyltrichlorosilanes and sufaces: Mononuclear and oligomeric covalently attached layers of dichloro- and trichloroalkylsilanes on silicon. *Langmuir* **2000**, *16*, 7268–7274. [CrossRef]
4. Fatunbi, H.O.; Bruch, M.D. Characterization of the structural morphology of chemically modified silica prepared by surface polymerization of a mixture of long and short alkyl chains using ^{13}C and ^{29}Si NMR spectroscopy. *Langmuir* **2013**, *29*, 4974–4987. [CrossRef] [PubMed]
5. Prado, A.G.S.; Miranda, B.S.; Dias, J.A. Attachment of two distinct humic acids into a silica gel surface. *Colloids Surf. A* **2004**, *242*, 137–143. [CrossRef]
6. Szabo, G.; Angelova, A.; Guczi, J.; Bulman, R.A. An examination of the uptake of radioiodine by chemically bound humic acid and by some solid phase of soil. *Sci. Total Environ.* **1993**, *130–131*, 375–382. [CrossRef]
7. Szabo, G.; Prosser, S.L.; Bulman, R.A. Determination of the adsorption coefficient (K_{OC}) of some aromatics for soil by RP-HPLC on two immobilized humic acid phases. *Chemosphere* **1990**, *21*, 777–788. [CrossRef]
8. Guczi, J.; Angelova, A.; Bulman, R.A.; Szabo, G. Investigation of the interactions of 110 mAg^+ and $125I^-$ with humic acid chemically immobilized on silica gel. *React. Polym.* **1992**, *17*, 61–68. [CrossRef]
9. Koopal, L.K.; Yang, Y.; Minnaard, A.J.; Theunissen, P.L.M.; van Riemsdijk, W.H. Chemical immobilization of humic acid on silica. *Colloids Surf. A* **1998**, *141*, 385–395. [CrossRef]
10. Yang, Y.H.; Koopal, L.K. Immobilisation of humic acids and binding of nitrophenol to immobilized humics. *Colloids Surf. A* **1999**, *151*, 201–212. [CrossRef]
11. Armon, R.; Zolkov, C.H.; Laor, Y. Entrapment of humic acid in a sol-gel matrix—A new tool for sorption studies. *J. Sol Gel Sci. Technol.* **2000**, *19*, 95–100. [CrossRef]
12. Klavins, M.; Eglite, L. Immobilisation of humic substances. *Colloids Surf. A* **2002**, *203*, 47–54. [CrossRef]
13. De la Rosa, G.; Gardea-Torresdey, J.L.; Peralta-Videa, J.R.; Herrera, I.; Contreras, C. Use of silica-immobilized humin for heavy metal removal from aqueous solution under flow conditions. *Bioresour. Technol.* **2003**, *90*, 11–17. [CrossRef]
14. Helles, R. Wastewater COD removal using sol-gel immobilized humicacid. *J. Toxicol. Environ. Health Sci.* **2010**, *2*, 7–10.
15. Rosa, A.H.; Vicente, A.A.; Rocha, J.C.; Trevisan, H.C. A new application of humic substances: Activation of supports for invertase immobilization. *Fresenius J. Anal. Chem.* **2000**, *368*, 730–733. [CrossRef] [PubMed]
16. Treccani, L.; Klein, T.Y.; Meder, F.; Pardun, K.; Rezwan, K. Functionalized ceramics for biomedical, biotechnological and environmental applications. *Acta Biomater.* **2013**, *9*, 7115–7150. [CrossRef] [PubMed]
17. Blaszykowski, C.; Sheikh, S.; Benvenuto, P.; Thomson, M. New functionalizable alkyltrichlorosiloxane surface modifiers for biosensor and biomedical applications. *Langmuir* **2012**, *28*, 2318–2322. [CrossRef] [PubMed]
18. Leal, O.; Bolivar, C.; Ovalles, C.; Garcia, J.J.; Espidel, Y. Reversible adsorption of carbon dioxide on amine surface-bonded silica gel. *Inorg. Chim. Acta* **1995**, *240*, 183–189. [CrossRef]
19. Liu, Y.; Liu, J.; Yao, W.; Cen, W.; Wang, H.; Weng, X.; Wu, Z. The effects of surface acidity on CO_2 adsorption over amine functionalized protonated titanate nanotubes. *RSC Adv.* **2013**, *3*, 18803–18810. [CrossRef]
20. Bacsik, Z.; Atluri, R.; Garcia-Bennett, A.E.; Hedin, N. Temperature-induced uptake of CO_2 and formation of carbamates in mesocaged silica modified with *n*-propylamines. *Langmuir* **2010**, *26*, 10013–10024. [CrossRef] [PubMed]
21. Engelhardt, G.; Michel, D. *High-Resolution Solid-State NMR of Silicates and Zeolites*; John Wiley and Sons: Chichester, UK, 1994.
22. Duer, M.J. *Introduction to Solis-State NMR Spectroscopy*; Blackwell Publishing: Oxford, UK, 2004.

23. Bonhomme, C.; Coelho, C.; Baccile, N.; Gervais, C.; Azais, T.; Babonneau, F. Advanced solid state NMR techniques for the characterization of sol-gel-derived materials. *Acc. Chem. Res.* **2007**, *40*, 738–746. [CrossRef] [PubMed]
24. Stejskal, E.O.; Memory, J. *High Resolution NMR in the Solid State*; Oxford University Press: Oxford, UK, 1994.
25. Pretsch, E.; Clerc, T.; Seibl, J.; Simon, W. *Tabellen Zur Strukturaufklärung Organischer Verbindungen Mit Spektroskopischen Methoden*; Springer-Verlag: Berlin, Germany, 1976.
26. Cestari, A.R.; Vieira, E.F.S.; Vieira, G.S.; Almeida, L.E. Aggregation and adsorption of reactive dyes in the presence of an anionic surfactant on mesoporous aminopropyl silica. *J. Colloid Interface Sci.* **2007**, *309*, 402–411. [CrossRef] [PubMed]
27. Janos, P. Separation methods in the chemistry of humic substances. *J. Chromatogr. A* **2003**, *983*, 1–18. [CrossRef]
28. Brambilla, R.; Pires, G.P.; dos Santos, J.H.Z.; Lacherda-Miranda, M.S.; Chornik, B. Octadecylsilane-modified silicas prepared by grafting and sol-gel methods. *J. Electron. Spectrosc.* **2007**, *156–158*, 413–420. [CrossRef]
29. Jung, H.S.; Moon, D.S.; Lee, J.K. Quantitative analysis and efficient surface modification of silica nanoparticles. *J. Nanomater.* **2012**, *2012*. [CrossRef]
30. Meera, K.M.S.; Sankar, R.M.; Paul, J.; Jaisankara, S.N.; Mandal, A.B. The influence of applied silica nanoparticles on a bio-renewable castor oil based polyurethane nanocomposite and its physicochemical properties. *Phys. Chem. Chem. Phys.* **2014**, *16*, 9276–9288. [CrossRef] [PubMed]

13

Theoretical and Experimental Studies on the Crystal Structure, Electronic Structure and Optical Properties of $SmTaO_4$

Song Wang, Miao Jiang, Lihong Gao *, Zhuang Ma and Fuchi Wang

Academic Editor: Lioz Etgar

School of Materials Science and Engineering, Beijing Institute of Technology, Beijing 100081, China; 2120131177@bit.edu.cn (S.W.); jiangmiao0226@gmail.com (M.J.); hstrong929@bit.edu.cn (Z.M.); wangfuchi@bit.edu.cn (F.W.)

* Correspondence: gaolihong@bit.edu.cn

Abstract: The crystal structure, electronic structure and optical properties of $SmTaO_4$ were identified through an experimental method and first principles calculation. X-ray powder diffraction (XRD) and a spectrophotometer were used to characterize the crystal structure, reflectivity and band gap of this material; furthermore, the electronic structure and optical properties were investigated according to three exchange-correlation potentials, LDA, GGA and GGA + *U*. Results show that the $SmTaO_4$ calcined at 1400 °C with the solid-state reaction method is in monoclinic phase in the space group I^2/a. In addition, the calculated lattice parameters are consistent with the experimental values. The electron transitions among the O 2*p* states, Sm 4*f* states and Ta 5*d* states play a key role in the dielectric function, refractive index, absorption coefficient and reflectivity of $SmTaO_4$. The calculation of first principles provides considerable insight into the relationship between the electronic structure and optical properties of this material.

Keywords: crystal structure; electronic structure; optical properties; first principles calculation

1. Introduction

Lanthanide orthotantalates ($LnTaO_4$; Ln = La, Ce, Pr, Nd and Sm, *etc.*) have recently been studied extensively because of their favorable dielectric property [1], luminescence [2,3], chemical and electrochemical stabilities [4], photo-electronic activity [5,6], and ion conductivity [7,8]. $LnTaO_4$ can be used in phosphors for solid-state lighting, in photocatalysts for both contaminant degeneration and H_2 production, in chemically robust hosts for nuclear materials and wastes, in solid-state laser materials, and in ion conductors for lithium batteries or solid-oxide fuel cells [9–12].

Electronic interactions, including the partly occupied Ln 4*f* levels, play an important role in the properties of $LnTaO_4$ [13,14]. Machida *et al.* [15,16] reported that the energy level of Ln 4*f* states can be lowered by increasing the number of 4*f* electrons; thus, the unoccupied 4*f* states are at the bottom of the conduction band, whereas the partially occupied 4*f* states drop with an increase in 4*f* electrons and finally fall into the valence band. Given that Sm is located in the middle position in the lanthanide series and that it contains six 4*f* electrons, the Sm 4*f* states may be located near the Fermi level, thereby increasing the probability that electron transition is enhanced. This phenomenon satisfies the requirements of high reflectivity, *i.e.*, the band gap is narrow, the conduction band moves to a low level energy position, and the Fermi level is located in the conduction band [17]. Therefore, $SmTaO_4$ may exhibit excellent optical properties and may be applied in protective or anti-radiation materials for optical use.

To our knowledge, a few structural and luminescent investigations have been conducted on $SmTaO_4$ [3,4,7], while its electronic structure and optical properties have rarely been reported. In this paper, $SmTaO_4$ was successfully synthesized through the solid-state reaction method, and its electronic structure and optical properties were investigated in detail. Its electronic structure and optical properties are correlated through first principles calculation.

2. Experimental Methods and Theoretical Calculations

2.1. Experimental Methods

Analytical-grade Ta_2O_5 and Sm_2O_3 powders were utilized as raw materials. The raw powders were preheated at 800 °C to remove the absorbed moisture and other gases; subsequently, these powders were weighed at a 1:1 stoichiometric ratio and then milled for 6 h in a ball-grinding mill with alcohol as a medium. The powders were then calcined at 1400 °C for 10 h. The phase composition of $SmTaO_4$ was determined through X-ray powder diffraction (XRD, X'Pert PRO MPD, PANalytical Inc., Almelo, The Netherlands) with monochromated CuKα radiation and a scintillation detector at a scanning rate of $4^{\circ}\cdot \mathrm{min}^{-1}$. The experimental values of lattice constants, unit cell volume, interatomic distances and atomic coordinates were calculated using Reflex Plus software packed in MS Modeling (Accelerys Inc., San Diego, CA, USA) with Rietveld refinement [18]. In addition, the calculated XRD pattern of structure model was also obtained via Reflex Plus. The reflectivity was measured by a spectrophotometer (Cary 5000, Varian Inc., Palo Alto, CA, USA) operating in the UV/VIS/NIR spectral range with a 110 mm integrating sphere. The band gap was determined from the diffuse reflectance spectrum as follows [19,20]:

Step 1. The relationship between band gap (E_g) and photon frequency is described in the Tauc equation:

$$(h\nu\alpha)^{\frac{1}{n}} = A\left(h\nu - E_g\right) \tag{1}$$

where h is the Planck's constant, ν is the photon frequency, α is the absorption coefficient, E_g is the band gap, and A is the proportional constant. The value of exponent n denotes the nature of sample transition; $n = 1/2$ for direct allowed transitions and $n = 3/2$ for direct forbidden transitions. Moreover, $n = 2$ for indirect allowed transitions and $n = 3$ for indirect forbidden transitions. Given that direct allowed transitions are considered in this study, $n = 1/2$ is chosen for this sample.

Step 2. The acquired diffuse reflectance spectrum is converted into a Kubelka–Munk function and the vertical axis is converted into quantity $F(R\infty)$, which is proportional to the absorption coefficient. Thus, α in Equation (1) is substituted with $F(R\infty)$ as follows:

$$(h\nu F\left(R\infty\right))^2 = A\left(h\nu - E_g\right) \tag{2}$$

Step 3. When the Kubelka–Munk function is employed, the $(h\nu F(R\infty))^2$ is plotted against $h\nu$. $h\nu$ is expressed in electron volts (eV), and its relationship with wavelength λ (nm) is written as $h\nu = 1239.7/\lambda$.

Step 4. A line is drawn tangent to the inflection point on the curve obtained in Step 3, and the $h\nu$ value at the intersection point of the tangent line and the horizontal axis corresponds to the band gap E_g value.

2.2. Computational Details

The theoretical calculations were performed with the pseudo-potential plane-wave total energy implemented in the Cambridge Sequential Total Energy Package code (CASTEP code). The interactions between electrons and ion cores were represented by the Vanderbilt-type ultra-soft pseudo-potential. Furthermore, the O $2s^2$, $2p^4$ electrons, Ta $5d^3$, $6s^2$ electrons and Sm $5s^2$, $5p^6$, $4f^6$, $6s^2$ electrons were explicitly regarded as valence electrons. The standard density functions describe poorly the electronic properties of these localized orbitals, but they perform very well in the total energy calculation which refers to structural properties. So to calculate the crystal structure, the exchange-correlation effects were described by the conventional local density approximation (LDA) of Ceperley-Alder-Perdew-Zunger (CA-PZ) [21] and generalized gradient approximation (GGA) of Perdew-Burke-Ernzerhof formula (PBE) [22]. To compute the electronic structure and optical properties, the GGA with additional Hubbard correlation terms, namely the GGA + U approach (where U is the Hubbard energy) [23,24] was applied in addition to LDA and GGA because the GGA + U approach effectively describes the on-site electron-electron repulsion associated with the 4f narrow bands of lanthanide atoms [25]. Hubbard U was set as 6.0 eV for the valence f shell of Sm atom which has already been successfully applied in lanthanide compounds [25–27]. A plane-wave cut-off energy of 600 eV was employed throughout the calculation process. To sample the Brillouin zone (BZ), the following k-point grid calculations were generated according to the Monkhorst–Pack scheme: 5 × 5 × 5 for structure optimization and the band structure [28]. In order to improve the calculation precision of properties, the k-point grids are promoted to 6 × 6 × 6 for electronic density of states and 8 × 8 × 8 for optical properties. Moreover, the Pulay scheme of density mixing [29] was applied to evaluate energy and stress levels. The Broyden-Fletcher-Goldfarb-Shanno (BFGS) minimization scheme [30] was also applied in geometry optimization. The tolerances for energy change, maximum force, maximum stress, and maximum displacement were set as 2 × 10^{-5} eV $atom^{-1}$, 0.05 eV $Å^{-1}$, 0.1 GPa, and 0.002 Å, respectively. The total energy converged to 1 × 10^{-6} eV $atom^{-1}$ in the self-consistent calculation process.

3. Results and Discussion

3.1. Crystal Structure

Figure 1 shows the XRD pattern of the sample calcined at 1400 °C. All of the peaks match those of the $SmTaO_4$ phase. The $SmTaO_4$ crystal structure belongs to the monoclinic system, and the coincident space group is I^2/a. The Sm^{3+} ion is surrounded by eight oxygen neighbors to form a dodecahedron and is located at the sites with C_2 symmetry; moreover, the Ta^{5+} ion is surrounded by six oxygen neighbors to form a distorted octahedron [31]. The experimental $SmTaO_4$ crystal structure is refined according to the XRD result using Reflex Plus software, and this refined structure is adopted as the structure model for use in theoretical calculation.

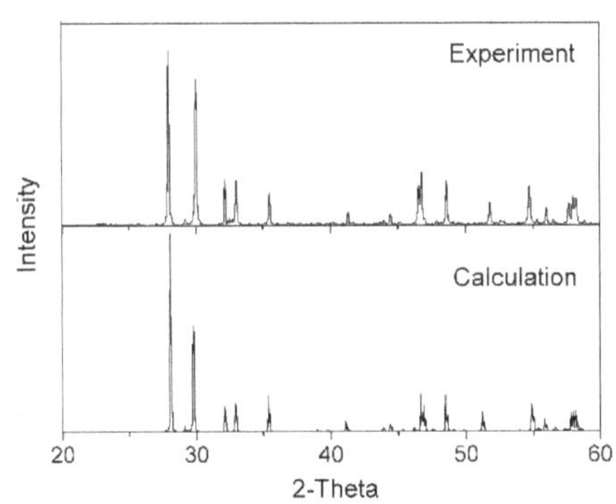

Figure 1. X-ray diffraction patterns of $SmTaO_4$: (**a**) Experimental result; (**b**) Calculated result.

Figure 2 depicts the refined crystal structure. The SmO_8 dodecahedron is generated, along with its two neighbors, along the b–c plane to form a general folding line shape, and the TaO_6 octahedrons connect these dodecahedrons with their vertices. Figure 1b presents the corresponding XRD pattern calculated based on this crystal structure; this result matches the experimental outcome, and this finding verifies the accuracy of the structure model.

The theoretical $SmTaO_4$ structure was investigated with both LDA and GGA exchange-correlation functions. Table 1 presents the experimental, calculated, and reported lattice constants (a, b, c, and β), the conventional cell volume (V), and the density of $SmTaO_4$. Tables 2 and 3 list the calculated and experimental interatomic distances and atomic coordinates of $SmTaO_4$ compared with reported results The computed findings show that the approximate density functional theory (DFT) executed by either LDA or GGA underestimated or overestimated the lattice constants and volume slightly more than the experimental values. Nonetheless, the calculated lattice constants differ from the experimental values by no more than 1.16%. Additionally, the calculated interatomic distances and atomic coordinates are very close to the experimental ones. So there is a good agreement between our experimental and calculated results, and all of these values match well with the available reported values. This agreement between calculation and experiment ensures the accuracy of the structure model used to derive physical properties.

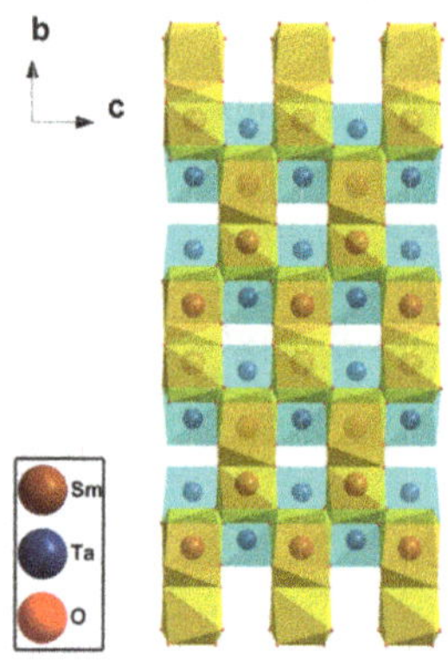

Figure 2. Crystal structure of $SmTaO_4$.

Table 1. Experimental, calculated and reported lattice constants (a, b, c, and β), conventional cell volume (V), and density of $SmTaO_4$.

Lattice Constant	Experimental	LDA	GGA	Reported
a (Å)	5.444 (±0.0085)	5.393	5.514	5.455 [a], 5.405 [b]
b (Å)	11.115 (±0.0072)	10.968	11.241	11.137 [a], 11.063 [b]
c (Å)	5.081 (±0.0079)	5.094	5.161	5.091 [a], 5.081 [b]
β (°)	95.74 (±0.0027)	96.06	95.47	94.73 [a], 95.62 [b]
V (Å^3)	305.93	299.63	318.43	308.24 [a]
Density (g/cm^3)	8.55	8.76	8.25	/

[a] Ref. [32]; [b] Ref. [10].

Table 2. Calculated and experimental interatomic distances (Å) in $SmTaO_4$.

Compound	Ta–O	Sm–O	O–O
LDA	1.9345 × 4 [a] 1.9955 × 4 2.2452 × 4	2.3002 × 4 2.3173 × 4 2.4774 × 4	2.6182 × 2 2.6952 × 4 2.7614 × 2 2.8789 × 4
GGA	1.9864 × 4 2.0482 × 4 2.3671 × 4	2.3570 × 4 2.3975 × 4 2.5567 × 4	2.7089 × 2 2.7779 × 4 2.8499 × 2 2.9822 × 4
Experimental	1.840 1.925 2.335	2.338 2.447 2.545	/

[a] × 4 and × 2 mean the number of the atomic bonds.

Table 3. Calculated and experimental atomic coordinates in $SmTaO_4$ compared with reported results.

Atom	LDA	GGA	Experimental	Reported [a]
Sm	(0.25, 0.116, 0)	(0.25, 0.115, 0)	(0.25, 0.112, 0)	(0.25, 0.118, 0)
Ta	(0.25, 0.652, 0)	(0.25, 0.651, 0)	(0.25, 0.649, 0)	(0.25, 0.651, 0)
O1	(0.010, 0.722, 0.231)	(0.009, 0.722, 0.227)	(0.006, 0.717, 0.205)	(0.015, 0.719, 0.216)
O2	(0.909, 0.465, 0.247)	(0.911, 0.453, 0.249)	(0.894, 0.454, 0.266)	(0.901, 0.456, 0.240)

[a] Ref. [32].

3.2. Electronic Structure

Since the electronic structure is crucial to understanding the optical properties of $SmTaO_4$, the energy band structure and density of states (DOS) of $SmTaO_4$ have been calculated with the calculated lattice constants. The GGA + U approach effectively describes the on-site electron-electron repulsion of the lanthanide atoms with 4f electrons; thus, this approach was employed in this part and was compared with LDA and GGA. The zero-point energy is set as the Fermi level.

Figure 3 exhibits the energy band structure of $SmTaO_4$ within the −4–6 eV region along the high-symmetry directions. The main similarity in the three images lies in the fact that the top of the valence bands appears flat, whereas the bottom of the conduction bands is slightly dispersed. In addition, a few extremely narrow bands are located in the forbidden gap that originates from the Sm 4f states; this outcome is consistent with the theoretical and experimental results of other lanthanide compounds [33,34]. The primary dissimilarity among the three images lies in the fact that as per the GGA + U, the Sm 4f states are observed above the Fermi level at approximately 2 eV, whereas the LDA and GGA results indicate that the Sm 4f states are positioned around the Fermi level. The $SmTaO_4$ band gap is indirect since the valence band maxima of LDA, GGA, and GGA + U lie at the G point, whereas their conduction band minima fall between the M and A points of the Brillouin zone. The values of band gaps between the valence and conduction bands, *i.e.*, between the O 2p and Ta 5d states, are 3.65, 3.49, and 3.57 eV for LDA, GGA, and GGA + U, respectively; these gaps are explained in the following studies. In addition, the experimental value determined based on the diffuse reflectance spectrum is 3.68 eV. Therefore, our calculated and experimental band gap values are in agreement with the reported value of 3.8 eV (Table 4).

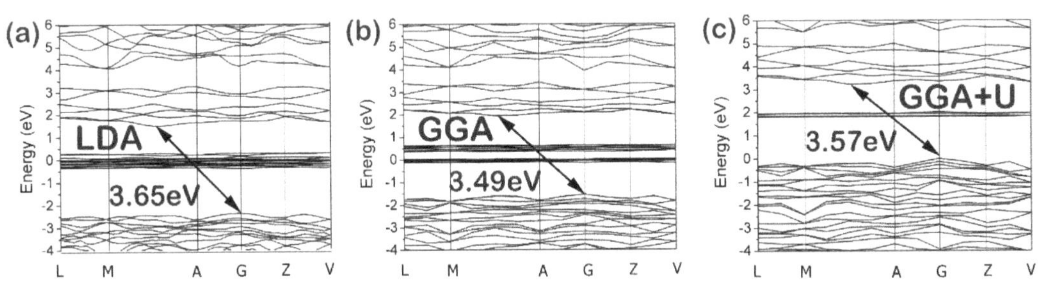

Figure 3. Band structure of $SmTaO_4$ as calculated with (**a**) local density approximation (LDA); (**b**) generalized gradient approximation (GGA) and (**c**) GGA with additional Hubbard correlation terms (GGA + U).

Table 4. Experimental, calculated and reported band gaps (eV) of $SmTaO_4$.

Band Gap	LDA	GGA	GGA + U	Experiment	Reported [16]
O 2p–Ta 5d	3.65	3.49	3.57	3.68	3.8
O 2p–Sm 4f	1.35	0.98	1.76	/	/
Sm 4f–Ta 5d	0.46	0.7	1.12	/	/

Figure 4 shows the total density of states (TDOS) and the partial density of states (PDOS) of Sm, Ta, and O. The top of the valence band is mainly dominated by O 2*p* states hybridized with Ta 5*d* states, which suggests the contributions of covalent bonding to $SmTaO_4$. The bottom of the conduction band is mainly dominated by very sharp and well-localized peaks of the Ta 5*d* states as well as slight contributions from the O 2*p* states; this phenomenon generates antibonding π^* and σ^* bands [35]. It can be seen from Figures 4a and 4b that the LDA and GGA results show that the Sm 4*f* electrons are localized at the Fermi level and are split into two peaks. The Sm 4*f* states spread from approximately −0.45 eV to 0.80 eV for LDA and from −0.41 eV to 0.91 eV for GGA with the weak overlap of the Ta 5*d* and O 2*p* states. In the case of GGA + *U*, the Sm 4*f* states are also divided into two peaks; however, one peak is located above the Fermi level and the other is positioned at the bottom of the valence band hybridized with the Ta 5*d* and O 2*p* orbits. Table 3 summarizes the band gaps of Sm 4*f*–O 2*p* and Sm 4*f*–Ta 5*d*. The band gaps between the Sm 4*f* states at the bottom and the top of the valence band (O 2*p*) are 1.35, 0.98, and 1.76 eV for LDA, GGA, and GGA + *U*, respectively. The band gaps between the Sm 4*f* states on the top and the Ta 5*d* states at the bottom of the conduction band are 0.46, 0.70, and 1.12 eV for LDA, GGA, and GGA + *U*.

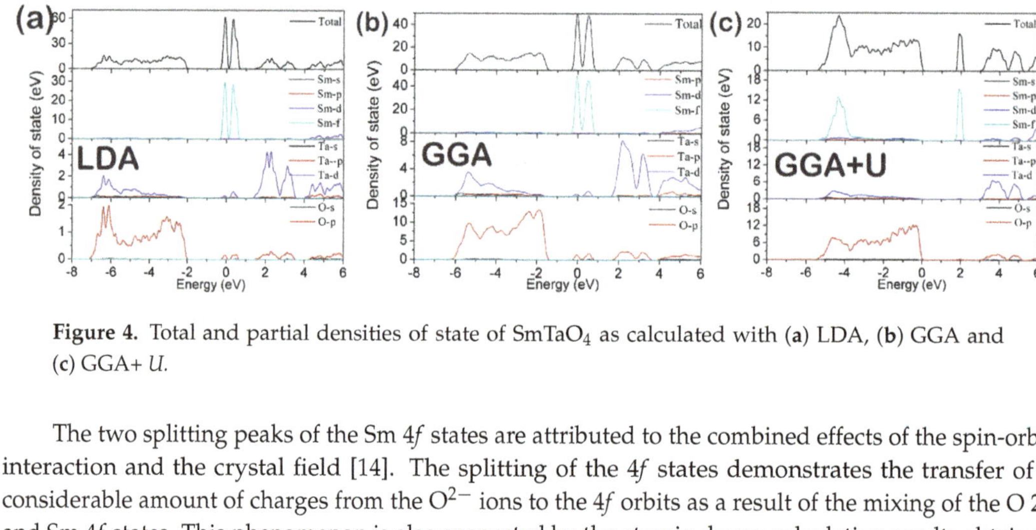

Figure 4. Total and partial densities of state of $SmTaO_4$ as calculated with (**a**) LDA, (**b**) GGA and (**c**) GGA+ *U*.

The two splitting peaks of the Sm 4*f* states are attributed to the combined effects of the spin-orbit interaction and the crystal field [14]. The splitting of the 4*f* states demonstrates the transfer of a considerable amount of charges from the O^{2-} ions to the 4*f* orbits as a result of the mixing of the O 2*p* and Sm 4*f* states. This phenomenon is also suggested by the atomic charge calculation results obtained from the Mulliken population analysis, as shown in Table 5. The atomic charges of the O atoms are −0.58 or −0.63 e (LDA), −0.63 or −0.67 e (GGA) and −0.64 or −0.68 e (GGA + *U*), while those of the Sm atoms are 1.39 e (LDA), 1.44 e (GGA) and 1.75 e (GGA + *U*).

Table 5. Calculated atomic charges determined by the Mulliken population analysis.

Atom	Ion	Charge (e)		
		LDA	GGA	GGA + *U*
O	1	−0.63	−0.67	−0.68
O	2	−0.58	−0.63	−0.64
O	3	−0.63	−0.67	−0.68
O	4	−0.58	−0.63	−0.64
O	5	−0.63	−0.67	−0.68
O	6	−0.58	−0.63	−0.64
O	7	−0.63	−0.67	−0.68
O	8	−0.58	−0.63	−0.64
Sm	1	1.39	1.44	1.75
Sm	2	1.39	1.44	1.75
Ta	1	1.02	1.17	0.90
Ta	2	1.02	1.17	0.90

As mentioned previously, the partially filled Sm 4*f* states are located at the Fermi level when a one-electron method with an orbital-independent potential is applied, such as LDA and GGA. This result is inconsistent with that observed in the experiments during the insulating state. In contrast, GGA + *U* facilitates the study of various strongly correlated compounds and improves considerably on LDA and GGA [36]. Therefore, GGA + *U* can provide appropriate results for $SmTaO_4$.

3.3. Optical Properties

The complete response to the applied electromagnetic radiation field is described by the dielectric function $\varepsilon(\omega) = \varepsilon_1(\omega) + i\varepsilon_2(\omega)$, where $\varepsilon_1(\omega)$ and $\varepsilon_2(\omega)$ are the real and imaginary parts of the dielectric function, respectively. The intraband and interband transitions of electrons mainly contribute to $\varepsilon(\omega)$ [37]. The former are important only for metals, whereas the interband transitions can be further divided into direct and indirect transitions. The direct interband contributions to $\varepsilon_2(\omega)$ are calculated by considering all the possible transitions from occupied to unoccupied states and taking the appropriate transition matrix element into account [38,39] as follows:

$$\varepsilon_2(q \to 0, h\omega) = \frac{2e^2\pi}{\Omega\varepsilon_0}\sum\left|\langle\psi_k^c\,|\hat{u}, r|\,\psi_k^v\rangle\right|^2 \delta\left(E_K^C - E_K^V - E\right) \tag{3}$$

where ω is the light frequency and *e* is the electronic charge; $\Psi_k{}^c$ and $\Psi_k{}^v$ are the conduction and valence band wave functions at *k*, respectively. Index *u* is the vector defining the polarization of the incident electric field, and index Ω is the primitive cell volume. Moreover, $\varepsilon_1(\omega)$ can be derived $\varepsilon_2(\omega)$ through the Kramers-Kronig dispersion equation:

$$\varepsilon_1(\omega) = 1 + \frac{2}{\pi}M\int_0^\infty \frac{\omega\prime\varepsilon_2(\omega\prime)}{\omega\prime^2 - \omega^2}d\omega\prime \tag{4}$$

where *M* is the principal value.

Optical constants, such as refractive index $n(\omega)$, extinction coefficient $k(\omega)$, absorption coefficient $I(\omega)$, and reflectivity $R(\omega)$, can be derived from $\varepsilon(\omega)$ and are given by:

$$n(\omega) = \left[\frac{\varepsilon_1(\omega)}{2} + \frac{\sqrt{\varepsilon_1^2(\omega) + \varepsilon_2^2(\omega)}}{2}\right]^{1/2} \tag{5}$$

$$k(\omega) = \left[\frac{\sqrt{\varepsilon_1^2(\omega) + \varepsilon_2^2(\omega)}}{2} - \frac{\varepsilon_1(\omega)}{2}\right]^{1/2} \tag{6}$$

$$I(\omega) = \sqrt{2}\omega\left[\sqrt{\varepsilon_1^2(\omega) + \varepsilon_2^2(\omega)} - \varepsilon_1(\omega)\right]^{1/2} \tag{7}$$

$$R(\omega) = \left|\frac{\sqrt{\varepsilon(\omega)} - 1}{\sqrt{\varepsilon(\omega)} + 1}\right|^2 \tag{8}$$

Since GGA + *U* accurately describes the electronic structure of $SmTaO_4$, the optical property is analyzed based on the results presented by GGA + *U*. Figure 5a shows the calculated $\varepsilon_2(\omega)$ and $\varepsilon_1(\omega)$ of the complex dielectric function in the energy range of 0–30 eV. Three conspicuous peaks labeled 1, 2 and 3 are centered at 2.36, 4.42 and 8.83 eV, respectively; these peaks are related to the electron transitions between different special *k* points at the first irreducible Brillouin zone. Peak 1 can be ascribed to the electron transitions from O 2*p* to the Sm 4*f* states located above the Fermi level. Peak 2 is attributed to the transitions from O 2*p* to the Ta 5*d* conduction bands. Peak 3 corresponds to the transitions from the Sm 4*f* located at the bottom of the valence bands to the Ta 5*d* states. Moreover, $\varepsilon_1(\omega)$ drops from the maximum to the local minimum in the range of 3.68–5.51 eV, and proceeds from this local maximum to another minimum in the range of 6.73–11.21 eV. This finding indicates that $SmTaO_4$ has two absorption peaks in the two regions. The static dielectric constant $\varepsilon(0)$ is approximately 5.36, as per the calculation at the equilibrium lattice constant.

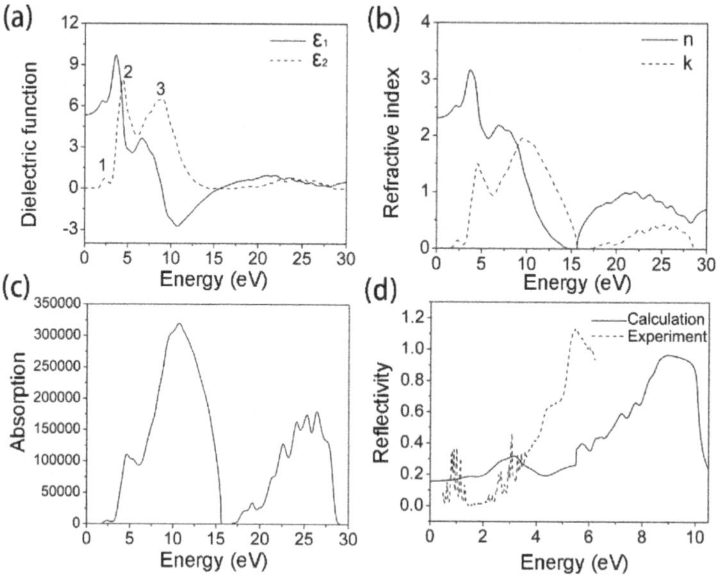

Figure 5. Optical properties of $SmTaO_4$ as calculated with GGA + U: (**a**) dielectric function; (**b**) refractive index; (**c**) absorption coefficient and (**d**) reflectivity.

Figure 5b shows the calculated refractive index. The static refractive index $n(0)$ value is 2.31, which increases with increasing the energy level in the transparency region, reaching a peak in the ultraviolet region at approximately 3.70 eV. It then decreases to a global minimum at 15.03 eV. The origin of the structures in $\varepsilon_2(\omega)$ also explains the structures in the refractive index. Furthermore, the local maxima of the extinction coefficient $k(\omega)$ correspond to the zero value of $\varepsilon_1(\omega)$.

Figure 5c shows the behavior of the absorption coefficient $I(\omega)$. The absorption edge begins from approximately 1.8 eV, which corresponds to the energy band gap O 2p–Sm 4f. The first small peak located at the low-energy region originates from the electron transition from O 2p to the Sm 4f states located above the Fermi level. The subsequent two peaks in the absorption spectrum positioned at 4.62 and 10.67 eV can be explained by the dielectric function analysis results. Other peaks are observed at 19.12, 22.61, 24.13, 25.27 and 26.41 eV.

Figure 5d presents the experimental and calculated reflectivity. Due to the limitation in the experimental measurement, we merely observed the reflectivity of $SmTaO_4$ at 0.5–6.2 eV. To facilitate the most accurate comparison, we conducted Savitzky-Golay smoothing by performing a local quartic polynomial regression on the experimental curve. It can be seen that the experimental reflectivity curve is reproduced properly, although the energetic positions in both spectra are not accurately consistent. This disparity is also reported in the literature [40].

4. Conclusions

In this work, we investigate the crystal structure, electronic structure, and optical properties of $SmTaO_4$ by using an experimental method and first principles. The results show that $SmTaO_4$ can be synthetized at 1400 °C, which is naturally observed with the monoclinic phase. The calculated lattice constants of $SmTaO_4$ are in good agreement with the experimental values. Moreover, the electronic structure and optical properties of $SmTaO_4$ were extensively studied according to three exchange-correlation potentials, namely LDA, GGA and GGA + U. The top of the valence band is dominated by the O 2p states hybridized with Ta 5d states, and the bottom of the conduction band is dominated by the Ta 5d states with slight contributions from the O 2p states. The two splitting peaks of the Sm 4f states are attributed to the combined effects of the spin-orbit interaction and the crystal field. In addition, the LDA and GGA results show that these two Sm 4f peaks are located around the Fermi level; in contrast, the GGA + U results indicate that one peak is located above the Fermi level at approximately 2 eV, whereas the other peak is positioned at the bottom of the valence band. The outcomes obtained with GGA + U should be more credible than those of the other two potentials because the former facilitates the study of various strongly correlated compounds and improves considerably on LDA and GGA. Furthermore, the optical properties of $SmTaO_4$ are mainly composed of contributions from three types of electron transitions: the electrons from O 2p to the Sm 4f states located above the Fermi level, the electrons from O 2p to the Ta 5d conduction bands, and the electrons from the Sm 4f located at the bottom of valence bands to the Ta 5d states. The experimental reflectivity curve is reproduced properly through calculation, although the energy positions in the spectra are slightly offset.

Acknowledgments: This study was funded by the National Natural Science Foundation of China (grant number 51302013). The authors are also very grateful to the Analytical and Testing Center of Beijing Normal University and to the Chinese Academy Of Sciences Physicochemical Technology Research Institute for providing support during our experiment.

Author Contributions: Song Wang and Lihong Gao had full access to all of the data in the study and take responsibility for the integrity of our data as well as the accuracy of the data analysis. Miao Jiang, Zhuang Ma, and Fuchi Wang contributed their ideas to this study and to the preparation of this manuscript.

Conflicts of Interest: The authors declare no conflict of interest.

References

1. Dorenbos, P.; Rogers, E.G. Vacuum referred binding energies of the lanthanides in transition metal oxide compounds. *J. Solid State Sci. Technol.* **2014**, *3*, R150–R158.
2. Porter, S.H. Perovskite and pyrochlore Tantalum Oxide Nitrides: Synthesis and characterization. Master's Thesis, Ohio State University, Columbus, OH, USA, December 2012.
3. Voloshyna, O.; Neicheva, S.V.; Starzhinskiy, N.G.; Zeny, I.M.; Gridin, S.S.; Baumer, V.N.; Sidletskiy, O.Ts. Luminescent and scintillation properties of orthotantalates with common formulae $RETaO_4$ (RE = Y, Sc, La, Lu and Gd). *Mater. Sci. Eng. B* **2013**, *178*, 1491–1496. [CrossRef]
4. Wu, M.; Liu, X.; Gu, M.; Li, C.; Liu, B.; Huang, S. Characterization and luminescence properties of sol-gel derived M′-type $LuTaO_4$:Ln^{3+} (Ln = Pr, Sm, Dy) phosphors. *Mater. Res. Bull.* **2014**, *60*, 652–658. [CrossRef]
5. Sych, A.M.; Golub, A.M. Niobates and tantalates of tervalent elements. *Russ. Chem. Rev.* **1977**, *46*, 210–2250. [CrossRef]
6. Shimizu, K.; Itoh, S.; Hatamachi, T.; Kodama, T.; Sato, M.; Toda, K. Photocatalytic water splitting on Ni-intercalated Ruddlesden-Popper Tantalate $H_2La_{2/3}Ta_2O_7$. *Chem. Mater.* **2005**, *17*, 5161–5166. [CrossRef]
7. Osterloh, F.E. Inorganic materials as catalysts for photochemical splitting of water. *Chem. Mater.* **2008**, *20*, 35–54. [CrossRef]
8. Mokkelbost, T.; Andersen, Ø.; Strøm, R.A.; Wiik, K.; Grande, T.; Einarsrud, M.-A. High-temperature proton-conducting $LaNbO_4$-based materials: Powder synthesis by spray pyrolysis. *J. Am. Ceram. Soc.* **2007**, *90*, 3395–3400. [CrossRef]
9. Shen, Y.; Leckie, R.M.; Levi, C.G.; Clarke, D.R. Low thermal conductivity without oxygen vacancies in equimolar $YO_{1.5}$ + $TaO_{2.5}$− and $YbO_{1.5}$ + $TaO_{2.5}$− stabilized tetragonal zirconia ceramics. *Acta Mater.* **2010**, *58*, 4424–4431.
10. Siqueira, K.P.F.; Carvalho, G.B.; Dias, A. Influence of the processing conditions and chemical environment on the crystal structures and phonon modes of lanthanide orthotantalates. *Dalton Trans.* **2011**, *40*, 9454–9460. [CrossRef] [PubMed]
11. Nyman, M.; Rodriguez, M.A.; Rohwer, L.E.S.; Martin, J.E.; Waller, M.; Osterloh, F.E. Unique $LaTaO_4$ polymorph for multiple energy applications. *Chem. Mater.* **2009**, *21*, 4731–4737. [CrossRef]

12. Forbes, T.Z.; Nyman, M.; Rodriguez, M.A.; Navrotsky, A. The energetics of lanthanum tantalate materials. *J. Solid State Chem.* **2010**, *183*, 2516–2521. [CrossRef]
13. Machida, M.; Yabunaka, J.; Kijima, T. Synthesis and photocatalytic property of layered perovskite tantalates, $RbLnTa_2O_7$ (Ln = La, Pr, Nd, and Sm). *Chem. Mater.* **2000**, *12*, 812–817. [CrossRef]
14. Gillen, R.; Clark, S.J.; Robertson, J. Nature of the electronic band gap in lanthanide oxides. *Phys. Rev. B* **2013**, *87*, 125116. [CrossRef]
15. Machida, M.; Yabunaka, J.; Kijima, T.; Matsushima, S.; Arai, M. Electronic structure of layered tantalates photocatalysts, $RbLnTa_2O_7$ (Ln = La, Pr, Nd, and Sm). *Int. J. Inorg. Mater.* **2001**, *3*, 545–550. [CrossRef]
16. Machida, M.; Murakami, S.; Kijima, T. Photocatalytic property and electronic structure of lanthanide tantalates, $LnTaO_4$ (Ln = La, Ce, Pr, Nd, and Sm). *J. Phys. Chem. B* **2001**, *105*, 3289–3294. [CrossRef]
17. Gao, L.H.; Ma, Z.; Fan, Q.B. First-principle studies of the electronic structure and reflectivity of $LaTiO_3$ and Sr doped $LaTiO_3$ ($La_{1-x}Sr_xTiO_3$). *J. Electroceram.* **2011**, *27*, 114–119. [CrossRef]
18. Mccusker, L.B.; Vonele, R.B.; Cox, D.E.; Louër, D.; Scardi, P. Rietveld refinement guidelines. *J. Appl. Cryst.* **1999**, *32*, 36–50. [CrossRef]
19. Tauc, J.; Grigorovici, R.; Vancu., A. Optical Properties and Electronic Structure of Amorphous Germanium. *Phys. Status Solidi* **1966**, *15*, 627–637. [CrossRef]
20. Liu, Y.J.; Zhu, N. Precise determination of band gap naturally via absorption/reflectance/transmission spectra. *Appl. Mech. Mater.* **2014**, *509*, 3–8. [CrossRef]
21. Ceperley, D.M.; Alder, B.J. Ground State of the Electron Gas by a Stochastic Method. *Phys. Rev. Lett.* **1980**, *45*, 566–569. [CrossRef]
22. Perdew, P.J.; Burke, K.; Ernzerhof, M. Generalized gradient approximation made simple. *Phys. Rev. Lett.* **1996**, *77*, 3865–3868. [CrossRef] [PubMed]
23. Clark, S.J.; Segall, M.D.; Pickard, C.J.; Hasnip, P.J.; Probert, M.J.; Refson, K.; Payne, M.C. First principles methods using CASTEP. *Z. Kristallogr* **2005**, *220*, 567–570. [CrossRef]
24. Dudarev, S.L.; Botton, G.A.; Savrasov, S.Y.; Humphreys, C.J.; Sutton, A.P. Electron-energy-loss spectra and the structural stability of nickel oxide: An LSDA+U study. *Phys. Rev. B* **1998**, *57*, 1505. [CrossRef]
25. Khadraoui, Z.; Horchani-Naifer, K.; Ferhi, M. Electronic structure and optical properties of $TbPO_4$: Experiment and density functional theory calculations. *Opt. Mater.* **2015**, *47*, 484–489. [CrossRef]
26. Pang, M.J.; Zhan, Y.Z.; Wang, H.Z.; Jiang, W.P.; Du, Y. Structural, electronic, elastic and thermodynamic properties of $AlSi_2RE$ (RE = La, Ce, Pr and Nd) from first-principle calculations. *Comput. Mater. Sci.* **2011**, *50*, 3303–3310.
27. Khadraoui, Z.; Horchani-Naifer, K.; Ferhi, M.; Ferid, M. Synthesis, characterization and DFT calculations of electronic and optical properties of $YbPO_4$. *Chem. Phys.* **2015**, *457*, 37–42. [CrossRef]
28. Payne, M.C.; Teter, M.P.; Allan, D.C.; Arias, T.A.; Joannopoulos, J.D. Iterative minimization techniques for ab initio total energy calculations: Molecular dynamics and conjugate gradients. *Rev. Mod. Phys.* **1992**, *64*, 1045–1097. [CrossRef]
29. Kresse, G.; Furthmuller, J. Efficient iterative schemes for ab initio total-energy calculations using a plane-wave basis set. *Phys. Rev. B* **1996**, *54*, 11169–11186. [CrossRef]
30. Pfrommer, B.G.; Côté, M.; Louie, S.G.; Cohen, M.L. Relaxation of crystals with the quasi-newton method. *J. Comput. Phys.* **1997**, *131*, 233–240. [CrossRef]
31. Zhou, P.; Zhang, Q.; Ning, K.; Yang, H.; Sun, D.; Luo, J.; Yin, S. Structural and spectral investigations on heavily Er^{3+} doped $RETaO_4$ (RE = Sc, Y, Gd, Lu) polycrystalline powders. In Proceedings of 1st Pacific Rim Laser Damage Symposium on Optical Materials for High Power Lasers (PLD), Shanghai, China, 6–9 November 2011.
32. Keller, C. Über ternäre Oxide des Niobs und Tantals vom Typ ABO_4. *Z. Anorg. Allg. Chem.* **1962**, *318*, 89–106. (In German). [CrossRef]
33. Wuilloud, E.; Delley, B.; Schneider, W.-D.; Bear, Y. Spectroscopic evidence for localized and extended f-symmetry States in CeO_2. *Phys. Rev. Lett.* **1984**, *53*, 202–205. [CrossRef]
34. Khadraoui, Z.; Bouzidi, C.; Horchani-Naifer, K.; Ferid, M. Crystal structure, energy band and optical properties of dysprosium monophosphate $DyPO_4$. *J. Alloys Compd.* **2014**, *617*, 281–286. [CrossRef]
35. Bouhemadou, A.; Djabi, F.; Khenata, R. First principles study of structural, elastic, electronic and optical properties of the cubic perovskite $BaHfO_3$. *Phys. Lett. A* **2008**, *372*, 4527–4531. [CrossRef]

36. Fujimori, A.; Minami, F. Valence-band photoemission and optical absorption in nickel compounds. *Phys. Rev. B* **1984**, *30*, 957–971. [CrossRef]
37. Smith, N.V. Photoelectron energy spectra and the band structures of the noble metals. *Phys. Rev. B* **1971**, *3*, 1862–1878. [CrossRef]
38. Gajdoš, M.; Hummer, K.; Kresse, G.; Furthmüller, J.; Bechstedt, F. Linear optical properties in the projector-augmented wave methodology. *Phys. Rev. B* **2006**, *73*, 045112. [CrossRef]
39. Hopfield, J.J. Theory of the contribution of excitons to the complex dielectric constant of crystals. *Phys. Rev.* **1958**, *112*, 1555. [CrossRef]
40. Pang, M.; Zhan, Y.; Ling, M.; Wei, S.; Liu, Y.; Du, Y. First-principles study on the crystal, electronic structure and mechanical properties of hexagonal Al_3Re (Re = La, Ce, Pr, Nd, Sm, Gd) intermetallic compounds. *Solid State Commun.* **2011**, *151*, 1135–1140. [CrossRef]

14

Fibre Morphological Characteristics of Kraft Pulps of *Acacia melanoxylon* Estimated by NIR-PLS-R Models

Helena Pereira [1,†], António José Alves Santos [1,†] and Ofélia Anjos [1,2,*,†]

Academic Editor: Maryam Tabrizian

1 Centro de Estudos Florestais, Instituto Superior de Agronomia, Universidade de Lisboa, Tapada da Ajuda, 1349-017 Lisboa, Portugal; hpereira@isa.ulisboa.pt (H.P.); antonioalvesantos@gmail.com (A.J.A.S.)
2 Instituto Politécnico de Castelo Branco, Escola Superior Agrária, Apartado 119, 6001-909 Castelo Branco, Portugal
* Correspondence: ofelia@ipcb.pt
† These authors contributed equally to this work.

Abstract: In this paper, the morphological properties of fiber length (weighted in length) and of fiber width of unbleached Kraft pulp of *Acacia melanoxylon* were determined using TECHPAP Morfi® equipment (Techpap SAS, Grenoble, France), and were used in the calibration development of Near Infrared (NIR) partial least squares regression (PLS-R) models based on the spectral data obtained for the wood. It is the first time that fiber length and width of pulp were predicted with NIR spectral data of the initial woodmeal, with high accuracy and precision, and with ratios of performance to deviation (RPD) fulfilling the requirements for screening in breeding programs. The selected models for fiber length and fiber width used the second derivative and first derivative + multiplicative scatter correction (2ndDer and 1stDer + MSC) pre-processed spectra, respectively, in the wavenumber ranges from 7506 to 5440 cm^{-1}. The statistical parameters of cross-validation (RMSECV (root mean square error of cross-validation) of 0.009 mm and 0.39 μm) and validation (RMSEP (root mean square error of prediction) of 0.007 mm and 0.36 μm) with RPD_{TS} (ratios of performance to deviation of test set) values of 3.9 and 3.3, respectively, confirmed that the models are robust and well qualified for prediction. This modeling approach shows a high potential to be used for tree breeding and improvement programs, providing a rapid screening for desired fiber morphological properties of pulp prediction.

Keywords: *Acacia melanoxylon*; Kraft pulps; fibre length and width; NIR

1. Introduction

Pulp is a major global commodity forest-based product that consists of the dissociated cells of the raw material, usually named pulp fibers. The fiber morphological properties are important quality parameters for pulp and paper properties and therefore they are measured both for research purposes and industrial quality control.

In fact, the fiber characteristics greatly influence the quality and properties of the final product, e.g., they are frequently correlated with the physical and mechanical properties of paper and paperboard [1–6]. For instance, strong correlation between fiber length and tear index was reported for both hardwoods and softwoods [7–11]. Tensile and tear strength of paper increase with fiber length, especially in weakly bonded sheets [12,13]. Fiber length influences the papersheet formation and its uniformity [13,14].

Fiber biometry has also been found correlated with other pulp variables. For instance, the pulp yield correlated positively with fiber length and negatively with fiber width [15] and

Wimmer *et al.* [16] reported that fiber length of *E. globulus* had a strong effect on pulp yield and freeness, as well as active alkali consumption in addition to tear index and bending stiffness.

Fiber morphological variables, namely length and width, are among the main fundamental characteristics of the pulp fibers. Therefore, methods for measuring fiber morphology are essential for determining pulp quality.

The traditional method involves classifying the pulp into screened fractions [17], measuring the weight and length of fibers in each fraction to calculate the weighted average length by weight [18]. Automated optical analyzers are now used [19] such as Kajaani FS300 (Metso Automation Inc., Helsinki, Finland), OpTest HiRes FQA (OpTest Equipment Inc., Hawkesbury, ON, Canada), Fiber Lab (Metso Automation Inc., Helsinki, Finland), TECHPAP Morfi lab (Techpap SAS, Grenoble, France), L & W STFI Fibermaster (Lorentzen & Wettre, Stockholm, Sweden). In spite of their relatively easy operation, these methods require the measurement of a large number of fibers from a pulp suspension, and are therefore not time- and cost-effective when a large number of wood samples have to be tested.

This is the case, for instance, in screening programs of pulping raw materials, namely in pulpwood improvement programs, for which fiber morphology is one of the quality traits. Given the high number of samples that have to be tested in such programs, it would be important to be able to predict fiber length and width in a faster way. The use of non-destructive measurements is tempting, especially the spectroscopic methodologies using the near infrared wavelength range (NIRS). The application of NIRS for predicting several properties of wood and pulps using statistical modeling tools, namely PLS models (Partial least squares), has already proved successful [20].

Comparatively few attempts were made to develop NIRS-based models for the morphological variables of wood. Schimleck *et al.* [21] examined the use of NIR spectroscopy for predicting tracheid length of loblolly pine wood and considered that the accuracy was sufficient for ranking purposes. NIRS-based predictions of tracheid coarseness and wall thickness were excellent [22]. Jones *et al.* [23] and Schimleck *et al.* [21] also reported the effect of green wood condition and a wide range of sites on the estimate of tracheid morphological characteristics of loblolly pine wood by NIR spectroscopy.

Via *et al.* [24] studied the variation of tracheid length in longleaf pine with tree age and height using NIR-based prediction. Wang *et al.* [25] established an NIR-based PLS-R model to predict fiber length of slash pine and poplar. Viana *et al.* [26] also predicted the morphological characteristics and basic density of eucalypt clones wood using NIR, but the best calibration correlations were obtained for basic density. Inagaki *et al.* [27] predicted fiber length for *Eucalyptus camaldulensis* with an RPD (ratios of performance to deviation) of 3.8 from solid wood samples using NIR-based PLS-R models. Sun *et al.* [28] estimated the MFA (microfibril angle) and fiber length of bamboo by NIRS, and the PLS models based on noise combined with orthogonal signal selection spectra gave the strongest correlations.

The prediction of pulp fiber morphology using NIRS of the wood has not been attempted. This is made in the present work, which focuses on the development of a NIR-PLS-R model for the prediction of fiber length and width in the unbleached Kraft pulps using *Acacia melanoxylon* wood meal as a case study. *A. melanoxylon* is a valuable timber species, but is also of interest for pulping as evaluated regarding pulp yield and properties [15,29–31]. Pulps produced from Acacia species are competitive in the world market of hardwood pulps now dominated by *Eucalyptus* [32].

In this work the pulps were produced under identical pulping conditions, as would be the case in a screening program or in pulp mill operation, and the measurements of fiber length weighted in length and fiber width used for the modeling were determined with TECHPAP Morfi® equipment (Techpap SAS, Grenoble, France). We aimed to obtain models performing better than those found in the literature, which are not precise enough for screening purposes, according to the AACC (American Association of Cereal Chemists) Method 39-00 [33], due to failing the ratios of performance to deviation (RPD) criteria.

2. Results

The fiber length and width of the unbleached pulps of *A. melanoxylon* obtained in this study ranged from 0.66 to 0.79 mm and 16.4 to 22.3 μm, respectively, with an average of 0.73 ± 0.03 mm and 18.8 ± 1.4 μm (Table 1). This variability represents a good data scattering and the two sets showed similar statistics.

Table 1. Number of samples and range of fibre length (weighted in length) and width in the calibration set and test set, with average, standard deviation (SD) and coefficient of variation (CV).

Parameter	Fiber Length (mm)		Fiber Width (μm)	
	Calibration Set	Test Set	Calibration Set	Test Set
Number of samples	45	15	45	15
Maximum	0.79	0.79	22.3	21.6
Minimum	0.66	0.69	16.4	16.7
Average	0.73	0.73	18.7	18.9
SD	0.03	0.03	1.4	1.2
CV	4.0	3.7	6.4	5.6

The near-infrared data from the 45 samples of the calibration set were regressed against their experimentally determined fibre length weighted in length and width, and the results obtained with the various pre-processing of the raw spectral data are summarized in Table 2. All pre-processing was reported in Table 2 in order to compare the choice model with the other tested models. In this study, the spectral information in the wavenumber range from 7506 to 5440 cm^{-1} was used for calibration, as found by automated optimization.

It is interesting to notice that the wavenumber range varies greatly with the parameter being studied. For instance, NIR-PLS-R models published for different parameters of *A. melanoxylon* used different wavenumber ranges, as summarized in Table 3.

The calibration for fiber length obtained a rank ranging from eight (2ndDer—seconded derivative, 1stDer + MSC—first derivative + multiplicative scatter correction, 1stDer + VecNor—first derivative + vector normalization) to 10 (ConOff—constant offset elimination, no spectral pre-processing), while the coefficients of determination (r^2) ranged from 80.2% to 93.4%, and RMSECV (root mean square error of cross-validation) from 0.008 to 0.014 mm. For fiber width, the obtained calibrations ranks ranged from six (1stDer + MSC) to 10 (no spectral pre-processing), with the coefficients of determination (r^2) ranging from 88.5% to 93.0%, and RMSECV from 0.37 to 0.45 μm. When using the test set validation for fiber length and fiber width, the coefficients of determination ranged from 62.4% to 93.5% and 72.3% to 93.7%, respectively.

The best model was selected by using the 2ndDer (second derivative) for fiber length and 1stDer + MSC for fiber width of the spectral data. The 2ndDer pre-processing for the classification of length was selected instead of 1stDer + SLS (first derivative + straight line subtraction) because the first one better corrects the effects of wood grain than the 1stDer + SLS. The capacity of the prediction by NIR of morphological properties of fibers is a consequence of the effect of the materials density which results from the arrangement of the particles in the structure, in this specific case the length and width of the fibers. Thus, when used in the diffuse reflectance mode in spectrum acquisition, those are quantified amounts of light unabsorbed by materials.

The corresponding plot of NIR-PLS-R predicted *versus* the laboratorial determined fiber length and fiber width is shown in Figure 1. Both the cross-validation and the validation showed high correlation between predicted and determined values, with a RMSEP of 0.007 mm and 0.36 μm, a rank of eight and six, and no outliers respectively for fiber length and fiber width. Figure 2 shows NIR diffuse reflectance spectra of the fiber length and fiber width for the 2ndDer and 1stDer + SLS preprocessing used, respectively.

Table 2. Results of the cross-validation of the calibration set (45 samples) and of the test set validation (15 samples) including percentage of outsiders (OS) and outliers (OL) of the test set obtained during prediction of the samples with unknown fiber length (weighted in length) and width, using various pre-processing methods of the raw spectral data. In bold: cross-validation for the best model with test set validation.

Parameter	Pre-Processing	Number of Samples	Cross-Validation (CV)				Test Set Validation (TS)					
			Rank	r^2 (%)	RMSECV	RPD	Rank	r^2 (%)	RMSEP	RPD	OS (%)	OL (%)
Fiber Length (mm)	1stDer	45	9	92.4	0.008	3.6	8	87.8	0.010	2.9	0.0	0.0
	2ndDer	**45**	**8**	**91.6**	**0.009**	**3.5**	**8**	**93.5**	**0.007**	**3.9**	**0.0**	**0.0**
	ConOff	44	10	90.6	0.009	3.3	9	76.9	0.014	2.2	0.0	0.0
	MinMax	45	9	80.2	0.014	2.3	8	76.6	0.014	2.1	0.0	0.0
	MSC	45	9	81.7	0.013	2.3	7	70.9	0.015	1.9	0.0	0.0
	None	42	10	88.5	0.010	3.0	9	62.4	0.017	1.8	0.0	0.0
	SLS	45	9	91.7	0.009	3.5	7	80.1	0.013	2.3	0.0	0.0
	VecNor	45	9	81.6	0.013	2.3	7	70.7	0.015	1.9	0.0	0.0
	1stDer + MSC	44	8	86.5	0.011	2.7	6	76.4	0.014	2.1	0.0	0.0
	1stDer + SLS	45	9	93.4	0.008	3.9	7	88.9	0.009	3.0	0.0	0.0
	1stDer + VecNor	44	8	86.4	0.011	2.7	6	76.4	0.014	2.1	0.0	0.0
Fiber Width (μm)	1stDer	45	7	91.8	0.41	3.5	6	85.0	0.45	2.6	0.0	0.0
	2ndDer	44	7	88.5	0.45	3.0	9	93.7	0.29	4.0	0.0	0.0
	ConOff	45	9	92.6	0.38	3.7	8	71.1	0.62	1.9	0.0	0.0
	MinMax	45	8	92.5	0.39	3.7	7	75.1	0.58	2.0	0.0	0.0
	MSC	45	8	93.0	0.37	3.8	6	75.0	0.58	2.0	0.0	0.0
	None	45	10	92.6	0.38	3.7	5	75.7	0.51	2.1	0.0	0.1
	SLS	45	8	92.5	0.37	3.7	7	72.3	0.61	1.9	0.0	0.0
	VecNor	45	8	93.0	0.37	3.8	6	74.7	0.58	2.0	0.0	0.0
	1stDer + MSC	**45**	**6**	**92.5**	**0.39**	**3.7**	**6**	**90.4**	**0.36**	**3.3**	**0.0**	**0.0**
	1stDer + SLS	44	7	90.2	0.41	3.2	7	88.2	0.40	2.9	0.0	0.0
	1stDer + VecNor	45	6	92.5	0.39	3.7	6	87.6	0.41	2.9	0.0	0.0

Rank—number of PLS components; r^2—coefficient of determination; RMSECV—root mean square error of cross-validation; RPD—ratios of performance to deviation; RMSEP—root mean square error of prediction; 1stDer—first derivative; 2ndDer—seconded derivative; ConOff—constant offset elimination; MinMax—minimum-maximum normalization; MSC—multiplicative scatter correction; None—no spectral data processing; SLS—straight line elimination; VecNor—vector normalization; 1stDer + MSC—first derivative + multiplicative scatter correction; 1stDer + SLS—first derivative + straight line subtraction; 1stDer + VecNor—first derivative + vector normalization.

Table 3. Wavenumber ranges and pre-processing used in the NIR-PLS-R (infrared (**NIR**) partial least squares regression (**PLSR**) models published in bibliography for different properties of *Acacia melanoxylon*.

Property	References	Material	Wavenumber Ranges (cm^{-1})	Pre-Processing
Wood basic density	[34]	Solid wood	8250–7494 + 5450–5304	1stDer + SLS
Pulp yield	[35]	Woodmeal	9087–5440 + 4605–4243	2ndDer
Kappa number	[36]	Kraft pulp	6110–5440	1stDer
ISO brightness of unbleached pulp	[37]		9404–7498 + 4605–4243	1stDer

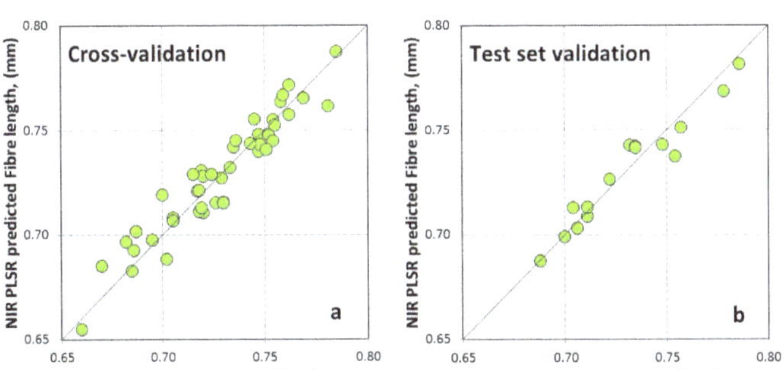

Figure 1. *Cont.*

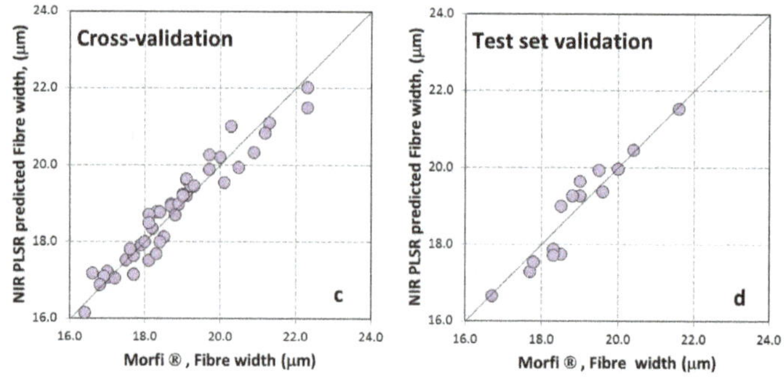

Figure 1. NIR-PLS-R predicted *versus* laboratory determined fiber length and fiber width for *Acacia melanoxylon* pulps for the cross-validation (**a**,**c**) and test set validation (**b**,**d**).

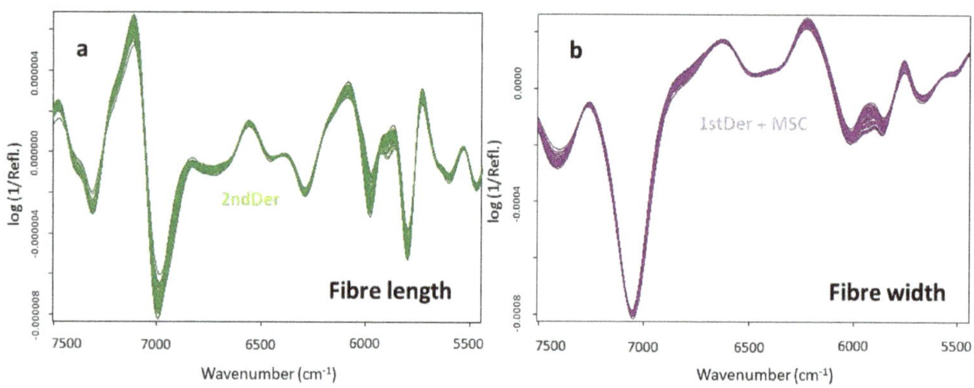

Figure 2. NIR diffuse reflectance for all spectra (45 samples—cross-validation) of the fiber length and fiber width for *Acacia melanoxylon* pulps (**a**,**b**, respectively).

3. Discussion

When considering all the samples of this study (60 samples), a good calibration for the fiber length and fiber width of the unbleached Kraft pulp was obtained with the spectral data of the woodmeal with r^2 = 96.5% and 95.0%, Rank = 8 and 6, root-mean-square error of estimation (RMSEE) = 0.006 mm and 0.32 μm and RPD = 5.3 and 4.5, respectively (Table 4). These results compare very favorably with the few data available for the estimation of morphological properties of wood using NIR spectroscopy, as shown in Table 5.

Table 4. Results of the calibration and cross-validation for all samples (60 samples) for fiber length (weighted in length) and width of *Acacia melanoxylon* pulps, using the best pre-processing methods of the raw spectral data.

Parameter	Pre-Processing	Calibration (C)				Cross-Validation (CV)			
		Rank	r^2 (%)	RMSEE	RPD	Rank	r^2 (%)	RMSECV	RPD
Fiber length (mm)	2ndDer	8	96.5	0.006	5.3	8	94.0	0.007	4.1
Fiber width (μm)	1stDer + MSC	6	95.0	0.32	4.5	6	92.8	0.37	3.7

Rank—number of PLS components; r^2—coefficient of determination; RMSECV—root mean square error of cross validation; RPD—ratios of performance to deviation; RMSEE—root mean square error of estimation; 2ndDer—seconded derivative; 1stDer + MSC—first derivative + multiplicative scatter correction.

For instance, the models for tracheid radial and tangential diameter of softwoods [21–23] were weaker with r^2 from 45.0% to 80.0%, standard error of calibration (SEC) from 0.60 to 1.60 μm, and the validation of the models showed higher standard error of prediction (SEP) from 1.04 to 2.70 μm.

Inagaki *et al.* [27] proposed (*Eucalyptus camaldulensis*) one model for the prediction of fiber length in wood for hardwoods, with a root mean square error of cross-validation (RMSECV) of 0.018 mm and 0.012 mm to a root mean square error of prediction (RMSEP), respectively, with r^2 87.0% and 93.0%.

The ratios of performance to deviation (RPD) may be used to evaluate if the prediction models fulfill the requirements of the AACC Method 39-00 for screening in breeding programs that require a RPD $\geqslant$ 2.5 [33]. The RPD was introduced by Williams and Norris [38] as the ratio between the standard deviation of the reference data of the validation set and the standard error of prediction of a cross-validation or of the test set validation. In the present case, the RPD for the validation of the NIR-PLS-R model was 3.9 and 3.3, respectively, for fiber length weighted in length and fiber width (Table 2), thereby allowing the conclusion that it is applicable for screening in breeding programs.

It should be stressed that the fiber biometric characteristics of the unbleached pulp can be accurately predicted by the spectral data of the unprocessed (*i.e.*, unpulped) wood material. The practical interest for improvement and breeding programs is obvious, and the application of NIRS-PLS-R models will allow a very high reduction in cost and time for the experimental evaluation.

Table 5. Model characteristics for some cellular morphological properties of wood using NIR spectroscopy published in the bibliography.

Species	Reference	Parameter	Pre-Processing	Rank	r^2	RMSECV	RMSEP
Eucalyptus camaldulensis	[27]	Fiber length (mm)	MSC	6	87.0/93.0	0.018 mm	0.012 mm
Pinus taeda	[23]	Tracheid radial diameter (μm)	1stDer (a)	10	46.0/42.0	1.58 * μm	1.77 ** μm
			1stDer (b)	10	45.0/46.0	1.59 * μm	1.60 ** μm
		Tracheid tangential diameter (μm)	1stDer (a)	6	60.0/63.0	1.00 * μm	1.09 ** μm
			1stDer (b)	6	60.0/62.0	1.02 * μm	1.04 ** μm
	[21]	Tracheid radial diameter (μm)	2ndDer (c)	6	70.0/36.0	1.10 * μm	1.50 ** μm
		Tracheid tangential diameter (μm)		6	80.0/61.0	0.60 * μm	1.30 ** μm
		Tracheid radial diameter (μm)	2ndDer (d)	5	59.0/16.0	1.30 * μm	1.70 ** μm
		Tracheid tangential diameter (μm)		4	73.0/16.0	0.70 * μm	1.70 ** μm
Pinus radiata	[22]	Tracheid radial diameter (μm)	2ndDer	6	65.0/28.0–30.0	1.60 * μm	1.90–2.70 ** μm
		Tracheid tangential diameter (μm)		4	69.0/69.0–79.0	1.00 * μm	1.20–1.70 ** μm

(a) original calibration; (b) original calibration with new data; (c) spectra on radial-longitudinal face of wood samples; (d) spectra on the transverse face of dry wood samples; * value corresponding to standard error of calibration (SEC); ** value corresponding to standard error of prediction (SEP).

4. Materials and Methods

4.1. Wood Samples

A total of 60 wood discs from *Acacia melanoxylon* R. Br., belonging to 20 trees from four sites in Portugal and collected at different stem height levels, were used in this study. Detailed information on samples, sites and stands is available elsewhere [39].

Wood meal samples were prepared by milling using a knife mill (Retsch) with a 1 mm output screen and the fraction coarser than 0.25 mm was retained for spectral acquisition.

4.2. Kraft Pulps

Samples of 25 g oven-dry woodmeal were Kraft pulped using a multi-batch digester system under the following reaction conditions that were set to obtain a target kappa number of 15: active alkali charge 21.3% (as NaOH); sulfidity 30%; liquor/wood ratio 4/1; time to temperature of 160 °C,

90 min; time at temperature of 160 °C, 90 min. The pulped samples were disintegrated, washed, and screened. Under these conditions the pulp yields ranged from 47.0% to 58.2% [15,35], as related to the variation of heartwood proportion in different wood discs [39].

4.3. Morphological Properties of Pulp Fibers

The morphological properties (fiber length—weighted in length—and width) of the unbleached pulps of *A. melanoxylon* were determined automatically by image analysis of a diluted suspension (20 mg L^{-1}) in the flow chamber of TECHPAP Morfi® Equipment, by measuring at least 5000 fibers. The morphological properties of the pulp fibers used in this study are the same samples determined by Santos *et al.* [15] according to TAPPI 271 pm-98 [19].

The fiber length weighted in length (L_w—(Equation (1))) and fiber width (l_N—(Equation (2))) are calculated by TECHPAP Morfi® as:

$$L_w = \frac{\sum_{i=0}^{N} (Li \times Li)}{\sum_{i=0}^{N} Li} \tag{1}$$

$$l_N = \frac{\sum_{i=0}^{N} li}{N} \tag{2}$$

where N is the number of fiber, i is fiber i, Li is the average length weighted in length, and li width of the fiber i.

The experimental error using the TECHPAP Morfi® in the measurements of the fiber length—weighted in length—and width was 0.5% and 1%, respectively.

4.4. Spectra Collection and Data Processing

The woodmeal samples were conditioned in a climatic chamber at 60 °C for a period of 48 h before spectral acquisition. NIR spectra were collected in the wavenumber range from 12,000 to 3800 cm^{-1} with a near infrared spectrometer (BRUKER, model Vector 22/N, Karlsruhe, Germany) in diffuse reflectance mode, using a spinning cup module. Each spectrum was obtained with 100 scans at a spectral resolution of 16 cm^{-1}. After collecting the spectra, the woodmeal samples were used for production of the Kraft pulps.

The samples were randomly divided into a calibration set containing 45 samples and a validation set (test set) containing 15 samples. The processing was done in two steps. First, the infrared data from the calibration samples were regressed against the measured fiber length and width, and by means of full cross-validation with one sample omitted a significant number of PLS components (rank) was obtained using OPUS/Quant 2 software (version 7.5.18 BRUKER, Bruker Corporation, Karlsruhe, Germany). Besides the raw spectra, also pre-processed spectra with 10 methods were used for PLS analysis [36,37,40]. In a second step, the validation of the PLS-R models was performed using the independent test set. The number of PLS factors was found by automated optimization.

The quality of the calibration models was assessed by means of cross-validation and by using the test set validation results by determining their coefficient of determination (r^2), root mean square error of cross-validation (RMSECV), root mean square error of prediction (RMSEP) and the residual prediction deviation or ratio of performance to deviation (RPD).

The selection of the final model was based on its predictive ability assessed by the least possible number of samples classified as outsiders and/or outliers.

5. Conclusions

NIRS-PLS-R models could be developed to predict the biometric characteristics of fiber length and width of unbleached Kraft pulps using the spectral data of the initial wood meal. The statistical parameters of cross-validation (RMSECV of 0.009 mm and 0.39 µm) and validation (RMSEP of

0.007 mm and 0.36 μm) with RPD_{TS} values of 3.9 and 3.3, respectively, confirm that the models are robust, stable, and well qualified for prediction.

This modeling approach using NIR spectral data of wood to predict pulp fiber dimensions was presented here for the first time. It has a high potential to be used for tree breeding and improvement programs by providing a rapid screening for desired fiber morphological properties of pulp.

Acknowledgments: The second author António José Alves Santos acknowledges a scholarship (SFRH/BD/42073/2007) given by Fundação para a Ciência e a Tecnologia (FCT), Portugal. Centro de Estudos Florestais is a research unit funded by FCT (UID/AGR/00239-2013). To Centro de Biotecnologia de Plantas da Beira Interior for the equipment and OPUS software availability. Rogério M. S. Simões (Universidade da Beira Interior—Faculdade de Ciências), in monitoring of the Morfi determination.

Author Contributions: These authors contributed equally to this work.

Conflicts of Interest: The authors declare no conflict of interest. The founding sponsors had no role in the design of the study; in the collection, analyses, or interpretation of data; in the writing of the manuscript, and in the decision to publish the results.

References

1. Anjos, O.; Santos, A.; Simões, R.; Pereira, H. Morphological, mechanical and optical properties of cypress. *Holzforschung* **2014**, *68*, 861–995. [CrossRef]
2. El-Hosseiny, F.; Anderson, D. Effect of fibre length and coarseness on the burst strength of paper. *Tappi J.* **1999**, *82*, 202–203.
3. Zobel, B.J.; Buijtenen, J.P. *Wood Variation: Its Causes and Control*; Springer: Heidelberg, Germany, 1989; p. 286.
4. Keays, J.L.; Hatton, J.V.; Bailey, G.R.; Neilson, R.W. Present and Future Uses of Canadian Poplars in Fibre and Wood Products. Available online: http://digitalcommons.usu.edu/aspen_bib/5234/ (accessed on 23 December 2015).
5. Dinwoodie, J.M. The relationship between fibre morphology and paper properties: A review of literature. *Tappi* **1965**, *48*, 440–447.
6. Van Buijtenen, J.P.; Einspahr, D.W.; Peckham, J.R. Natural variation in *Populus tremuloides* Michx. II. Variation in pulp and papermaking properties. *Tappi* **1962**, *45*, 58–61.
7. Malan, F.S.; Male, J.R.; Venter, J.S. Relationship between the properties of eucalyptus wood and some chemical, pulp and paper properties. *Paper South. Africa* **1994**, *2*, 6–16.
8. Labosky, P., Jr.; Ifju, G. A study of loblolly pine growth increments. Part IV. Papermaking properties. *Wood Fiber* **1981**, *13*, 120–137.
9. DuPlooy, A.B.J. The relationship between wood and pulp properties of *E. grandis* (Hill ex-Maiden) grown in South Africa. *Appita* **1980**, *33*, 257–264.
10. Wandgaard, F.F.; Williams, D.L. Fiber length and fiber strength in relation to tearing resistance of hardwood pulps. *TAPPI* **1970**, *53*, 2153–2154.
11. Watson, A.J.; Dadswell, H.E. Influence on fibre morphology on paper properties. Part 1. Fibre length. *Appita J.* **1961**, *14*, 168–176.
12. Seth, R.S.; Page, D.H. Fiber properties and tearing resistance. *Tappi J.* **1988**, *71*, 103–107.
13. Niskanen, K. *Paper Physics*; Fapet Oy: Helsinki, Finland, 1998.
14. Paavilainen, L. Quality-competitiveness of Asian short-fibre raw materials in different paper grades. *Paperi Ja Puu* **2000**, *82*, 156–161.
15. Santos, A.J.A.; Anjos, O.; Amaral, M.E.; Gil, N.; Pereira, H.; Simões, R. Influence on pulping yield and pulp properties of wood density of *Acacia melanoxylon*. *J. Wood Sci.* **2012**, *58*, 479–486. [CrossRef]
16. Wimmer, R.; Downes, G.M.; Evans, R.; Rasmussen, G.; French, J. Direct effects of wood characteristics on pulp and handsheet properties of *Eucalyptus globulus*. *Holzforschung* **2002**, *56*, 244–252. [CrossRef]
17. Technical Association of the Pulp & Paper Industry (TAPPI). *Fiber Length of Pulp by Classification, Test Method T 233 cm-06, TAPPI Standard Test Methods*; TAPPI: Atlanta, GA, USA.
18. Technical Association of the Pulp & Paper Industry (TAPPI). *Fiber Length of Pulp by Projection, Test Method T 232 cm-01, TAPPI Standard Test Methods*; TAPPI: Atlanta, GA, USA.
19. Technical Association of the Pulp & Paper Industry (TAPPI). *Fiber Length of Pulp and Paper by Automated Optical Analyser Using Polarized Light, Test Method T 271 om-12, TAPPI Standard Test Methods*; TAPPI: Atlanta, GA, USA.

20. Tsuchikawa, S.; Kobori, H. A review of recent application of near infrared spectroscopy to wood science and technology. *J. Wood Sci.* **2015**, *61*, 213–220. [CrossRef]
21. Schimleck, L.R.; Mora, C.; Daniels, R.F. Estimation of tracheid morphological characteristics of green *Pinus taeda* L. radial strips by near infrared spectroscopy. *Wood Fiber Sci.* **2004**, *36*, 527–535.
22. Schimleck, L.R.; Evans, R. Estimation of *Pinus radiata* D. Don tracheid morphological characteristics by near infrared spectroscopy. *Holzforschung* **2004**, *58*, 66–73. [CrossRef]
23. Jones, P.D.; Schimleck, L.R.; Peter, G.F.; Daniels, R.F.; Clark, A. Non-destructive estimation of *Pinus taeda* L. tracheid morphological characteristics for samples from a wide range of sites in Georgia. *Wood Sci. Technol.* **2005**, *39*, 529–545. [CrossRef]
24. Via, B.K.; Shupe, T.F.; Stine, M.; So, C.L.; Groom, L.H. Tracheid length prediction in *Pinus palustris* by means of near infrared spectroscopy: The influence of age. *Holz als Roh-Und Werkstoff* **2005**, *63*, 231–236. [CrossRef]
25. Wang, Y.H.; Feim, B.H.; Fu, F.; Jiang, Z.H.; Qin, D.C.; Yang, Z.A. Novel method for estimating wood fiber length using near infrared spectroscopy. *China Pulp Paper* **2008**, *27*, 6–9.
26. Viana, L.C.; Trugilho, P.F.; Hein, P.R.; Lima, J.T.; da Silva, J.R. Predicting the morphological characteristics and basic density of Eucalyptus wood using the NIRS technique. *Cerne* **2009**, *15*, 421–429.
27. Inagaki, T.; Schwanninger, M.; Kato, R.; Kurata, Y.; Thanapase, W.; Puthson, P.; Tsuchikawa, S. *Eucalyptus camaldulensis* density and fiber length estimated by near-infrared spectroscopy. *Wood Sci. Technol.* **2012**, *46*, 143–155. [CrossRef]
28. Sun, B.L.; Chai, Y.B.; Huang, A.M.; Liu, J.H. Application of NIR spectroscopy to estimate of MFA and fiber length of *Neosinocalamus affinis*. *Spectrosc. Spectr. Anal.* **2011**, *31*, 3251–3255.
29. Santos, A.J.A.; Anjos, O.; Simões, R. Papermaking potential of *Acacia dealbata* and *Acacia melanoxylon*. *Appita J.* **2006**, *59*, 58–64.
30. Anjos, O.; Santos, A.J.A.; Simões, R. Effect of *Acacia melanoxylon* fibre morphology on papermaking potential. *Appita J.* **2011**, *64*, 185–191.
31. Lourenço, A.; Baptista, I.; Gominho, J.; Pereira, H. The influence of heartwood on the pulping properties of *Acacia melanoxylon* wood. *J. Wood Sci.* **2008**, *54*, 464–469. [CrossRef]
32. Griffin, A.R.; Twayi, H.; Braunstein, R.; Downes, G.M.; Son, D.H.; Harwood, C.E. A comparison of fibre and pulp properties of diploid and tetraploid *Acacia mangium* grown in Vietnam. *Appita J.* **2014**, *67*, 43–49.
33. American Association of Cereal Chemists (AACC) International. *Near-Infrared Methods-Guidelines for Model Development and Maintenance, AACC-Standards Method 39-00:15*; AACC International: St. Paul, MN, USA.
34. Santos, A.J.A.; Alves, A.M.; Simões, R.M.; Pereira, H.; Rodrigues, J.; Schwanninger, M. Estimation of wood basic density of *Acacia melanoxylon* (R. Br.) by near infrared spectroscopy. *J. Near Infrared Spec.* **2012**, *20*, 267–274. [CrossRef]
35. Santos, A.J.A.; Anjos, O.; Pereira, H. Prediction of blackwood Kraft pulps yields with wood NIR-PLS-R models. *Wood Sci. Technol.* **2015**. under review.
36. Santos, A.J.A.; Anjos, O.; Simões, R.; Rodrigues, J.; Pereira, H. Kappa number prediction of *Acacia melanoxylon* unbleached Kraft pulps using NIR-PLSR models with a narrow interval of variation. *BioResources* **2014**, *9*, 6735–6744. [CrossRef]
37. Santos, A.J.A.; Anjos, O.; Pereira, H. Estimation of *Acacia melanoxylon* unbleached Kraft pulp brightness by NIR spectroscopy. *Forest Syst.* **2015**, *24*. [CrossRef]
38. Williams, P.; Norris, K. *Near-Infrared Technology in the Agricultural and Food Industries*, 2nd ed.; American Association of Cereal Chemists: St. Paul, MN, USA, 2004; p. 296.
39. Santos, A.J.A.; Simões, R.; Tavares, M. Variation of some wood macroscopic properties along the stem of *Acacia melanoxylon* R. Br. adult trees in Portugal. *Forest Syst.* **2013**, *22*, 463–470. [CrossRef]
40. Alves, A.; Santos, A.J.A.; da Silva Perez, D.; Rodrigues, J.; Pereira, H.; Simões, R.; Schwanninger, M. NIR PLSR model selection for Kappa number prediction of maritime pine Kraft pulps. *Wood Sci. Technol.* **2007**, *41*, 491–499. [CrossRef]

15

A Critical Evaluation of the Influence of the Dark Exchange Current on the Performance of Dye-Sensitized Solar Cells

Rodrigo García-Rodríguez [1,*], Julio Villanueva-Cab [2], Juan A. Anta [3] and Gerko Oskam [1,*]

Academic Editor: Federico Bella

1 Departamento de Física Aplicada, Centro de Investigación y de Estudios Avanzados del Instituto Politécnico Nacional (CINVESTAV-IPN), Mérida, Yucatán 97310, Mexico
2 Instituto de Física, Benemérita Universidad Autónoma de Puebla (BUAP), Puebla, Puebla 72570, Mexico; juliovc@ifuap.buap.mx
3 Nanostructured Solar Cells Group, Department of Physical, Chemical and Natural Systems, Universidad Pablo de Olavide, Seville ES-41013, Spain; anta@upo.es
* Correspondence: rgarcia@mda.cinvestav.mx (R.G.-R.); oskam@mda.cinvestav.mx (G.O.)

Abstract: The influence of the thickness of the nanostructured, mesoporous TiO_2 film on several parameters determining the performance of a dye-sensitized solar cell is investigated both experimentally and theoretically. We pay special attention to the effect of the exchange current density in the dark, and we compare the values obtained by steady state measurements with values extracted from small perturbation techniques. We also evaluate the influence of exchange current density, the solar cell ideality factor, and the effective absorption coefficient of the cell on the optimal film thickness. The results show that the exchange current density in the dark is proportional to the TiO_2 film thickness, however, the effective absorption coefficient is the parameter that ultimately defines the ideal thickness. We illustrate the importance of the exchange current density in the dark on the determination of the current–voltage characteristics and we show how an important improvement of the cell performance can be achieved by decreasing values of the total series resistance and the exchange current density in the dark.

Keywords: dye-sensitized solar cells; continuity equation; dark current; exchange current density; optimal film thickness; super-position principle

1. Introduction

The development of new sources of energy that can be economically viable and environmentally friendly is one of the most important goals in current scientific research. Solar energy is one of the most promising alternatives as a clean energy source as the Sun provides the Earth in one hour with as much energy as all mankind uses in one year [1,2]. Development of this solar energy includes the development of third generation solar cells. This type of solar cell includes dye-sensitized solar cells (DSCs) [3–6] and the recently discovered perovskite-based solar cells [7–10]. Dye-sensitized solar cells are photoelectrochemical systems that represent an interesting option in the development of new technology in the photovoltaics field related to the use of earth abundant materials, potentially low cost, and simplicity of manufacturing techniques. The DSC consists of a transparent conductive oxide (TCO) on which a layer of a nanostructured, mesoporous semiconductor, usually TiO_2, is deposited. TiO_2 has a large bandgap (3.2 eV) and it does not absorb visible light, hence, the semiconductor is sensitized with a molecular dye, which can be an organometallic ruthenium complex or a purely organic dye. The solar cell is sealed with a thermoplastic polymer and another TCO acting as counter electrode, which is usually activated with an electrocatalyst consisting of small Pt islands or carbon materials. The solar cell is filled with an electrolyte solution based on an organic low-viscosity solvent and a redox couple, classically I^-/I_3^- or the newer, higher-performance $Co^{2+/3+}(bpy)_3$ and related couples [11–14]. Recent advances in dye-sensitized solar cells include the development of flexible fiber/wire-shaped solar cells [7,15], co-sensitization for increasing light absorption, and performance [11,16] and copper-based redox couples [17–19]. This photoelectrochemical system is very complex, combining physical and chemical properties of inorganic nanomaterials, optoelectronic properties of molecular organic dyes, and non-aqueous electrochemistry, and in order to understand the phenomena that take place in the solar cell and to achieve improved performance, the study of interrelated parameters using advanced characterization techniques is essential.

One of the factors that affect DSC performance is the thickness of the nanostructured, mesoporous, TiO_2 film [20–27]. The exposed TiO_2 surface area is proportional to the thickness, hence, a thicker TiO_2 film allows for more dye to be adsorbed resulting in an increase in light harvesting efficiency; however, the surface is also the main source of recombination sites, which directly affects the charge collection efficiency. Various studies have shown that the efficiency of the DSC increases with thickness of the film until it reaches a critical value, and then decreases, thus defining an optimal film thickness. The optimal thickness is generally reported to be around 10 μm [20,23,26], although 15–20 μm has also been reported as the ideal thickness [21,22,25,27]. It has been proposed theoretically that this dependence on thickness is a consequence of a balance between absorption of light and photogeneration of electrons and dark electron transfer processes in the cell [28]. In describing dark processes, we need to take into account a variety of parameters. One of these parameters is the exchange current density in the dark, J_0, and several groups have analyzed the influence of J_0 on the performance of the solar cell [27–31]. Assuming that $J_0 \propto d$, where d is the semiconductor film thickness, Zhu and coworkers [28] determined theoretically the effect of the film thickness dependence of J_0 on the performance of DSCs. Using this assumption they concluded that, as the semiconductor film thickness increases, the solar cell becomes more efficient due to better absorption of light and a larger photogenerated current, but after reaching a thickness of 10 μm or more, the processes in the dark have more relevance in the performance of the cell and ultimately limit the efficiency of the device; the optimal film thickness was found to be around 10 μm. In spite of this, the relationship between J_0 and d has not been yet unambiguously demonstrated and has not been verified experimentally.

In this work, we have determined values of J_0 by means of both steady state current–voltage measurements analyzed using the diode equation and small perturbation techniques, such as electrochemical impedance spectroscopy (EIS), intensity-modulated photovoltage spectroscopy (IMVS) and intensity-modulated photocurrent spectroscopy (IMPS). Using a continuity equation model for the total electron density, we analytically derive a relationship between J_0 and the recombination time constant and electron density in the dark, which are parameters that can be determined experimentally using small perturbation measurements; the relationship was corroborated experimentally. We also describe the influence of other parameters on the determination of the optimal film thickness in DSCs according to the diode equation.

2. Theoretical Background

2.1. Determination of J_0 from Steady State Measurements

The current voltage response of a solar cell can be approximated as the sum of the short circuit photocurrent and the dark current as a function of applied voltage; this method is known as the superposition approximation. Although the reverse current which flows in response to voltage in an illuminated cell is not formally equal to the current which flows in the dark, the approximation is reasonable for many photovoltaic materials. The sign convention for current and voltage in photovoltaics is such that the photocurrent is positive. With this sign convention and for a non-ideal diode, the net current density in the cell is [32–34]:

$$J = J_{SC} - J_0 \left[exp\left(\frac{qV}{mk_BT}\right) - 1 \right] \tag{1}$$

where J_{SC} is the short-circuit photocurrent density, J_0 is the exchange current density in the dark, k_B is the Boltzmann constant, T is temperature, q is the elementary charge, and m is the ideality factor of the diode equation. The value of m depends on the order of the recombination reaction rate in electrons and oxidized species in the electrolyte, and the distribution of recombination sites [35,36]. In addition, the value of J_0 is related to the rate constant of the recombination reaction, the film thickness, the order of reactions of electrons and oxidized redox species, and the background concentrations of electrons in the TiO_2 and oxidized redox species in the electrolyte [28]. The second part of Equation (1) represents the dark current density. At sufficiently high bias, the two currents cancel, corresponding to open-circuit conditions, while at zero bias the cell produces its maximum current, corresponding to short-circuit conditions. Together with series resistance losses, the balance between transport and recombination determines the shape of the current–voltage curve of the device [33,37].

If only the resistance losses relating to current are considered, the expression for the net current density (J_{net}) of a DSC can be expressed as:

$$J_{net}(V) = J_{SC} - J_0 \left\{ exp\left[\frac{q\left(V + J_{net}R_S^{total}\right)}{mk_BT} \right] - 1 \right\} \tag{2}$$

where J_{net} is the net photocurrent density, $R_s{}^{total}$ is the total series resistance of the solar cell and V is the applied voltage. From the last equation we can obtain:

$$\frac{dV}{dJ_{net}} = -R_S^{total} + \frac{mk_BT}{q} \cdot \frac{1}{J_{net} - J_{sc} - J_0} \tag{3}$$

The dark exchange current density, J_0, is normally orders of magnitude smaller than the sum of J_{net} and J_{SC} and, therefore, can be neglected in this equation [28,38]. Isolating the J_0 term in Equation (2) and under open circuit conditions ($J_{net} = 0$), we obtain the following expression:

$$J_0 = \frac{J_{sc}}{exp\left(\frac{qV_{oc}}{mk_BT}\right) - 1} \tag{4}$$

From Equation (4) we can obtain a value for J_0. Considering no recombination losses at short-circuit conditions, the value of J_{SC} will be equal to the light absorbed:

$$J_{SC} = q\phi T_{FTO}\left[1 - exp\left(-\alpha_{abs}d\right)\right] \tag{5}$$

where ϕ is the incident photon flux, T_{FTO} is the effective transmittance of the FTO substrate, and α_{abs} is the effective absorption coefficient of the solar cell. Assuming (i) a collection efficiency of over 99% ($\eta_{cc} \geqslant 99\%$), (ii) $J_0 \propto d$, (iii) ohmic losses have been corrected for and light scattering, (iv) reflection at the counter electrode, and (v) light absorption by the iodide electrolyte and TiO_2 are negligible [28], the diode equation can be expressed as:

$$J(V) = -J_{00}d\left[exp\left(\frac{qV}{mk_BT}\right) - 1 \right] + q\phi T_{FTO}\left[1 - exp\left(-\alpha_{abs}d\right)\right] \tag{6}$$

where J_{00} is the exchange current density per unit film thickness. The first and second terms on the right side of the last equation denote the contributions to $J(V)$ in the dark and in the light, respectively [28].

2.2. Determination of J_0 from Small Perturbation Techniques

The electron transport and recombination properties of the DSC have been investigated using a continuity equation for electrons, which can either be expressed in terms of the free electron density in the conduction band, including a separate term for the trapping and detrapping processes, or in terms of the total electron density; it has been shown that these two approaches are mathematically equivalent under "quasi-static conditions" [33]. It is convenient to express the continuity equation in terms of the total electron density in the photoanode, n, as this is an experimentally accessible quantity: the total electron density includes both free and trapped electrons and can be determined experimentally by charge extraction measurements. Assuming 1-dimensional geometry for electronic processes in the photoanode, the continuity equation can be written as [33]:

$$\frac{\partial n(x,t)}{\partial t} = \frac{\partial}{\partial x}\left[D_n(n(x))\frac{\partial n(x)}{\partial x}\right] + G(x) - k_n(n(x))\,n(x,t) \tag{7}$$

which is the continuity equation in terms of the total electron density, n, taking the approximation that a quasi-static equilibrium holds between free and trapped electrons. Here, x is the distance to the working electrode, $G(x)$ is the volume generation rate and the trapping-detrapping dynamics are implicitly considered in the form of an effective diffusion coefficient, D_n, and an effective recombination constant, k_n, which are a function of the total electron density. Another advantage of using Equation (7) is that the transient and stationary behaviors of the device are described with a single continuity equation [33], so this equation can also be used in the time-dependent analysis.

The photovoltage is related to the total density accumulated in the semiconductor as [33]:

$$n = n_0 exp\left[\frac{\alpha q V}{k_B T}\right] \tag{8}$$

where n_0 is the total electron density in the dark and α is a parameter that reflects the average energy of the exponential distribution of trap states below the conduction band. This parameter can be estimated from capacitance data obtained from impedance measurements. In the continuity equation (Equation (7)), the last term represents the recombination term, in which k_n stands as a small-perturbation recombination constant, which can be obtained directly by small perturbation techniques. The recombination constant represents all possible recombination processes, including recombination with the dye, and can be expressed as [33]:

$$k_n(n) = k_0\left(\frac{n}{n_0}\right)^{\frac{\beta-\alpha}{\alpha}} \tag{9}$$

where k_0 corresponds to the recombination constant in the dark and β is the recombination reaction order with respect to free electrons, corresponding to the inverse of the ideality factor, m. From Equations (8) and (9), we obtain an expression for the recombination term in the continuity equation:

$$k_n n = k_0 n_0 exp\left[\frac{\beta q V}{k_B T}\right] = U \tag{10}$$

The recombination current density (J_{rec}) at the open circuit can be approximated as [30,37]:

$$J_{rec} = Uqd = J_0 exp\left[\frac{\beta q V}{k_B T}\right] \tag{11}$$

where U is the recombination term of the continuity equation. The recombination current density, J_{rec}, is a useful parameter to describe the flux of electrons lost from the TiO_2. It gives the total rate of loss over the full TiO_2 film thickness. As the cell voltage increases, the recombination current also increases as expected due to the increasing total charge concentration, n, in the TiO_2 [39]. Combining Equations (10) and (11), we obtain an expression for the exchange current density in the dark:

$$J_0 = k_0 n_0 q d \quad (12)$$

The values of k_0 and n_0 can be obtained from small perturbation techniques (Figures S1 and S2, Supplementary Materials), which allows us to determine the value of J_0 and to compare its value with the value obtained from steady state measurements, in particular, the current–voltage curve.

3. Experimental Section

Elaboration and Characterization of Dye-Sensitized Solar Cells

The procedure to prepare a TiO_2 screen printing paste was a variation of a standard procedure. An amount of 0.2 mol of acetic acid was added all at once to 0.2 mol of titanium iso-propoxide and vigorously stirred for 15 min at room temperature. The mixture of titanium iso-propoxide and acetic acid was added drop-wise into a flask with 290 mL water with vigorous stirring. The final mixture was stirred at 700 rpm for 40 min. After adding 4 mL of nitric acid, the mixture was heated from room temperature to 80 °C within 40 min and subsequently peptized for 75 min. Water was then added to the cooling liquid mixture to adjust the volume to 370 mL. Aliquots from the resultant mixture were hydrothermally treated at 200 °C for 12 h in a Teflon-lined titanium autoclave (Parr Instruments, Moline, IL, USA). After the heating process, the mixture was sonicated for 1 h, centrifuged, and washed with ethanol. The final material was a colloid with 40 wt% TiO_2. An amount of 1.8 g of the colloid was dissolved in 9 ml ethanol and mixed with 7.3 g terpineol; the resulting mixture was sonicated for 1 h. This mixture was added drop-wise to a solution of 0.9 g ethyl cellulose in 10 mL ethanol. The final mixture was sonicated for 1 h and the excess ethanol was removed with the aid of a rotary evaporator [20].

To prepare the DSC working electrodes, fluorine-doped tin oxide (FTO) glass is used as current collector (Pilkington TEC 15, Xop Glass, Castellón, Spain). The FTO glass was cleaned in a detergent solution (Alconox) using an ultrasonic bath for 10 min, and then rinsed with deionized water. The FTO glass was then submerged sequentially in deionized water, ethanol, and iso-propanol and treated in an ultrasonic bath for 10 min. The FTO substrate glasses were heated up to 450 °C for 30 min in order to eliminate all the organic residues. A layer of TiO_2 paste was coated on the FTO by screen printing and then dried for 10 min at 120 °C; five layers of TiO_2 were deposited using this method, after which the films were heated following the temperature ramp described in Figure S3 in the Supplementary Materials. After the heat treatment, another five layers were deposited onto the previously deposited films in order to increase the thickness of the film. After deposition, the films underwent the previously mentioned heat treatment. This procedure was repeated until the film reaches the desired thickness. Films of 5, 10, 15, 20, 25, and 30 layers were elaborated. The application of the temperature ramp is necessary every five depositions in order to prevent failures and cracks in the semiconductor film. All the electrodes were, once again, heated according to the temperature ramp described in Supplementary Materials, but they were taken to a final constant temperature of 80 °C. The electrodes at 80 °C were immersed in a solution of 0.5 mM N-719 dye in ethanol and kept at room temperature for 24 h in the dark to assure complete sensitizer uptake. To prepare the counter electrodes, two holes were drilled in FTO-covered glass electrodes (Pilkington TEC 8, Xop Glass). The perforated counter electrodes were cleaned following the same procedure as used for the working electrodes. The counter electrode was covered with Pt electrocatalyst by spreading a drop of Platisol (Solaronix, Aubonne, Switzerland) and sintering at 400 °C for 5 min. The sensitized TiO_2 electrode and Pt-catalyzed counter electrode were assembled into a sandwich-type cell and sealed with Surlyn polymer of 60 µm thickness by applying mechanical pressure and a temperature of 215 °C for 15 min. The electrolyte solution was introduced trough the two small holes in the back of the counter electrode. The electrolyte composition was 0.6 M DMII (1,2-dimethyl-3-propylimidazolium iodide), 0.1 M LiI, 0.1 M GuSCN (guanidinium thiocyanate),

0.05 M I_2, and 0.5 M TBP (tert-butyl pyridine) in a mixture of acetonitrile and valeronitrile (volume ratio 85:15). In order to have good electrical contact for the connections to the measurement set-up, the edges of the FTO outside the cell were covered with a conductive silver paint (SPI-High Purity Silver Paint) [20]. A picture of the DSCs with a different number of coatings is presented in Figure S4 in Supplementary Materials.

The thickness of the TiO_2 nanoparticulate film was measured using a KLA Tencor AlphaStep D-120 profilometer (KLA Tencor, Milpitas, CA, USA). The crystal structure and average particle size of the TiO_2 films were determined by X-ray diffraction (XRD) using a Siemens D-5000 (Siemens, Munich, Germany) with CuKα radiation. Photovoltaic characterization was performed using a set-up consisting of a 450 W ozone-free Xe-lamp (Oriel, Newport Corporation, Santa Clara, CA, USA) with a water filter, calibrated to an irradiance of 100 mW cm^{-2} on the surface of the solar cell using an Air Mass 1.5 Global (AM 1.5G) optical filter (Newport Corporation); the intensity was calibrated using a certified 4 cm^2 monocrystalline silicon reference cell with incorporated KG-5 filter. Photovoltaic characterization with blue, green, and red light LED illumination (467, 525, and 625 nm wavelength, respectively) was also employed. The scan rate for all the current–voltage curves was 0.01 V/s. Current–voltage curves (I–V), electrochemical impedance spectroscopy (EIS), intensity modulated photocurrent spectroscopy (IMPS) and intensity modulated photovoltage spectroscopy (IMVS) were recorded with an Autolab PGSTAT302N/FRA2 set-up (Metrohm Autolab, Utrecht, The Netherlands). The evolution of the electron transport process in the cell was investigated using EIS, both in the dark and under red light LED illumination (625 nm). An AC amplitude of 10 mV was applied and the frequency range was from 0.1 Hz to 100 kHz. The electrochemical impedance spectrum was analyzed using Z-View software with the aid of the equivalent circuit proposed by Fabregat-Santiago *et al*, and is presented in Figure S5 in the Supplementary Materials [20,31]. IMPS and IMVS measurements were performed at modulation frequencies between 1 mHz and 10 kHz. A red LED (625 nm) was used to illuminate the sample from the substrate side and it served both as the bias illumination and the small sinusoidally modulated probe beam.

4. Results and Discussion

A series of solar cells was fabricated using screen printed TiO_2 films of different thicknesses, and the films and the solar cells were characterized using a variety of techniques. The corresponding experimental results of profilometry, Scanning Electron Microscopy (SEM), and X-ray diffraction measurements are shown in the Supplementary Materials in Figures S6–S8. Crack free films were obtained up to a thickness of 32 μm and the TiO_2 film was crystalline and consisted of pure anatase with an average nanoparticle size of approximately 15 nm (diameter). Figure 1 shows representative J–V curves for dye-sensitized solar cells, illustrating that the current at short circuit tends to increase with film thickness while the open circuit potential decreases. The characteristics of the film and solar cell are summarized in Table S1 in the Supplementary Materials.

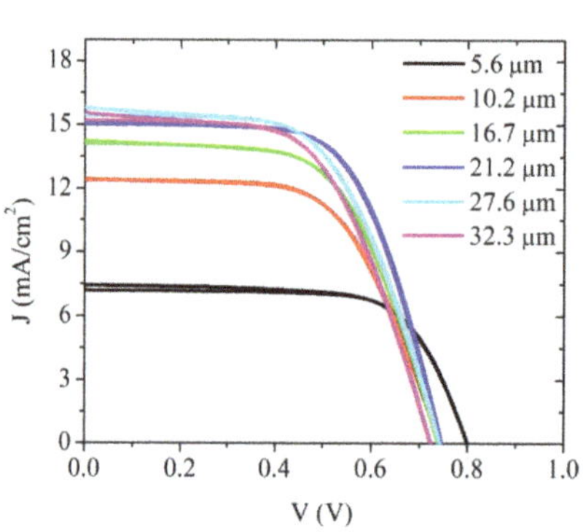

Figure 1. Current–voltage curves for a selection of DSCs as a function of TiO_2 film thickness.

Figure 2 shows the dependencies of the short circuit current density, J_{SC}, open circuit photovoltage, V_{oc}, fill factor, *FF*, and cell efficiency as a function of the film thickness. Upon increasing the film thickness, the exposed TiO_2 surface area increases, which results in the adsorption of more dye on the film [21,27]. In turn, this leads to an increase of the photogenerated electron flux, which is reflected in the increase of J_{SC}. Another important feature is that the photocurrent increases with thickness until it reaches a saturation value. This is related to the very limited increase of light absorption with increasing thickness for thick films as most light is already absorbed [24]. We can also observe that as the thickness increases, V_{oc} decreases. The V_{oc} is a parameter strongly related with recombination in the cell [20,21,26]. The main loss of current is due to a recombination process between photogenerated electrons in the TiO_2 and the oxidized redox species in the electrolytic solution [40]. As the surface area of the semiconductor film and contact area with the electrolyte solution increases, the number of recombination pathways increases thus facilitating recombination, resulting in a lower V_{oc} [21,26,27]. It is important, however, to take other factors into consideration such as dye regeneration, especially for solar cells with large film thickness; although frequently omitted, electron transfer back to the oxidized dye could become an important recombination processes that should not be ignored [39]. The fill factor also diminishes with the increase of thickness. This could be due to inefficient electron injection and/or dye regeneration, especially at high voltages, where regeneration of the dye becomes voltage dependent [41], but also to increasing series resistance losses. At high voltages, the electron concentration in the TiO_2 can be several times larger than that at short circuit, and this could also increase the probability of undesirable electron transfer to the oxidized sensitizer (S^+) [42]. As the electron Fermi level is raised, the probability of back reaction between electrons and dye cations increases, resulting in, or as a result of, poor dye-regeneration by the electron donor in the solution [33,39]. In addition, it has been argued that the regeneration of the dye under illumination leads to a higher I_3^- concentration, which should increase the rate of recombination in DSCs [28], although it has been pointed out that poor dye regeneration could ultimately limit the DSCs performances [21]. As mentioned, an increase in series resistance will also decrease the value of fill factor, but, as shown in Figure 4, this parameter varies only slightly with the film thickness in the solar cells. The solar cell efficiency increases with thickness due to a better absorption of light and an increase in photocurrent, until it reaches a maximum value after which it does not change significantly. If the thickness is further increased, recombination will become more important resulting in a decreased solar energy conversion efficiency [40].

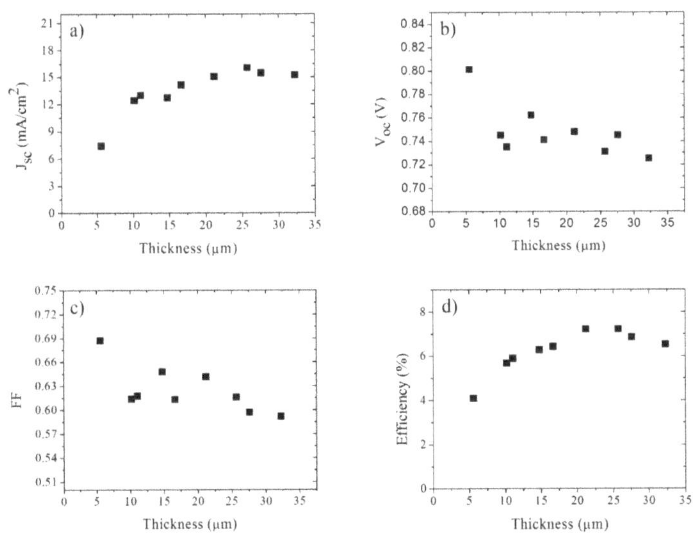

Figure 2. Main parameters of the solar cells as a function of TiO_2 film thickness; (**a**) J_{SC}, (**b**) V_{oc}, (**c**) Fill factor, and (**d**) Overall efficiency.

Figure 3 shows the charge transfer resistance, chemical capacitance and electronic lifetime values obtained by EIS as a function of the TiO_2 film thickness. The comparison is made at 0.74 V, although the tendency is the same at other voltages. The insets show the dependence of R_{ct} and C_μ on the open circuit potential under illumination (at different light intensities) for the 21.2 μm cell. The values of the trap distribution parameter α (Equations (8) and (9)) and the recombination reaction order, β, (Equation (9)) are also displayed. Nyquist plots for this cell and the corresponding fitting are illustrated in Figure S9 in Supplementary Materials.

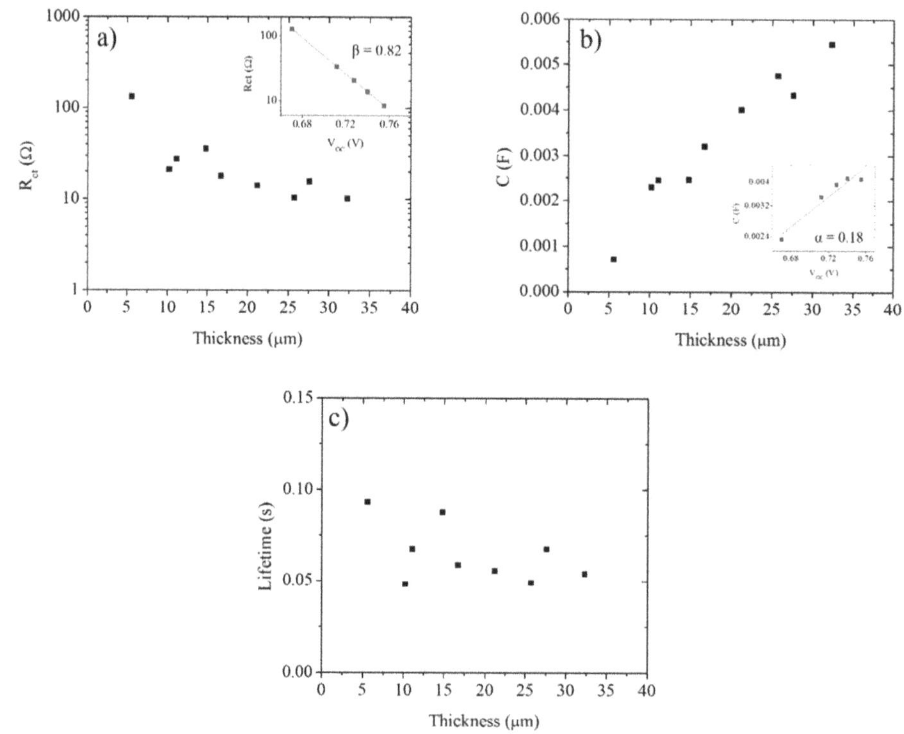

Figure 3. (**a**) Charge transfer resistance, (**b**) chemical capacitance and (**c**) electron lifetime obtained from EIS as a function of TiO_2 film thickness. The comparison was made at 0.74 V. The insets in (**a**) and (**b**) show the open circuit photovoltage dependence of the charge transfer resistance and chemical capacitance for the 21.2 μm solar cell, respectively; the corresponding values of the parameters α and β are shown.

The results show a decrease in R_{ct} value with increasing thickness, which means that the recombination processes between electrons in the TiO_2 film and the oxidized species in the electrolyte or dye become more significant as the thickness increases. This result is consistent with the V_{oc} thickness dependence observed in Figure 2. Since the surface area is proportional to the thickness (if the morphology is independent of the number of layers deposited), R_{ct} is expected to be proportional to $1/d$; Figure 2a illustrates that this is the case. The chemical capacitance increases linearly with film thickness and is proportional to the increase in surface area of the mesoporous layer. Although the chemical capacitance increases with thickness, the value of α for the cell remained constant ($\alpha \approx 0.18$), indicating that the energetic distribution of trap states stays the same. The electron lifetime is a function of the charge transfer resistance and the chemical capacitance [31,35,43]. Since both have opposite behaviors with respect to increasing electrode thickness, the electron lifetime is essentially independent of electrode thickness [44], as illustrated in Figure 3c.

In Figure 4, the series resistances of the cells is shown versus film thickness. The values of R_s were determined both by means of the diode equation and by EIS. From Figure 4 it can be concluded that as the thickness increases, the series resistance of the cell increases slightly as well. An increase of the series resistance is expected for thicker cells, hence reducing the efficiency somewhat [21,23,33]. Both experimental methods illustrate that the series resistance does not increase substantially, hence it is not a crucial parameter to determine an ideal film thickness for optimal cell performance. A small but significant difference is observed between the series resistance obtained by the diode equation and by EIS measurements. The series resistance obtained by means of Equation (3) includes the contributions from the active layer corresponding to r_{tr} and the contribution from the elements connected to the active layer, such as the FTO of the working and counter electrodes, contacts, and wires, and the electrolyte solution; the series resistance obtained by EIS is related only with the series resistance of the FTO [31]. The difference between both values can be observed in Figure 4.

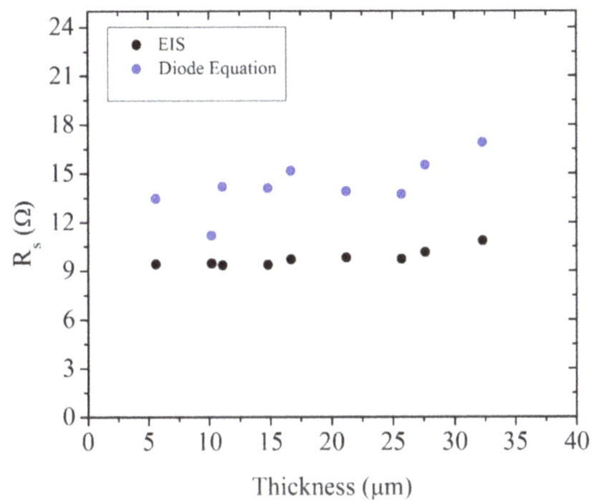

Figure 4. Comparison of the series resistances of the solar cells as a function of TiO_2 film thickness obtained by two diferent methods: steady state (blue dots) and small perturbation (dark dots) techniques.

As the thickness of the semiconductor film increases, the electrochemical system that is the cell becomes more complex and different parameters can have important influence on its performance. As mentioned in the introduction, one of those important parameters is the exchange current density in the dark, J_0, which is directly related with recombination as shown in Equation (11). Using Equation (4), the values of J_0 were determined for all the cells as a function of TiO_2 film thickness. The experiments were performed under 1 sun illumination and the results are presented in Figure 5.

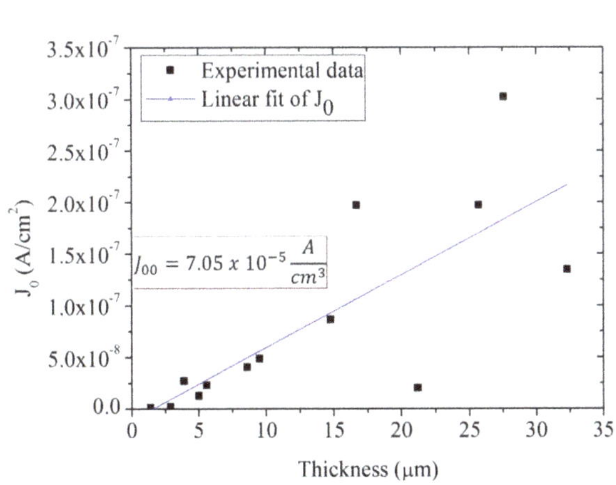

Figure 5. J_0 as a function of film thickness obtained by steady state measurements and the diode Equation (6) under 1 sun illumination. The slope represents J_{00}.

Figure 5 shows that the exchange current density in the dark, J_0, is approximately linear with respect to thickness, in agreement with what would be expected from Equation (12). However, it has to be taken into account that the step-wise process that has been used to prepare the films (alternating layer deposition and calcination) can lead to a non-negligible variation of the porosity of the film [45], which could explain the increased scattering at larger thickness. Note, however, that the dependence of R_{ct}, C, and the lifetime on thickness imply that the morphology is essentially independent of thickness.

According to Equation (12), the value of J_0 depends on the rate constant, k_0, for electron transfer from the TiO_2 to oxidized redox species in solution (recombination) or the oxidized dye, the intrinsic electron density in the film, n_0, and the electrode thickness, d; the value of J_0 should become larger with increasing values of these parameters [28]. In a previous study of the influence of thickness in the distributed parameters of the cell by EIS, the results of Jennings *et al.* [44] imply that the electron lifetime and, as a consequence, τ_0 are independent of electrode thickness. From Equation (S1) in the Supplementary Materials [30,39,40], taking a constant value of porosity and assuming that $Jsc \propto d$, the value of n_0 will be independent of the electrode thickness. Since the value of k_0 and n_0 are independent of electrode thickness, then their product will also be independent of d. According to Equation (12), J_0 should follow a linear behavior with film thickness, as was also established by Zhu and coworkers [28]. The results in Figure 5 are in agreement with Zhu's approach, indicating that from the slope of Figure 5 we can obtain the value of J_{00}; this value is independent of TiO_2 film thickness.

In order to determine the influence of J_0 on the performance of a DSC and using the diode equation (Equation (2)), J–V curves for different *hypothetical* cells are presented in Figure 6 and compared to experimental results.

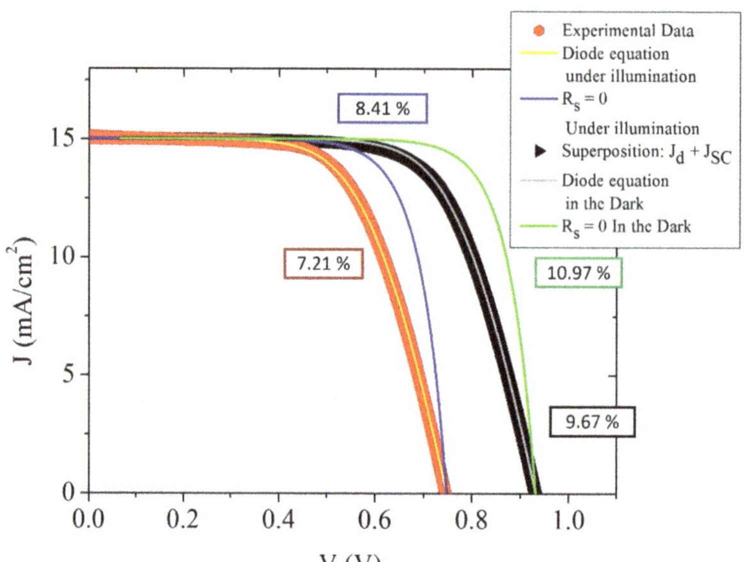

Figure 6. Comparison of the experimental J–V curve (Red dots) under 1 sun illumination obtained for the cell with thickness of 21.2 μm with calculated curves. The dark dots represent the J–V curve for this cell based on the super-position approximation, where the experimental J–V curve measured in the dark in is shifted by the J_{SC} value obtained under illumination.

In Figure 6, experimental data obtained for the cell with thickness of 21.2 μm are compared with theoretical J–V curves. The red dots represent the experimental J–V curve for this cell under 1 sun illumination. The dark dots represent the J–V curve for this cell based on the super-position approximation, where the experimental J–V curve measured in the dark in is shifted by the J_{SC} value obtained under illumination. Using Equation (4), we obtained the values for J_0 for these two curves, assuming the same values for the series resistance and β. Using these parameters and Equation (2), we simulated two J–V curves (yellow and gray lines in Figure 6). Although the conditions in the dark and under illumination are different and this could lead to different values of R_s and β, we kept them constant to evaluate the influence of J_0 in the performance of the cell. We also simulated these two curves with the same values of J_0 and β but with a series resistances equal to zero. For these 4 simulations we calculated the theoretical efficiency of the cells, which is presented in Table 1.

Table 1. Comparison of experimental and simulated J–V curves with different values of J_0 and R_s.

21.2 μm Solar Cell Data	Cell under Illumination	Cell under Illumination; $R_s = 0\ \Omega$	Superposition Approximation	Superposition Approximation; $R_s = 0\ \Omega$
J_0 (A/cm^2)	2.01×10^{-8}	2.01×10^{-8}	7.2×10^{-10}	7.2×10^{-10}
J_{00} (A/cm^3)	9.47×10^{-6}	9.47×10^{-6}	3.4×10^{-7}	3.4×10^{-7}
R_s (Ω)	13.90	0	13.90	0
β	0.47	0.47	0.47	0.47
Efficiency (%)	7.21	8.41	9.67	10.97

Table 1 shows that the value of J_0 for the cell under dark conditions is lower than under illumination, assuming the same values for R_s and β in both cases. This difference in J_0 reflects the difference in recombination processes under light and dark conditions. The main difference between the conditions in the dark and under illumination is that under illumination, a variety of processes occur involving the dye and the redox couple, thus possibly changing the recombination dynamics. Upon illumination the dye is brought into an excited state, and subsequent ultrafast electron injection results in the generation of oxidized dye molecules. These oxidized dye molecules can be reduced back to their original state by either accepting an electron from the TiO_2 via the back reaction, which is a recombination process [46], or by accepting an electron from the electron donor in the electrolyte solution, creating an increased local concentration of the electron acceptor. This, in turn, may also affect the recombination kinetics via the electron transfer from the TiO_2 to the electron acceptor in the solution [47]. The results in Table 1 suggest that these processes could be significant, which is reflected in an increase of the J_0 value under illumination. In the total density approximation (Equation (7)), recombination with the dye is implicitly considered when calculating the recombination constant. In contrast, in the diode equation, this effect is not explicitly considered but leads to a difference in J_0 between dark and illumination conditions. Figure 6 illustrates that this difference in J_0 values leads to a loss of photovoltage of almost 200 mV, which translates to a significant difference in the efficiency of the cell on the order of 2%–3%. This result strongly suggests that dye regeneration is a very important process in the determination of the final performance of the cell, as has been argued by Jennings *et al.* [42]. In addition, the results clearly lead to the conclusion that the superposition approximation does not hold. It is in fact well-known that the light J–V characteristics of DSCs cannot usually be predicted from the sum of the dark J–V and the short-circuit photocurrent (J_{sc}) [39,42]. Barnes *et al.* examined differences between dark and light J–V curves of DSCs; they found that non-negligible recombination with the oxidized dye, together with local changes in electrolyte concentrations due to current flow, could explain their data [39]. If we take into consideration the influence of the series resistance of the solar cell we can observe in the J–V curves that decreasing R_s results in an increase of the fill factor and the efficiency of the cell. Even for the idealized curve under dark conditions, we can achieve a significant improvement on the efficiency of over 1%. This implies that if we can achieve a very small value of R_s together with a decrease in the value of J_0, the performance of the solar cell could be improved significantly.

In order to compare the different methods proposed to obtain the exchange current density in the dark, the values of J_0 were also obtained from small perturbation techniques in the dark (EIS) and under illumination (EIS, IMVS, and IMPS), in accordance with Equation (12). The results are presented in Table 2. We also present the values of J_0 obtained from Equation (4) for a solar cell measured under illumination and in the dark, applying the superposition approximation. The most efficient cell (21.2 μm TiO_2 thickness) was employed for this analysis. From Table 2 we can conclude that the values of J_0 obtained under illumination conditions via the small perturbation methods IMVS-IMPS and the steady-state J–V measurements according to Equation (4) are very similar. We can also observe that the values for J_0 obtained under dark conditions via Equation (4) and the superposition approximation or EIS in the dark lead to practically the same value of J_0, however, J_0 is a factor 300 lower. These results imply that the approximations made with the continuity equation as a function of the total electron density is equivalent to the diode equation with respect to the determination of the J_0 value, and either of the two methods may be employed. Small perturbation methods therefore adequately show that the value of J_0 increases significantly under illumination, since the recombination processes are more significant under these conditions as discussed in the previous paragraph. Note that the results obtained by impedance measurements under illumination led to a different value of J_0 of about one order of magnitude lower, implying that the analysis method may not be accurate or that there are differences in conditions.

Table 2. Comparison between the values obtained for J_0 by different methods, both in the dark and under illumination.

21.2 μm Solar Cell Data	Diode Equation under Illumination	Diode Equation with Superposition Approximation	IMVS and IMPS	EIS under Illumination	EIS in the Dark
J_0 (A/cm^2)	2.01×10^{-8}	6.75×10^{-11}	1.17×10^{-8}	2.44×10^{-9}	9.44×10^{-11}
J_{00} (A/cm^3)	9.47×10^{-6}	3.19×10^{-8}	5.51×10^{-6}	1.15×10^{-6}	4.45×10^{-8}
R_s (Ω)	13.90	15.45	–	9.80	9.75
β	0.47	0.53	0.79	0.82	0.55

It has been suggested that J_0 ultimately determines the optimal film thickness for the best solar cell performance, because J_0 is an approximately linear function of film thickness [28]. In order to determine the parameters that can influence the optimum film thickness, and using Equation (6) and the experimental values obtained and presented in Figure 5, we have calculated simulated power versus voltage plots as a function of film thickness with different values of J_{00}, α_{abs}, m, φ, and T (Table S2 and Figure S10, Supplementary Materials). From the power density *vs.* voltage plots, the maximum power point was obtained as a function of these parameters, and the results are shown in Figure 7a–d.

Figure 7a shows the maximum power output of solar cells for three values of J_{00}. It can be observed that the optimum film thickness is around 10 μm, regardless of the value of J_{00}, implying that, although J_{00} is an important parameter for the performance of the DSCs, its value does not determine the optimum film thickness. From Figure 7b, it can be observed that regardless of the value of the ideality coefficient m, the behavior of the cells is the analogous to that of Figure 7a. Once again, the optimum film thickness does not change if the value of m is modified, and the optimal thickness remains at values of about 10 μm. When analyzing Figure 7b, it is very important to take into consideration that changing the value of β (or m) will lead to a change of the J_0 value, as can be observed in Equation 4 in agreement with the results reported by Fabregat-Santiago *et al.* [31]. They found that a decrease in β value would lead to an increase on J_0 and a decrease in the performance and the fill factor of DSCs. In Figure 7c, cells with different values of α_{abs} are presented. The values of α_{abs} were taken from the absorption coefficient of the Z907 dye adsorbed to the TiO_2 film for blue, green, and red light (wavelengths of 467, 525, and 625 nm, respectively). These results were compared with experimental data obtained from J–V curves for the cells under blue, green, and red LED illumination (Figure 7d). It can be observed that the theoretical (Figure 7c) and experimental (Figure 7d) data show a different optimal film thickness for different illumination wavelengths. In spite of the differences in the actual value of optimal film thickness between theoretical and experimental data, the same tendency is observed: changing the value of α_{abs} changes the ideal thickness of the semiconductor film of TiO_2. The absorption length, which is the distance for which the light intensity decreases by a factor of $1/e$, is inversely proportional to the value of α_{abs}. The value of α_{abs} for the green and blue wavelengths for the Z907 dye is larger than that for red wavelengths. Therefore, the absorption length for blue and green light is smaller, leading to an optimum film thickness of around 10–15 μm, while for a value of α_{abs} for red light, the optimum film thickness is around 20–30 μm, approximately 2–3 times larger. This implies that the absorption coefficient of the cell and the irradiation wavelength that is applied to the solar cell ultimately determines the ideal thickness for DSCs. Gonzalez-Vázquez *et al.* established via random walk numerical simulation (RWNS) that the collection efficiency depends on the characteristic optical absorption length in the cells [48]. This result could have implications for the measurements of DSCs, especially using small perturbation techniques, which are often performed with a red LED light illumination. Since white light illumination (operating conditions) corresponds to smaller shorter absorption lengths than red light, the actual diffusion length of the cell under operating conditions could be overestimated when experimentally determined using red LED light illumination.

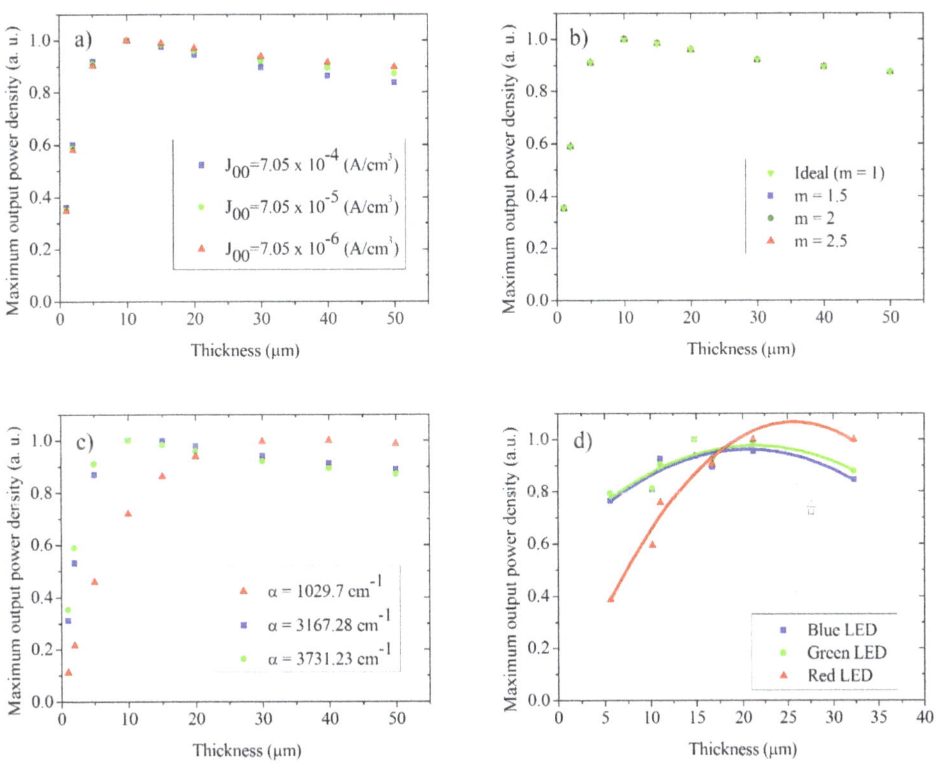

Figure 7. Maximum power density versus film thickness as a function of: (**a**) the value of J_{00} ($\alpha = 3731.23$ cm^{-1}; $m = 2$; $\Phi = 2.62 \times 10^{17}$ cm$^{-2}\cdot$s^{-1}; $T = 0.9$; $R_s = 0$ Ω); (**b**) the value of m ($\alpha = 3731.23$ cm^{-1}; $J_{00} = 7.05 \times 10^{-5}$; $\Phi = 2.62 \times 10^{17}$ cm$^{-2}\cdot$s^{-1}; $T = 0.9$; $R_s = 0$ Ω); (**c**) the value of α_{abs} ($m = 2$; $J_{00} = 7.05 \times 10^{-5}$; $T = 0.9$; $R_s = 0$; $\Phi_{C1} = 2.35 \times 10^{17}$ cm$^{-2}\cdot$s^{-1}, $\Phi_{C2} = 2.62 \times 10^{17}$ cm$^{-2}\cdot$s^{-1}, $\Phi_{C3} = 3.14 \times 10^{17}$ cm$^{-2}\cdot$s^{-1}); (**d**) Wavelength of light; experimental data obtained for the cells under blue, green, and red LED illumination. Tendency lines are presented and the open dots were not taken into account in the calculation of the fit. Note that the value of α was taken according to the absorption coefficient of the Z907 dye adsorbed onto films at 525 nm (green).

5. Conclusions

The exchange current density in the dark reflects the performance of DSCs, and decreasing its value led to an increase of 200 mV in theoretical cells. Although its value increases linearly with electrode thickness, is very susceptible to changes in porosity, increasing the difficulty to determine it for thick films ($d \geqslant 15$ μm). Differences in J_0 values between illumination and dark conditions were obtained, showing the fundamental differences in recombination processes under illumination and dark conditions. This result implies that the superposition approximation will not properly describe the J–V characteristics of the DSCs. Starting with the continuity equation as a function of total electron density, we obtained an expression of J_0 equivalent to the one deduced from the diode equation, and that allows us to determine J_0 via small perturbation techniques. Although J_0 depends on film thickness, it was shown that it does not determine the optimal film thickness of the DSCs, the value of α_{abs} being the one that determines the optimal film thickness in DSCs.

Acknowledgments: Consejo Nacional de Ciencia y Tecnología (CONACyT) is gratefully acknowledged for funding under grant CB-2012-178510, and CONACyT/ Secretaría de Energía (SENER) and Instituto de Energías Renovables de la Universidad Nacional Autónoma de México (IER-UNAM) for funding through the Mexican Center for Innovation in Solar Energy (CeMIE-Sol), Project P-27. We would like to acknowledge the invaluable help of Dora Huerta, José Bante, Beatriz Heredia, Leny Pinzón, Daniel Aguilar, and the National Laboratory for Nano and Biomaterials, with support from grants FOMIX-Yucatan 2008-108160, CONACYT LAB-2009-01-123913, INFRA 204822 & 188345, for access to materials characterization equipment. Juan A. Anta thanks Ministerio de Economía y Competitividad of Spain under grant MAT2013-47192-C3-3-R. Julio Villanueva-Cab thanks Programa para el Desarrollo Profesional Docente (PRODEP) under grant DSA/103.5/15/7449.

Author Contributions: Julio Villanueva-Cab suggested the analytical approach to relate the diode equation and continuity equation to obtain values for the dark exchange current density. Rodrigo García-Rodríguez performed the experimental work, acquired and analyzed the data, and wrote the manuscript under the supervision and expertise of Juan A. Anta, Gerko Oskam and Julio Villanueva-Cab. All authors contributed equally to the interpretation of results and preparation of the published article.

Conflicts of Interest: The authors declare no conflicts of interest.

References

1. Reddy, K.G.; Deepak, T.G.; Anjusree, G.S.; Thomas, S.; Vadukumpully, S.; Subramanian, K.R.V.; Nair, S.V.; Nair, A.S. On global energy scenario, dye-sensitized solar cells and the promise of nanotechnology. *Phys. Chem. Chem. Phys.* **2014**, *16*, 6838–6858. [CrossRef] [PubMed]
2. Litzov, I.; Brabec, C.J. Development of efficient and stable inverted bulk heterojunction (BHJ) solar cells using different metal oxide interfaces. *Materials* **2013**, *6*, 5796–5820. [CrossRef]
3. Oprea, C.I.; Panait, P.; Cimpoesu, F.; Ferbinteanu, M.; Gîrţu, M.A. Density functional theory (DFT) study of coumarin-based dyes adsorbed on TiO_2 nanoclusters—Applications to dye-sensitized solar cells. *Materials* **2013**, *6*, 2372–2392. [CrossRef]
4. Yang, J.; Bark, C.; Kim, K.; Choi, H. Characteristics of the dye-sensitized solar cells using TiO_2 nanotubes treated with $TiCl_4$. *Materials* **2014**, *7*, 3522–3532. [CrossRef]
5. Hardin, B.E.; Snaith, H.J.; McGehee, M.D. The renaissance of dye-sensitized solar cells. *Nat. Photon.* **2012**, *6*, 162–169. [CrossRef]
6. Nazeeruddin, M.K.; Baranoff, E.; Grätzel, M. Dye-sensitized solar cells: A brief overview. *Sol. Energy* **2011**, *85*, 1172–1178. [CrossRef]
7. Bella, F. Polymer electrolytes and perovskites: Lights and shadows in photovoltaic devices. *Electrochimica Acta* **2015**, *175*, 151–161. [CrossRef]
8. Gonzalez-Pedro, V.; Juarez-Perez, E.J.; Arsyad, W.-S.; Barea, E.M.; Fabregat-Santiago, F.; Mora-Sero, I.; Bisquert, J. General working principles of $CH_3NH_3PbX_3$ perovskite solar cells. *Nano Lett.* **2014**, *14*, 888–893. [CrossRef] [PubMed]
9. Kojima, A.; Teshima, K.; Shirai, Y.; Miyasaka, T. Organometal halide perovskites as visible-light sensitizers for photovoltaic cells. *J. Am. Chem. Soc.* **2009**, *131*, 6050–6051. [CrossRef] [PubMed]
10. Bi, D.; Boschloo, G.; Hagfeldt, A. High-efficient solid-state perovskite solar cell without lithium salt in the hole transport material. *Nano* **2014**, *9*, 1440001. [CrossRef]
11. Yella, A.; Lee, H.-W.; Tsao, H.N.; Yi, C.; Chandiran, A.K.; Nazeeruddin, M.K.; Diau, E.W.-G.; Yeh, C.-Y.; Zakeeruddin, S.M.; Grätzel, M. Porphyrin-sensitized solar cells with cobalt (ii/iii)–based redox electrolyte exceed 12 percent efficiency. *Science* **2011**, *334*, 629–634. [CrossRef] [PubMed]

12. Nusbaumer, H.; Moser, J.-E.; Zakeeruddin, S.M.; Nazeeruddin, M.K.; Grätzel, M. $Co^{II}(dbbip)_2^{2+}$ complex rivals tri-iodide/iodide redox mediator in dye-sensitized photovoltaic cells. *J. Phys. Chem. B* **2001**, *105*, 10461–10464. [CrossRef]
13. Feldt, S.M.; Gibson, E.A.; Gabrielsson, E.; Sun, L.; Boschloo, G.; Hagfeldt, A. Design of organic dyes and cobalt polypyridine redox mediators for high-efficiency dye-sensitized solar cells. *J. Am. Chem. Soc.* **2010**, *132*, 16714–16724. [CrossRef] [PubMed]
14. Hao, Y.; Tian, H.; Cong, J.; Yang, W.; Bora, I.; Sun, L.; Boschloo, G.; Hagfeldt, A. Triphenylamine groups improve blocking behavior of phenoxazine dyes in cobalt-electrolyte-based dye-sensitized solar cells. *Chem. Phys. Chem.* **2014**, *15*, 3476–3483. [CrossRef] [PubMed]
15. Peng, M.; Zou, D. Flexible fiber/wire-shaped solar cells in progress: Properties, materials, and designs. *J. Mater. Chem. A* **2015**, *3*, 20435–20458. [CrossRef]
16. Cid, J.-J.; Yum, J.-H.; Jang, S.-R.; Nazeeruddin, M.K.; Martínez-Ferrero, E.; Palomares, E.; Ko, J.; Grätzel, M.; Torres, T. Molecular cosensitization for efficient panchromatic dye-sensitized solar cells. *Angew. Chem. Int. Ed.* **2007**, *119*, 8510–8514. [CrossRef]
17. Hattori, S.; Wada, Y.; Yanagida, S.; Fukuzumi, S. Blue copper model complexes with distorted tetragonal geometry acting as effective electron-transfer mediators in dye-sensitized solar cells. *J. Am. Chem. Soc.* **2005**, *127*, 9648–9654. [CrossRef] [PubMed]
18. Freitag, M.; Daniel, Q.; Pazoki, M.; Sveinbjornsson, K.; Zhang, J.; Sun, L.; Hagfeldt, A.; Boschloo, G. High-efficiency dye-sensitized solar cells with molecular copper phenanthroline as solid hole conductor. *Energy Environ. Sci.* **2015**, *8*, 2634–2637. [CrossRef]
19. Bai, Y.; Yu, Q.; Cai, N.; Wang, Y.; Zhang, M.; Wang, P. High-efficiency organic dye-sensitized mesoscopic solar cells with a copper redox shuttle. *Chem. Commun.* **2011**, *47*, 4376–4378. [CrossRef] [PubMed]
20. Escalante, R.; García-Rodríguez, R.; Heredia-Cervera, B.E.; Aguilar-Frutis, M.A.; Gómez-Ortíz, N.; Villanueva-Cab, J.; Oskam, G. Influence of TiO_2 film thickness on the performance of dye-sensitized solar cells: Relation between optimum film thickness and electron diffusion length. *Energy Environ. Focus* **2013**, *2*, 280–286. [CrossRef]
21. Hsu, C.-P.; Lee, K.-M.; Huang, J.T.-W.; Lin, C.-Y.; Lee, C.-H.; Wang, L.-P.; Tsai, S.-Y.; Ho, K.-C. EIS analysis on low temperature fabrication of TiO_2 porous films for dye-sensitized solar cells. *Electrochimica Acta* **2008**, *53*, 7514–7522. [CrossRef]
22. Huang, C.-Y.; Hsu, Y.-C.; Chen, J.-G.; Suryanarayanan, V.; Lee, K.-M.; Ho, K.-C. The effects of hydrothermal temperature and thickness of TiO_2 film on the performance of a dye-sensitized solar cell. *Sol. Energy Mater. Sol. Cells* **2006**, *90*, 2391–2397. [CrossRef]
23. Zhang, H.; Wang, W.; Liu, H.; Wang, R.; Chen, Y.; Wang, Z. Effects of TiO_2 film thickness on photovoltaic properties of dye-sensitized solar cell and its enhanced performance by graphene combination. *Mater. Res. Bull.* **2014**, *49*, 126–131. [CrossRef]
24. Kao, M.C.; Chen, H.Z.; Young, S.L.; Kung, C.Y.; Lin, C.C. The effects of the thickness of TiO_2 films on the performance of dye-sensitized solar cells. *Thin Solid Films* **2009**, *517*, 5096–5099. [CrossRef]
25. Shin, I.; Seo, H.; Son, M.-K.; Kim, J.-K.; Prabakar, K.; Kim, H.-J. Analysis of TiO_2 thickness effect on characteristic of a dye-sensitized solar cell by using electrochemical impedance spectroscopy. *Curr. Appl. Phys.* **2010**, *10*, S422–S424. [CrossRef]
26. Park, K.-H.; Kim, T.-Y.; Kim, J.-H.; Kim, H.J.; Hong, C.K.; Lee, J.-W. Adsorption and electrochemical properties of photoelectrodes depending on TiO_2 film thickness for dye-sensitized solar cells. *J. Electroanal. Chem.* **2013**, *708*, 39–45. [CrossRef]
27. Wang, Z.-S.; Kawauchi, H.; Kashima, T.; Arakawa, H. Significant influence of TiO_2 photoelectrode morphology on the energy conversion efficiency of N719 dye-sensitized solar cell. *Coord. Chem. Rev.* **2004**, *248*, 1381–1389. [CrossRef]
28. Zhu, K.; Jang, S.-R.; Frank, A.J. Impact of high charge-collection efficiencies and dark energy-loss processes on transport, recombination, and photovoltaic properties of dye-sensitized solar cells. *J. Phys. Chem. Lett.* **2011**, *2*, 1070–1076. [CrossRef]
29. Ito, S.; Liska, P.; Comte, P.; Charvet, R.; Pechy, P.; Bach, U.; Schmidt-Mende, L.; Zakeeruddin, S.M.; Kay, A.; Nazeeruddin, M.K.; *et al.* Control of dark current in photoelectrochemical ($TiO_2/I^-–I_3^-$) and dye-sensitized solar cells. *Chem. Commun.* **2005**, 4351–4353. [CrossRef] [PubMed]

30. Shi, Y.; Dong, X. Coupled analysis of steady-state and dynamic characteristics of dye-sensitized solar cells for determination of conduction band movement and recombination parameters. *Phys. Chem. Chem. Phys.* **2013**, *15*, 299–306. [CrossRef] [PubMed]
31. Fabregat-Santiago, F.; Garcia-Belmonte, G.; Mora-Sero, I.; Bisquert, J. Characterization of nanostructured hybrid and organic solar cells by impedance spectroscopy. *Phys. Chem. Chem. Phys.* **2011**, *13*, 9083–9118. [CrossRef] [PubMed]
32. Nelson, J. *The Physics of Solar Cells*; Imperial College Press: London, UK, 2007; p. 363.
33. Anta, J.A.; Idigoras, J.; Guillen, E.; Villanueva-Cab, J.; Mandujano-Ramirez, H.J.; Oskam, G.; Pelleja, L.; Palomares, E. A continuity equation for the simulation of the current-voltage curve and the time-dependent properties of dye-sensitized solar cells. *Phys. Chem. Chem. Phys.* **2012**, *14*, 10285–10299. [CrossRef] [PubMed]
34. Gonzalez-Vazquez, J.P.; Oskam, G.; Anta, J.A. Origin of nonlinear recombination in dye-sensitized solar cells: Interplay between charge transport and charge transfer. *J. Phys. Chem. C* **2012**, *116*, 22687–22697. [CrossRef]
35. Bisquert, J.; Fabregat-Santiago, F.; Mora-Seró, I.; Garcia-Belmonte, G.; Giménez, S. Electron lifetime in dye-sensitized solar cells: Theory and interpretation of measurements. *J. Phys. Chem. C* **2009**, *113*, 17278–17290. [CrossRef]
36. Frank, J.A.; Kopidakis, N.; Lagemaat, J. Electrons in nanostructured TiO_2 solar cells: Transport, recombination and photovoltaic properties. *Coord. Chem. Rev.* **2004**, *248*, 1165–1179. [CrossRef]
37. Bisquert, J.; Mora-Seró, I. Simulation of steady-state characteristics of dye-sensitized solar cells and the interpretation of the diffusion length. *J. Phys. Chem. Lett.* **2010**, *1*, 450–456. [CrossRef]
38. Hegedus, S.S.; Shafarman, W.N. Thin-film solar cells: Device measurements and analysis. *Prog. Photovolt. Res. Appl.* **2004**, *12*, 155–176. [CrossRef]
39. Barnes, P.R.F.; Anderson, A.Y.; Juozapavicius, M.; Liu, L.; Li, X.; Palomares, E.; Forneli, A.; O'Regan, B.C. Factors controlling charge recombination under dark and light conditions in dye sensitised solar cells. *Phys. Chem. Chem. Phys.* **2011**, *13*, 3547–3558. [CrossRef] [PubMed]
40. Zhu, K.; Kopidakis, N.; Neale, N.R.; van de Lagemaat, J.; Frank, A.J. Influence of surface area on charge transport and recombination in dye-sensitized TiO_2 solar cells. *J. Phys. Chem. B* **2006**, *110*, 25174–25180. [CrossRef] [PubMed]
41. Li, F.; Jennings, J.R.; Wang, Q. Determination of sensitizer regeneration efficiency in dye-sensitized solar cells. *ACS Nano* **2013**, *7*, 8233–8242. [CrossRef] [PubMed]
42. Jennings, J.R.; Liu, Y.; Wang, Q. Efficiency limitations in dye-sensitized solar cells caused by inefficient sensitizer regeneration. *J. Phys. Chem. C* **2011**, *115*, 15109–15120. [CrossRef]
43. Halme, J.; Vahermaa, P.; Miettunen, K.; Lund, P. Device physics of dye solar cells. *Adv. Mater.* **2010**, *22*, E210–E234. [CrossRef] [PubMed]
44. Jennings, J.R.; Liu, Y.; Safari-Alamuti, F.; Wang, Q. Dependence of dye-sensitized solar cell impedance on photoelectrode thickness. *J. Phys. Chem. C* **2011**, *116*, 1556–1562. [CrossRef]
45. Guldin, S.; Docampo, P.; Stefik, M.; Kamita, G.; Wiesner, U.; Snaith, H.J.; Steiner, U. Layer-by-layer formation of block-copolymer-derived TiO_2 for solid-state dye-sensitized solar cells. *Small* **2012**, *8*, 432–440. [CrossRef] [PubMed]
46. Barnes, P.R.F.; Anderson, A.Y.; Durrant, J.R.; O'Regan, B.C. Simulation and measurement of complete dye sensitised solar cells: Including the influence of trapping, electrolyte, oxidised dyes and light intensity on steady state and transient device behaviour. *Phys. Chem. Chem. Phys.* **2011**, *13*, 5798–5816. [CrossRef] [PubMed]
47. Fabregat-Santiago, F.; Bisquert, J.; Garcia-Belmonte, G.; Boschloo, G.; Hagfeldt, A. Influence of electrolyte in transport and recombination in dye-sensitized solar cells studied by impedance spectroscopy. *Sol. Energ. Mater. Sol. Cells* **2005**, *87*, 117–131. [CrossRef]
48. Gonzalez-Vazquez, J.P.; Morales-Flórez, V.; Anta, J.A. How important is working with an ordered electrode to improve the charge collection efficiency in nanostructured solar cells? *J. Phys. Chem. Lett.* **2012**, *3*, 386–393. [CrossRef] [PubMed]

16

Biodegradable Nanocomposite Films Based on Sodium Alginate and Cellulose Nanofibrils

B. Deepa [1,2], Eldho Abraham [3], Laly A. Pothan [1,*], Nereida Cordeiro [4], Marisa Faria [4] and Sabu Thomas [5]

Academic Editor: Marco Morreale

1 Department of Chemistry, Bishop Moore College, Mavelikara-690101, Kerala, India; deepabkrishnan@gmail.com
2 Department of Chemistry, Church Missionary Society(C.M.S) College, Kottayam-686001, Kerala, India
3 Robert H. Smith Faculty of Agriculture, Food and Environment, Hebrew University, Jerusalem, POB 12, Rehovot 76100, Israel; eldhoabraham@gmail.com
4 Faculty of Exact Science and Engineering, University of Madeira, 9000-390 Funchal, Portugal; ncordeiro@uma.pt (N.C.); marisa.faria@uma.pt (M.F.)
5 School of Chemical Sciences, Mahatma Gandhi University, Kottayam-686560, Kerala, India; sabuchathukulam@yahoo.co.uk
* Correspondence: lalybmc@gmail.com

Abstract: Biodegradable nanocomposite films were prepared by incorporation of cellulose nanofibrils (CNF) into alginate biopolymer using the solution casting method. The effects of CNF content (2.5, 5, 7.5, 10 and 15 wt %) on mechanical, biodegradability and swelling behavior of the nanocomposite films were determined. The results showed that the tensile modulus value of the nanocomposite films increased from 308 to 1403 MPa with increasing CNF content from 0% to 10%; however, it decreased with further increase of the filler content. Incorporation of CNF also significantly reduced the swelling percentage and water solubility of alginate-based films, with the lower values found for 10 wt % in CNF. Biodegradation studies of the films in soil confirmed that the biodegradation time of alginate/CNF films greatly depends on the CNF content. The results evidence that the stronger intermolecular interaction and molecular compatibility between alginate and CNF components was at 10 wt % in CNF alginate films.

Keywords: cellulose nanofibrils; sodium alginate; nanocomposite films; solvent casting; mechanical properties

---

1. Introduction

Recently, bio-nanocomposite films developed from biopolymers have gained considerable attention due to their renewability, biodegradability, biocompatibility, low toxicity and their potential use in the packaging industry. In this context, studies based on two such biopolymers, namely cellulose and sodium alginate, have received renewed interest. Cellulose is the most abundant renewable biopolymer on earth and it is a polydispersed linear polymer of poly-β(1,4)-D-glucose units [1]. The two novel forms of cellulose, namely cellulose nanofibrils (CNF) and cellulose nanocrystals (CNC), appear to be attractive building blocks to produce high performance bionanomaterials. They exhibit outstanding properties such as low density, low thermal expansion, high aspect ratio and important mechanical properties, which make them relevant in various applications [2–6].

Alginate is another natural biopolymer, composed mainly of (1,4)-linked β-D-mannuronic acid units and α-L-guluronic acid units [7–9]. The main source of alginate is the cell wall of brown algae, and for commercial purposes alginate is extracted from seaweed. Alginates and alginate-based biocomposites have been used in food packaging, tissue engineering, biomedicine and pharmaceutical fields due to their non-toxicity, biodegradability, biocompatibility and unique gel-forming characteristics [8]. It has also been used for the preparation of biodegradable films. However, poor mechanical and weak water resistance properties limit its application, particularly in the presence of water and humidity [10,11]. The present study attempted to overcome these problems by developing nanocomposite films with the incorporation of cellulose nanofibrils into the alginate matrix. Cellulose nanofibrils used for this study were extracted from sisal fibers by steam explosion coupled with the acid hydrolysis process. The objective of this study was to investigate the effect of cellulose nanofibrils (CNF) at various loadings on the mechanical, biodegradability and swelling behavior of the resulting nanocomposite films.

2. Results and Discussion

2.1. Mechanical Properties

The effect of CNF loading on the tensile properties of alginate films is presented in Figure 1. In general, we observed an improvement of the mechanical properties, attributed to the good interfacial interaction between CNF and the alginate matrix due to the similar polysaccharide structures [10] and to the surface acid-base character (discussed below) of cellulose and alginate matrix.

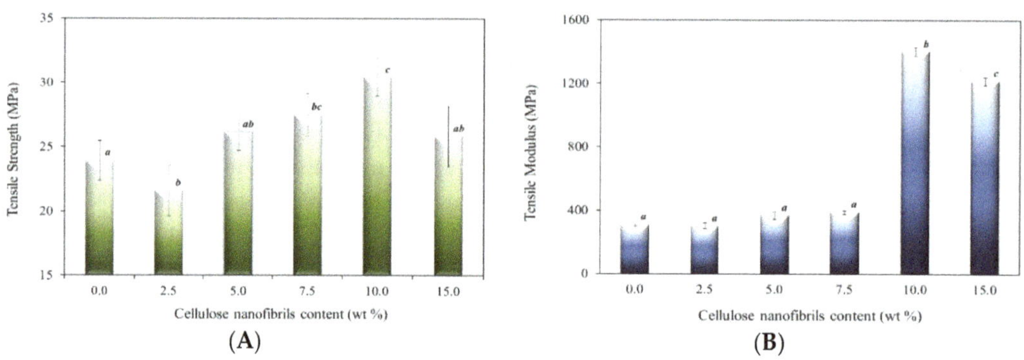

Figure 1. (**A**) Tensile strength and (**B**) Tensile modulus of cellulose nanofibril (CNF)-reinforced alginate films as the function of the CNF content (lower case letters (a, b, c) show Duncan grouping; distinct letters represent means significantly different ($p < 0.05$)).

It has been noted that the incorporation of 2.5 wt % of CNF negatively influences the mechanical properties of the alginate films. This can be attributed to the (i) destruction of the alginate chain organization, which weakens the composite and was not counterbalanced by the few links stabilized between the CNF and the alginate chains, or to the (ii) failure strain difference between the CNF and the matrix [12]. In other words, the reinforcement of CNF does not have an effect when the failure strain of the matrix polymer is much greater than that of the nanocellulose. So, the nanocomposite shows a failure before the stress is transferred to the CNF from the matrix [12]. At lower CNF content (≤10 wt %), CNF could disperse well in the alginate matrix, which increases the mechanical properties (tensile strength (TS) and tensile modulus(TM)) of the nanocomposite. However, at higher content (>10 wt %), the aggregation of CNF (observed by the Scanning electron microscopy (SEM) images, Figure 2) causes a poor dispersion of CNF in the alginate matrix [9]. As a result, both the tensile strength and tensile modulus of the films decrease for CNF additions higher than 10 wt %.

The tensile strength decreases at 2.5 wt %, and then increases with an increase in the CNF content up to 10 wt %, followed by a decrease at higher CNF content (Figure 1a). Similar behavior was observed in the tensile modulus of CNF-reinforced alginate films (Figure 1b). The tensile modulus (TM) value of the pure alginate film was found to be around 308 MPa. The incorporation of CNF into the alginate matrix caused a significant ($p < 0.05$) increase of TM to 1403 MPa, which is around 356% higher than that of the pure alginate film. The highest TM was observed for 10% CNF loading. The increased TM values of the alginate films obtained after the incorporation of CNF may be attributed to the increased stiffness of the films by the addition of nanocellulose. It also reveals the high reinforcing efficiency of CNF material in the alginate matrix.

For the tensile modulus, it was not observed the expected linear increase with the increase of CNF content; instead we observed a decrease at 2.5 wt % (above-mentioned) and a very high increase at 10 wt % of CNF incorporation. Given the surface nature of the CNF and alginate chain (later debated in the Inverse Gas Chromatography (IGC) analysis, this may be due to an optimization of the interactions between the chains of the CNF and alginate matrix, with the formation of a network-like structure increasing the TM significantlyto the 10 wt % of CNF incorporation.

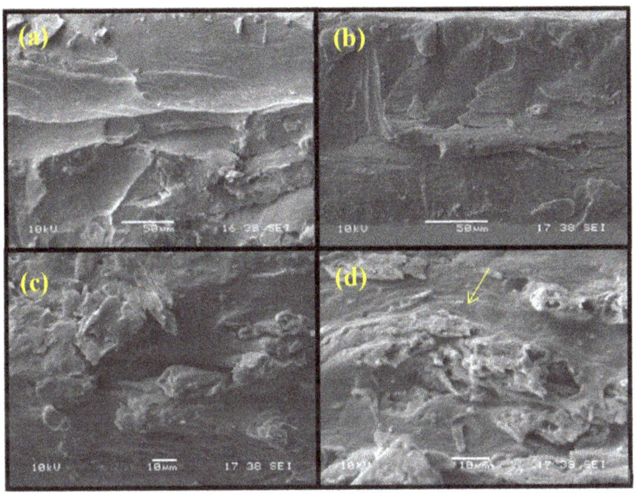

Figure 2. Scanning electron microscopy images of cross-section of the (**a**) alginate; (**b**) alginate with 5 wt % CNF; (**c**) alginate with 10 wt % CNF; and (**d**) alginate with 15 wt % CNF films (arrow indicates the CNF agglomeration in alginate matrix).

2.2. Fourier Transform Infrared Spectroscopy (FTIR) Analysis

Fourier Transform Infrared Spectroscopy (FTIR) analysis of films attempted to characterize the incorporation of CNF into the alginate-based film matrix by distinguishing the infrared (IR) bands and vibration shifts related to the CNF-alginate interactions (Figure 3). The characteristic peak of alginate is located at 1602 cm^{-1}, corresponding to the carbonyl bond (–C=O). The peaks at 1424 and 1024 cm^{-1} in both alginate and cellulose were attributed to carboxyl stretching bands (–COO and –C–O) [13]. In addition, peaks at 3334 and 2930 cm^{-1} were ascribed to the–O–H and –C–H of aliphatic chain stretching vibrations, respectively [13,14]. Some differences can be observed after CNF addition into the alginate matrix. A slight increase of the intensity and width of the overall –O–H band was observed as the CNF was incorporated into the alginate matrix. This can indicate an increase of hydrogen bonding between the alginate and CNF [15]. Overall, FTIR spectra of cellulose-loaded alginate films provided qualitative insights into the effect of cellulose concentration on the position, width and intensity of IR vibrations related to alginate-cellulose interactions, although many bands from the alginate spectra masked typical cellulose vibrations, especially in the 1800–1270 cm^{-1} region (implying bands related to the degree of order of cellulosic materials) [9]. All these changes support the evidence of intermolecular interactions and molecular compatibility between alginate and cellulose components.

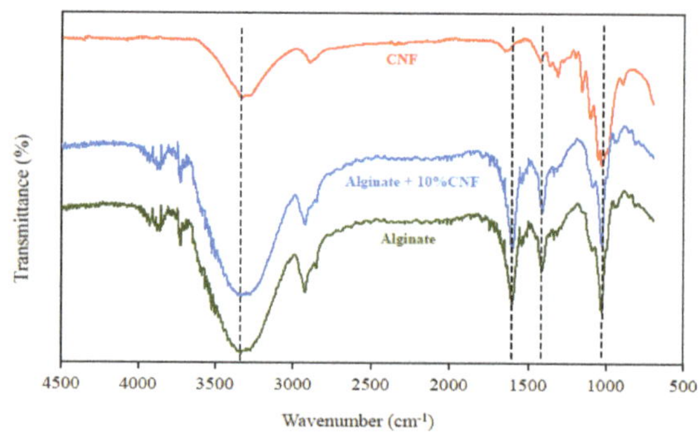

Figure 3. Fourier Transform Infrared (FTIR) spectra of cellulose nanofibrils (CNF), alginate film, alginate with 10 wt % of CNF film.

2.3. Inverse Gas Chromatography (IGC) Analysis

In order to understand the nature of alginate/CNF interactions and their effect on surface properties, the films were analyzed by Inverse Gas Chromatography (IGC). Through non-polar and polar probe molecules, the dispersive component of surface energy (γ_S^d) and the specific component of surface energy (G_S^{sp}) were determined, respectively. Table 1 includes the obtained IGC results. The γ_S^d found shows that the interaction between alginate/CNF makes a slightly less dispersive film surface.

Table 1. Dispersive component of the surface energy (γ_S^d), acid (K_a) and base (K_b) constants and acid/base ratio (K_a/K_b) of the cellulose nanofibrils (CNF) and alginate films at 25 °C.

Sample	γ_S^d (mJ/m^2) *	K_a	K_b	K_a/K_b
CNF	39.79	0.08	0.20	0.4
Alginate	38.66	0.09	0.05	1.8
Alginate with 10 wt % CNF	36.87	0.09	0.04	2.3

* Correlation coefficient between 0.998 and 0.999.

The G_S^{sp} was determined by IGC (Figure 4) using four polar probes: tetrahydrofuran (THF), ethanol (EtOH), ethyl acetate (EtAc) and acetonitrile (AcN). A more significant increase in the G_S^{sp} was observed for the THF probe (base probe) due to CNF incorporation in the alginate matrix. This can indicate a more acidic surface in the alginate/CNF composites.

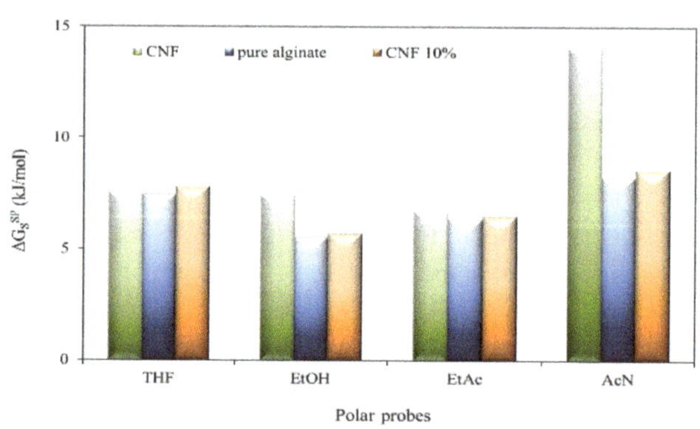

Figure 4. Specific surface free energy (G_S^{sp}) obtained for cellulose nanofibrils (CNF) and alginate films at 25 °C.

The obtained values of G_S^{sp} were converted into acid-base constants (K_a and K_b) and the results (Table 1) lead to the conclusion that the interaction between CNF and alginate takes place preferentially via the hydrogen bond between the oxygen carboxylic (basic group) and the OH from the CNF cellulose. Furthermore, this interaction probably causes a reorganization in the alginate chain and exposes more acid groups (OH) to the surface, as schematized in Figure 5.

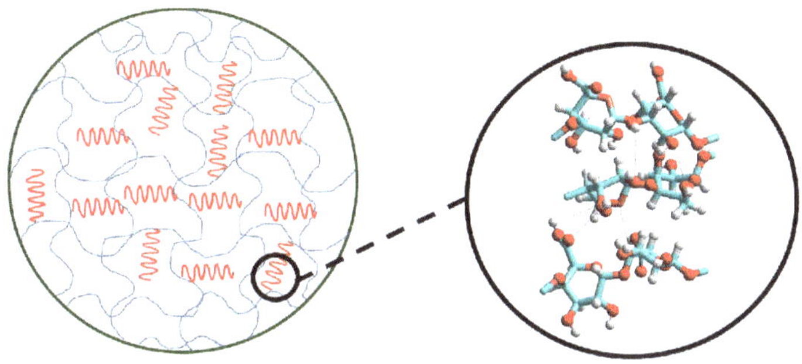

Figure 5. Schematic representation of the interaction between cellulose nanofibrils (CNF) and alginate matrix.

2.4. Water Solubility

Solubility in water is an important property of films for food packaging applications. Some potential uses may require water insolubility to enhance product integrity and water resistance. However, in other cases, water solubility (WS) of the film before product consumption might be useful as in the encapsulation of food or additives.

WS of alginate-based films as a function of CNF addition is given in Figure 6. The addition of CNF significantly reduces ($p < 0.05$) the film water solubility from 8.6% to 1.7% (10 wt % CNF). It is indicative of a strong hydrogen bond formation between the CNF and the film matrix. The hydroxyl groups of CNF can form strong interactions through hydrogen bonds with the hydroxyl and carboxyl groups on alginate and improve the cohesiveness of the biopolymer matrix while decreasing the water sensitivity. These interactions provide stability and resistance to alginate/CNF films. Inother words, water molecules cannot break these strong bonds sufficiently, thus the WS is decreased. Rhim [16] reported that the reduction observed in solubility of films reinforced with different nanofillers is mainly related to the strong hydrogen bond formation between hydroxyl groups of the biopolymer and the nanoparticles. It has also been reported that the high crystallinity of CNF (in this study, 91.3%) and their strong hydrogen-bonded networks within the polymer matrix were responsible for the water resistance improvement [17].

2.5. Moisture Absorption

Water sensitivity of the alginate/CNF films is also evaluated by monitoring the moisture absorption at different time intervals (Figure 6). The experimental results obtained indicate that the incorporation of CNF reduces moisture absorption, demonstrating that water has less affinity with the film. These results corroborate the behavior observed in water solubility tests. CNF presents a stronger interaction with alginate than with water. The number of active sites for water binding decreases due to the formation of strong bonds between the alginate matrix and CNF, reducing the moisture content of alginate/CNF films.

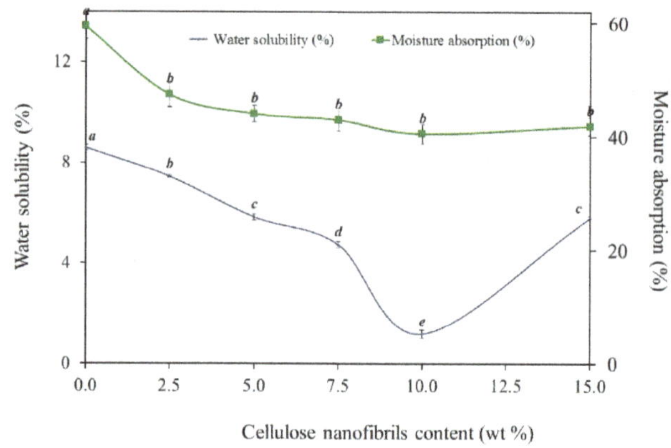

Figure 6. Effect of CNF content on the moisture absorption and water solubility of alginate/CNF films (lower case letters (a, b, c, d, e) show Duncan grouping; distinct letters represent means significantly different ($p < 0.05$)).

2.6. Swelling Behaviour

The swelling behavior of CNF-reinforced alginate-based films is presented in Figure 7. Incorporation of CNF significantly reduced the swelling percentage (SW) of alginate-based films, with the most significant decrease at 10 wt % CNF incorporation, decreasing about 55% after 6 h of immersion in water when compared with pure alginate films. After 1 h, the SW value of the pure alginate films was found to be 124%, whereas due to the incorporation of 10% and 15% CNF, the SW values of the nanocomposite films were found to be 65% and 77%, respectively.

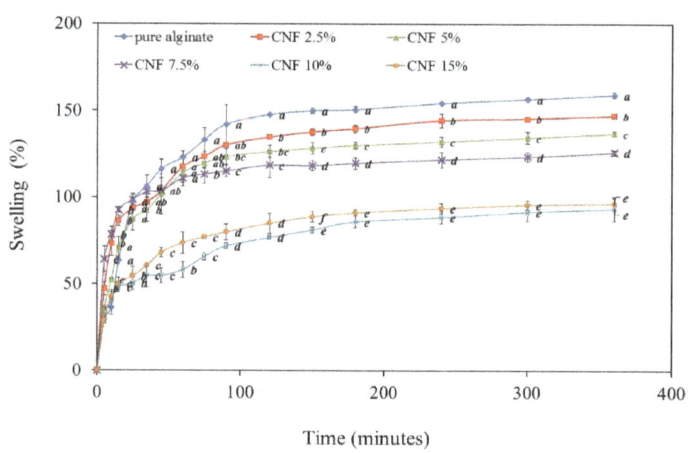

Figure 7. Effect of CNF content on the swelling ratio of alginate/CNFfilms (lower case letters (a, b, c, d, e) show Duncan grouping; distinct letters represent means significantly different ($p < 0.05$)).

2.7. Biodegradability

Biodegradability is another important property for films in food packaging. Biodegradability of pure alginate and CNF-reinforced alginate films were studied. Figure 8 shows the effect of CNF content on the rate of degradation of alginate/CNF films in compost. It is observed that the biodegradability of pure alginate films increased up to 90% as the burial time increased in the compost for five weeks. However, in the case of CNF-reinforced alginate films, the rate of degradation changes with varying amounts of CNF added to the alginate matrix.

When CNF is added to the alginate matrix, there arises a strong interaction between the matrix and the filler due to the homogeneous dispersion of CNF particles in the alginate matrix. Thus, the degradation of films becomes difficult due to difficulty in breaking the strong bonds between the alginate matrix and nanocellulose. Therefore, it can be concluded from the results that the CNF incorporation in alginate matrix decreases the degradation percentage of films up to 10 wt % of CNF. However, at higher concentrations of CNF (15 wt % of CNF), a reverse effect is observed due to agglomeration of CNF particles in the alginate matrix and the decrease in the alginate/CNF interaction.

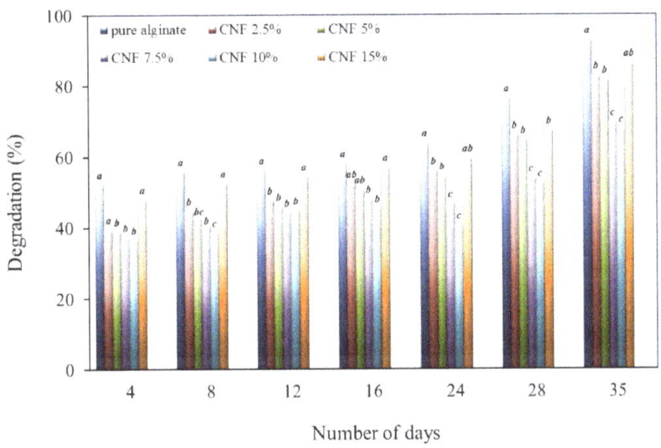

Figure 8. Effect of CNF content on the rate of degradation of alginate/CNF films (lower case letters (a, b, c) show Duncan grouping; distinct letters represent means significantly different ($p < 0.05$)).

3. Experimental Section

3.1. Materials

Sodium alginate used in this study was obtained from LobaChemie Pvt. Ltd. (Mumbai, India). Glycerol, sodium hydroxide, acetic acid, sodium chlorite and oxalic acid were purchased from Merck Pvt. Ltd. (Mumbai, India). Cellulose nanofibrils (CNF) used were isolated from sisal fibers by steam explosion process (percentage yield of 39%) as described and characterized in Deepa *et al.* (2015) [18]. Atomic force microscopy images (AFM) revealed a high value (5.3 nm) of the surface roughness, which predict a good interfacial adhesion between the CNF and the matrix in the composites preparation. The diameters of the nanocellulose structures were found to be between 20–80 nm with an average width of 13 ± 4 nm. Transmission electron microscopy (TEM) images showed that CNF were long, flexible, entangled, with several micrometers in length. A very high aspect ratio (length/diameter) was found as important parameter to obtain composites with good mechanical properties. By X-ray Diffraction (XRD) analysis, the crystallinity index and the crystal size of CNF were calculated to be 81.6% and 3.7 nm, respectively.

3.2. Films Preparation

A solution of sodium alginate was prepared by dissolving the sodium alginate (3% w/w) in distilled water at 60–70 °C under stirring for 30 min. After complete dissolution, 0.5 g glycerol/1 g alginate was added as a plasticizer to enhance film flexibility, decrease brittleness and facilitate their detachment from the petri dishes. The film-forming solution was cast onto petri dishes which were 14cm in diameter and oven dried at 40 °C for approximately 24 h. The dried films were then removed from the petri dishes and stored in polyethylene bags prior to characterization.

In order to prepare the nanocomposite films, different amounts of the cellulose nanofibrils (CNF), 2.5%, 5%, 7.5%, 10% and 15% (w/w) on solid sodium alginate, were added to 30 mL of deionized water and dispersed/homogenized using a homogenizer at 10,000 rpm for 30 min at room temperature. Afterwards, the CNF solution was added slowly to the alginate solutions (made as described above). The resulting mixture was stirred for 1 h. Finally, the films were prepared and stored, as described previously.

3.3. Mechanical Properties

Mechanical properties of the film samples including tensile strength (TS) and tensile modulus (TM) were measured according to standard method ASTM D882-02 [19] using an Instron Universal Testing Machine equipped with a 100 N-load cell. Prior to the test, the film samples were cut into rectangular strips (15 mm × 70 mm × 0.2 mm). The initial grip separation was set at 50 mm and crosshead speed at 50 mm/min. An average value was taken from the measurements of five films under the same conditions for each specimen.

3.4. Scanning Electron Microscopy (SEM)

Scanning electron microscopy (SEM) images of film samples were obtained using a Scanning Electron Microscope model JEOL JSM-6390LV (JEOL Ltd., Tokyo, Japan) with an accelerating voltage of 10–20 kV. Prior to SEM examination, the samples were oven dried at 60 °C for 8 h and then a fine layer of gold was deposited on samples by means of a plasma sputtering apparatus.

3.5. Fourier Transform Infrared Spectroscopy (FTIR)

Infrared spectra (4000–640 cm^{-1}) of the film samples were recorded on a Fourier Transform Infrared Spectrometer Perkin-Elmer (PC1600, Perkin-Elmer, Waltham, MA, USA) equipped with an Attenuated Total Reflectance (ATR) device with a spectral resolution of 4 cm^{-1} and after 32 scans.

3.6. Inverse Gas Chromatography (IGC)

Surface properties were analyzed using a commercial inverse gas chromatography (Surface Measurements Systems, London, UK). The measurements and theory applied were described exhaustively by Cordeiro [20]. In this study, all the measurements were carried out at 0% relative humidity (RH), a helium flow rate of 10 mL/min and in duplicate producing an experimental variation below 4%. Methane was used as reference molecule. Heptane, octane, nonane and decane were the four *n*-alkanes used for determination of the dispersive component of the surface free energy at 25 °C. Acetonitrile (ACN), ethyl acetate (EtOAc), ethanol (EtOH) and tetrahydrofuran (THF) were used for determination of the Gibbs specific free energy (G_S^{SP}) and acid-base surface character (K_A and K_B) at 25 °C.

3.7. Water Solubility

The solubility of the films in water was determined as percentage of film dry matter solubilized in water during a period of 24 h was described by Gontard [21]. The oven-dried film samples were cut (20 mm × 20 mm × 0.2 mm), weighed, immersed into 20 mL distilled water and slowly agitated. The amount of dry matter in the initial (W_i) and final (W_f) samples was determined by drying the samples at 105 °C. The film water solubility (WS) was calculated using the Equation (1):

$$WS\,(\%) = \frac{W_i - W_f}{W_i} \times 100 \quad (1)$$

3.8. Moisture Absorption

The moisture absorption of the films was determined after drying in an oven at 105 °C for 24 h, according to ASTM D5229 standard methodology [19]. Briefly, the dry films were conditioned at 25 °C in desiccators containing oversaturated solutions of sodium sulfate in order to ensure a relative humidity (RH) of 95%. The samples were removed at specific time and the rate of moisture absorption (MA) was calculated by the Equation (2):

$$MA\,(\%) = \frac{W_t - W_i}{W_i} \times 100 \tag{2}$$

where W_t and W_i represents the film weight at time t and before exposure to 95% RH, respectively.

3.9. Swelling Behavior

The swelling behavior of the films was described by the swelling ratio (SW) determined according to standard method ASTM D2765-95C [19]. The swelling ratio (SW) of the films was calculated using the Equation (3):

$$SW\,(\%) = \frac{W_s - W_d}{W_d} \times 100 \tag{3}$$

where W_s is the samples weight after swelling, and W_d is the weight of dry sample.

3.10. Biodegradation Analysis

Soil burial tests for biodegradation of the film samples were carried out by a method described by Martucci [22]. The samples were cut into rectangular pieces (20 mm × 20 mm), dried in an oven at 105 °C for 24 h and weighed (W_i). The films were then buried under compost in plastic boxes (120 mm × 60 mm × 1.5 mm) at a depth of 8 cm from the soil surface in order to ensure aerobic degradation conditions. The assay was performed at 32 °C and 35% relative humidity (RH) by adding water periodically. Fluctuations in soil moisture were followed gravimetrically using the standard method of oven drying (ASTM D2216) [19]. Samples were taken from the soil at different times and cleaned by wiping gently with a tissue paper. Subsequently, they were dried in an oven at 105 °C for 12 h and weighed (W_t) to determine the % of Weight Loss (WL) using the Equation (4):

$$WL\,(\%) = \frac{W_i - W_t}{W_i} \times 100 \tag{4}$$

3.11. Statistical Analysis

All experiments were carried at least in duplicate. Analysis of variance (ANOVA) in SPSS statistics software (Version 23; SPSS Inc., Chicago, IL, USA) followed by Duncan's test were used for statistical analysis of data. p-values < 0.05 were considered statistically significant.

4. Conclusions

Cellulose nanofibrils (CNF) isolated from sisal fiber by steam explosion coupled with the acid hydrolysis process was used to prepare nanocomposite films from alginate biopolymer. The similar chemical structures of the alginate matrix and the CNF led to strong adhesion between them through hydrogen bonding. In general, the results indicate that the biofilm alginate with 10% CNF presents the best properties. The incorporation of CNF into the alginate matrix improved its water resistance and mechanical properties. Swelling and water solubility measurements indicated that CNF-reinforced alginate films have less affinity towards water molecules. The biodegradability of the films investigated by the soil burial method demonstrated that the biodegradation time of alginate/CNF films was longer than that of pure alginate films. These nanocomposite films present excellent potential as a new biomaterial for application in food packaging and conservation.

Acknowledgments: Laly A. Pothan is grateful to the University Grants Commission (UGC), Government of India, for the Emeritus fellowship and B. Deepa is grateful to the Council of Scientific & Industrial Research (CSIR), New Delhi, India, for the CSIR-Senior Research Fellowship.

Author Contributions: B. Deepa and Laly A. Pothan conceived and designed the experiments; B. Deepa prepared the samples and performed the experiments; Nereida Cordeiro and Marisa Faria performed IGC analysis; B. Deepa,

Eldho Abraham, Laly A. Pothan, Nereida Cordeiro, Marisa Faria and Sabu Thomas worked together on the development of the ideas presented in this paper, and contributed to the data analysis and manuscript writing.

Conflicts of Interest: The authors declare no conflict of interest.

References

1. Klemm, D.; Kramer, F.; Moritz, S.; Lindstrom, T.; Ankerfors, M.; Gray, D.; Dorris, A. Nanocelluloses: A new family of naturebased materials. *Angew. Chem. Int. Ed.* **2011**, *50*, 5438–5466. [CrossRef] [PubMed]
2. Abdul Khalil, H.P.S.; Davoudpour, Y.; Islam, M.N.; Mustapha, A.; Sudesh, K.; Dungani, R.; Jawaid, M. Production and modification of nanofibrillated cellulose using various mechanical processes: A rewiew. *Carbohydr. Polym.* **2014**, *99*, 649–665. [CrossRef] [PubMed]
3. Sacui, I.A.; Nieuwendaal, R.C.; Burnett, D.J.; Stranick, S.J.; Jorfi, M.; Weder, C.; Foster, E.J.; Olsson, R.T.; Gilman, J.W. Comparison of the properties of cellulose nanocrystals and cellulose nanofibrils isolated from bacteria, tunicate, and wood processed using acid, enzymatic, mechanical, and oxidative methods. *ACS Appl. Mater. Interfaces* **2014**, *6*, 6127–6138. [CrossRef] [PubMed]
4. Habibi, Y.; Lucia, L.A.; Rojas, O.J. Cellulose nanocrystals: Chemistry, self-assembly, and applications. *Chem. Rev.* **2010**, *110*, 3479–3500. [CrossRef] [PubMed]
5. Eichhorn, S.J.; Dufresne, A.; Aranguren, M.; Marcovich, N.E.; Capadona, J.R.; Rowan, S.J.; Weder, C.; Thielemans, W.; Roman, M.; Renneckar, S. Review: Current international research into cellulose nanofibresand nanocomposites. *J. Mater. Sci.* **2010**, *45*, 1–33. [CrossRef]
6. Azzam, F.; Moreau, C.; Cousin, F.; Menelle, A.; Bizot, H.; Cathala, B. Cellulose nanofibril-based multilayered thin films: Effect of ionic strength on porosity, swelling, and optical properties. *Langmuir* **2014**, *30*, 8091–8100. [CrossRef] [PubMed]
7. Norajit, K.; Kim, K.M.; Ryu, G.H. Comparative studies on the characterization and antioxidant properties of biodegradable alginate films containing ginseng extract. *J. Food Eng.* **2010**, *98*, 377–384. [CrossRef]
8. Abdollahi, M.; Alboofetileh, M.; Rezaei, M.; Behrooz, R. Comparing physico-mechanical and thermal properties of alginate nanocomposite films reinforced with organic and/or inorganic nanofillers. *Food Hydrocoll.* **2013**, *32*, 416–424. [CrossRef]
9. Huq, T.; Salmieri, S.; Khan, A.; Khan, R.A.; Le Tien, C.; Riedl, B.; Fraschini, C.; Bouchard, J.; Uribe-Calderon, J.; Kamal, M.R.; *et al.* Nanocrystalline cellulose (NCC) reinforced alginate based biodegradable nanocomposite film. *Carbohydr. Polym.* **2012**, *90*, 1757–1763. [CrossRef] [PubMed]
10. Abdollahi, M.; Alboofetileh, M.; Behrooz, R.; Rezaei, M.; Miraki, R. Reducing water sensitivity of alginate bio-nanocomposite film using cellulose nanoparticles. *Int. J. Biol. Macromol.* **2013**, *54*, 166–173. [CrossRef] [PubMed]
11. Rhim, J.W. Physical and mechanical properties of water resistant sodium alginate films. *LWT Food Sci. Technol.* **2004**, *37*, 323–330. [CrossRef]
12. Cho, M.J.; Park, B.D. Tensile and thermal properties of nanocellulose-reinforced poly (vinyl alcohol) nanocomposites. *J. Ind. Eng. Chem.* **2011**, *17*, 36–40. [CrossRef]
13. Lin, N.; Bruzzese, C.; Dufresne, A. Tempo-oxidized nanocellulose participating as crosslinking aid for alginate-based sponges. *ACS Appl. Mater. Inter.* **2012**, *4*, 4948–4959. [CrossRef] [PubMed]
14. Han, J.; Guenier, A.S.; Salmieri, S.; Lacroix, M. Alginate and chitosan functionalization for micronutrient encapsulation. *J. Agric. Food Chem.* **2008**, *56*, 2528–2535. [CrossRef] [PubMed]
15. Khan, R.A.; Salmieri, S.; Dussault, D.; Uribe-Calderon, J.; Kamal, M.R.; Safrany, A.; Lacroix, M. Production and properties of nanocellulose-reinforced methylcellulose-based biodegradable films. *J. Agric. Food Chem.* **2010**, *58*, 7878–7885. [CrossRef] [PubMed]
16. Rhim, J.W.; Ng, P.K.W. Natural biopolymer-based nanocomposite films for packaging applications. *Crit. Rev. Food Sci.* **2007**, *47*, 411–433. [CrossRef] [PubMed]
17. De Paula, E.L.; Mano, V.; Pereira, F.V. Influence of cellulose nanowhiskers on the hydrolytic degradation behavior of poly (d, l-lactide). *Polym. Degrad. Stabil.* **2011**, *96*, 1631–1638. [CrossRef]
18. Deepa, B.; Abraham, E.; Cordeiro, N.; Mozetic, M.; Mathew, A.P.; Oksman, K.; Faria, M.; Thomas, S.; Pothan, L.A. Utilization of various lignocellulosic biomass for the production of nanocellulose: A comparative study. *Cellulose* **2015**, *22*, 1075–1090. [CrossRef]

19. *ASHRAE Handbook, Fundamentals*; American Society of Heating and Refrigerating and Air conditioning Engineers, Inc.: New York, NY, USA, 1997.
20. Cordeiro, N.; Gouveia, C.; Moraes, A.G.O.; Amico, S.C. Natural fibres characterization by inverse gas chromatography. *Carbohydr. Polym.* **2011**, *84*, 110–117. [CrossRef]
21. Gontard, N.; Guilbert, S.; Cuq, J.L. Edible wheat gluten films: Influence of the main process variables on film properties using response surface methodology. *J. Food Sci.* **1992**, *57*, 190–195. [CrossRef]
22. Martucci, J.F.; Ruseckaite, R.A. Biodegradation of three-layer laminate films based on gelatin under indoor soil conditions. *Polym. Degrad. Stabil.* **2009**, *94*, 1307–1313. [CrossRef]

17

Influence of Different Post-Plasma Treatment Storage Conditions on the Shear Bond Strength of Veneering Porcelain to Zirconia

Mun-Hwan Lee [1,†], Bong Ki Min [2,†], Jun Sik Son [3] and Tae-Yub Kwon [4,*]

Academic Editor: Jonathan Phillips

1 Department of Medical & Biological Engineering, Graduate School, Kyungpook National University, 2-188-1 Samduk-dong, Jung-gu, Daegu 700-412, Korea; leemunhwan@knu.ac.kr
2 Center for Research Facilities, Yeungnam University, 214-1 Dae-dong, Gyeongsan 712-749, Korea; bkmin@ynu.ac.kr
3 Korea Textile Development Institute, 1083 Jungri-dong, Seo-gu, Daegu 703-712, Korea; sonjk1@empas.com
4 Department of Dental Biomaterials, School of Dentistry, Kyungpook National University, 2-188-1 Samduk-dong, Jung-gu, Daegu 700-412, Korea
* Correspondence: tykwon@knu.ac.kr
† These authors contributed equally to this work.

Abstract: This *in vitro* study investigated whether different storage conditions of plasma-treated zirconia specimens affect the shear bond strength of veneering porcelain. Zirconia plates were treated with a non-thermal atmospheric argon plasma (200 W, 600 s). Porcelain veneering (2.38 mm in diameter) was performed immediately (P-I) or after 24 h storage in water (P-W) or air (P-A) on the treated surfaces ($n = 10$). Untreated plates were used as the control. Each group was further divided into two subgroups according to the application of a ceramic liner. All veneered specimens underwent a shear bond strength (SBS) test. In the X-ray photoelectron spectroscopy (XPS) analysis, the oxygen/carbon ratios of the plasma-treated groups increased in comparison with those of the control group. When a liner was not used, the three plasma-treated groups showed significantly higher SBS values than the control group ($p < 0.001$), although group P-A exhibited a significantly lower value than the other two groups ($p < 0.05$). The liner application negatively affected bonding in groups P-I and P-W ($p < 0.05$). When the veneering step was delayed after plasma treatment of zirconia, storage of the specimens in water was effective in maintaining the cleaned surfaces for optimal bonding with the veneering porcelain.

Keywords: zirconia; veneering porcelain; bond strength; plasma treatment; storage condition

1. Introduction

Recently, zirconia, specifically yttria-stabilized tetragonal zirconia polycrystal (Y-TZP), has become favored as a core material for crowns and fixed dental prostheses (FDPs) mainly due to its improved fracture resistance by transformation toughening [1,2]. The zirconia-based core structure is veneered with zirconia veneering ceramic, which generally consists of an amorphous and glassy silica matrix embedded in varying amounts of feldspar and leucite crystals [3].

In zirconia-based restorations, however, chipping and/or delamination of veneering ceramic has been reported at higher rates compared with metal-ceramic systems [4,5]. Ceramic chipping caused by an inherent weakness of the veneering porcelain is one of the most common failures of zirconia-based restorations [6–8]. Another clinical problem is delamination at the zirconia-veneer interface, probably caused by the weak interfacial adhesion and large difference in mechanical properties of the zirconia and veneer materials [7,9]. Thus, metal frameworks veneered with ceramics still represent the standard for ceramic FDPs [5]. Enhancing bonding between zirconia and veneering porcelain is required to ensure the long-term clinical success of zirconia-based restorations.

Various surface conditioning methods of the zirconia substructure have been researched to achieve enhanced zirconia/porcelain bonding. Among these methods, airborne-particle abrasion and ceramic liner application are the most commonly studied [7]. Air-abrasion of zirconia with alumina particles may increase the surface roughness and provide undercuts for bonding with veneering porcelain [5]. Some manufacturers recommend the application of a liner on the zirconia core surface to improve bonding and to mask the opaqueness of the zirconia [7]. However, the effect of the liner on the bond strength of the zirconia-veneer interface remains controversial. Kim *et al.* [5] suggested that the application of a liner increased the possibility of interfacial failure of the veneering ceramic from the zirconia core and that airborne-particle abrasion would be more useful than a liner application. Moreover, Wang *et al.* [7] showed that that the application of a liner before veneering reduces the bond strength of the interface.

The chemical composition of commercial ceramic liners varies depending on the manufacturer; however, the primary component is silica, indicating a similar composition to veneering porcelain [5,7]. Thus, the properties of a ceramic liner resemble those of veneering porcelain, and a stronger bond is expected at the porcelain/liner interface than at the liner/zirconia interface [7]. The wetting behavior is of primary importance for good adhesion of veneering porcelain or ceramic liner onto the zirconia substructure [10]. The wetting behavior is dependent not only on the porcelain or liner composition but also on the morphology and surface energy of the zirconia material [11].

Non-thermal plasmas are partially ionized "cold" gases that contain highly reactive particles [12]. Due to the relatively mild, non-destructive character of the gas phase, non-thermal plasma surface treatment effectively alters the surface characteristics, leaving the bulk properties of the materials un-altered after treatment [13]. Plasma surface treatment is generally used for cleaning and surface activation, which raise the "wettability" and surface energy of the substrate and lead to an improvement of the bonding properties [14,15]. A recent study [4] showed that argon plasma treatment improved bonding between zirconia and veneering porcelain. Although such plasma cleaning may be simple and useful for enhancing zirconia-veneer bonding, plasma-cleaned zirconia surfaces may be easily re-contaminated once exposed to air [16]. In a dental laboratory, the porcelain veneering step may be delayed for various reasons, allowing re-contamination, which may affect the bonding of veneered porcelain to the plasma-treated zirconia surfaces. For the practical application of this technology, the surface properties of plasma-treated substrates during storage should be maintained [17].

The purpose of this *in vitro* study was to investigate the influence of different post-plasma treatment storage conditions on the shear bond strength (SBS) between zirconia and veneered porcelain. To investigate the combined effect of the plasma and liner, specimens coated with a liner prior to plasma treatment were also prepared (Figure 1). The null hypothesis tested was that different storage conditions after plasma treatment do not affect the SBS between zirconia and veneered porcelain.

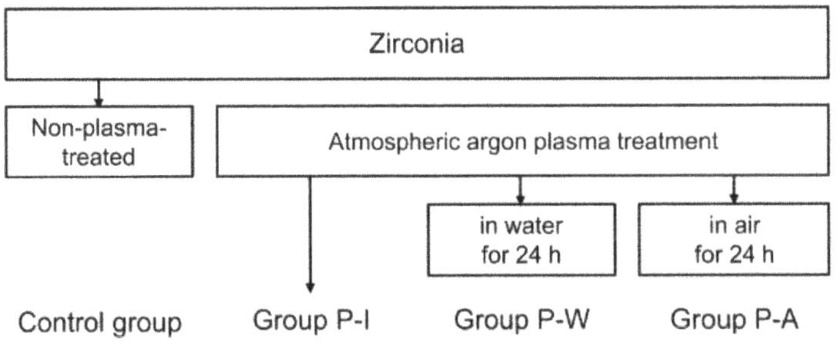

Figure 1. Study design for one untreated (control) and three plasma-treated groups. In the shear bond strength test, each group was further divided into two subgroups according to the use of a liner prior to veneering.

2. Results

2.1. Scanning Electron Microscopy and Atomic Force Microscopy

Representative scanning electron microscopy (SEM) and atomic force microscopy (AFM) images before and after the plasma treatment are shown in Figure 2. No morphological changes were apparent in the images following treatment. In addition, there was no significant difference in average surface roughness (R_a) between the untreated and the plasma-treated zirconia surfaces ($p = 0.828$).

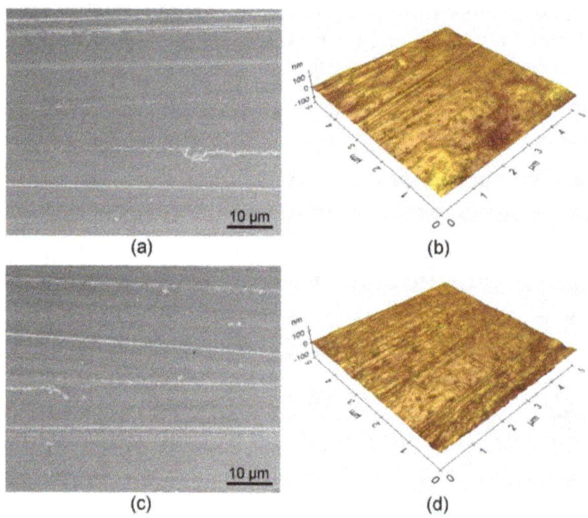

Figure 2. Scanning electron microscopy (SEM) (left, magnification: 2000×, bar = 10 μm) and atomic force microscopy (AFM) (right, 5 μm × 5 μm) images of the zirconia surfaces: (**a**,**b**) untreated; (**c**,**d**) plasma-treated.

2.2. Water Contact Angles

Figure 3 shows the water contact angle (CA) values for the four experimental groups. The control (untreated) group showed the highest CA among the groups, the difference being statistically significant ($p < 0.001$). The plasma treatment significantly lowered the CA values in comparison with the control group, although group P-A showed a significantly higher value than groups P-I and P-W ($p < 0.001$).

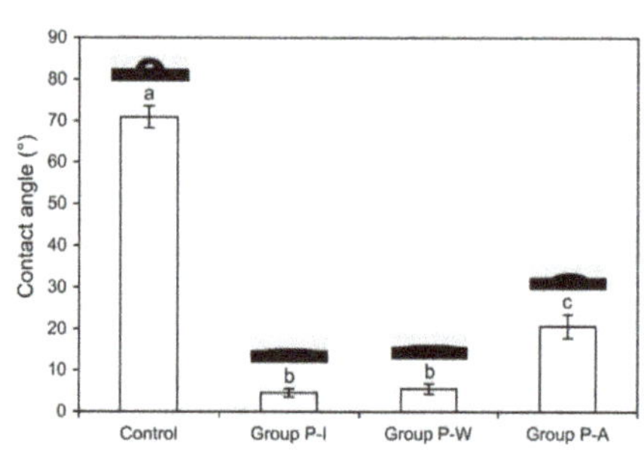

Figure 3. Contact angles of water droplets on the zirconia surfaces ($n = 6$). Different lowercase letters indicate significant differences ($p < 0.05$).

2.3. X-ray Photoelectron Spectroscopy

The results of the X-ray photoelectron spectroscopy (XPS) analysis are shown in Figures 4 and 5 and Table 1. The oxygen/carbon (O/C) ratios of the plasma-treated groups were considerably increased compared to that of the control group due to an increase in oxygen content (at %) and a decrease in carbon content. When the O 1s peak was deconvoluted into three components (ZrO_2 and acidic/basic hydroxyl groups), an increased oxygen content in the plasma-treated groups was found. The amount of basic hydroxyl groups was substantially decreased in the plasma-treated groups when compared to the control group, being lower in groups P-I and P-W than in group P-A.

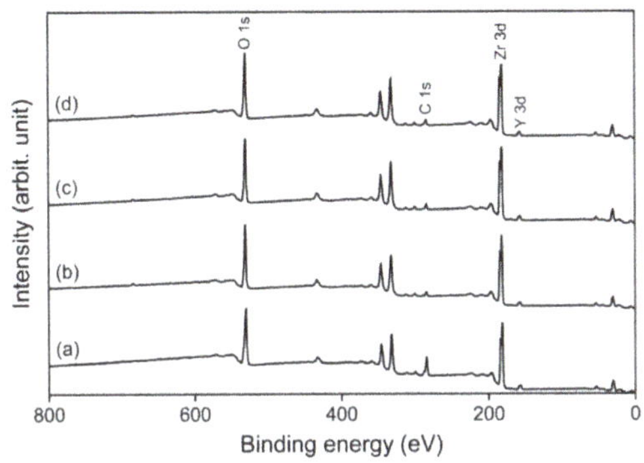

Figure 4. Survey X-ray photoelectron spectroscopy (XPS) spectra for untreated (**a**) and plasma-treated (**b–d**) zirconia surfaces: groups P-I (b); P-W (c); and P-A (d).

Table 1. Atomic percent of each element and oxygen/carbon (O/C) ratio for each group, calculated from the survey X-ray photoelectron spectroscopy (XPS) spectra (Figure 4).

Group	Zr 3d	O 1s	Y 3d	C 1s	O/C
Control	15.05	48.32	1.32	35.31	1.37
P-I	20.73	64.93	1.71	12.63	5.14
P-W	22.31	62.04	1.66	13.99	4.43
P-A	22.11	61.23	1.65	15.01	4.08

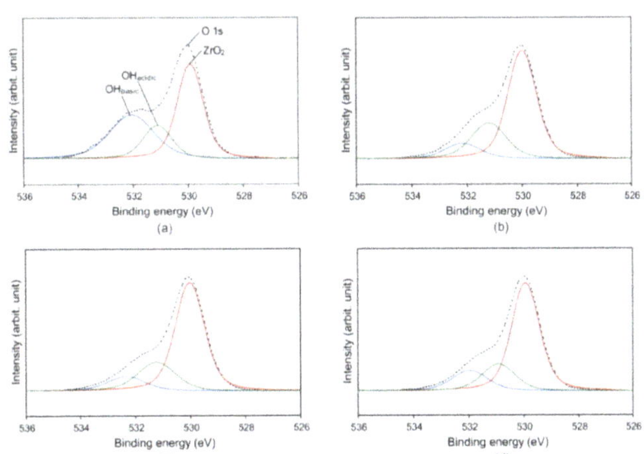

Figure 5. High-resolution X-ray photoelectron spectroscopy (XPS) spectra of O 1s region of untreated (**a**) and plasma-treated (**b–d**) zirconia surfaces: groups P-I (b); P-W (c); and P-A (d). Deconvolution of O 1s high resolution spectrum resulted in three components: basic hydroxyl groups at 532.6 eV, acidic hydroxyl groups at 531.8 eV, and ZrO_2 at 530.2 eV.

2.4. Shear Bond Strength and Failure Pattern

Table 2 shows the shear bond strength (SBS) value for each experimental group. Two-way analysis of variance (ANOVA) showed statistically significant differences based on the storage condition (including control) and for the interaction between the two main effects ($p < 0.001$). Within the no-liner subgroups, the lowest SBS was obtained in the control group, the difference being statistically significant ($p < 0.001$). Of the plasma-treated groups, groups P-I and P-W, which were statistically similar to each other ($p = 0.311$), exhibited significantly higher SBS values than group P-A ($p < 0.05$). On the other hand, there were no significant differences in the SBS values among the four liner subgroups ($p = 0.742$). Within the control (untreated) group, liner application significantly increased the SBS ($p < 0.001$). In groups P-I and P-W, the no-liner subgroups showed significantly higher SBS values than the liner subgroups ($p < 0.05$). In group P-A, on the other hand, there was no significant difference in the value between the two subgroups ($p = 0.461$).

Table 2. Mean shear bond strength (MPa) (standard deviation) for the experimental groups ($n = 10$).

Group	No-Liner Subgroups	Liner Subgroups
Control	17.65 (2.98) Aa	26.99 (2.93) b
P-I	32.59 (3.58) Ba	27.80 (3.00) b
P-W	30.42 (2.27) Ba	27.84 (2.51) b
P-A	26.07 (1.93) Ca	26.78 (2.25) a

Within the no-liner subgroups, the same uppercase letter indicates lack of statistically significant difference among the groups ($p > 0.05$). Within the liner subgroups, one-way ANOVA revealed no significant differences in the values among the four groups ($p = 0.742$). For each row (each group), the same lowercase letter indicates lack of statistically significant difference between the two subgroups ($p > 0.05$).

After SBS testing, all the veneered zirconia specimens exhibited a combination of cohesive and adhesive failure modes (*i.e.*, a mixed failure) (Figure 6). The fractures were found to start from the veneering porcelain and to propagate along the zirconia-porcelain interface. Within the no-liner subgroups, the control group showed primarily adhesive failure, whereas the three plasma-treated groups exhibited an increased area of cohesive failure (Figure 6a). In the four liner-applied groups, complete delamination of the porcelain or liner was frequently observed (Figure 6b).

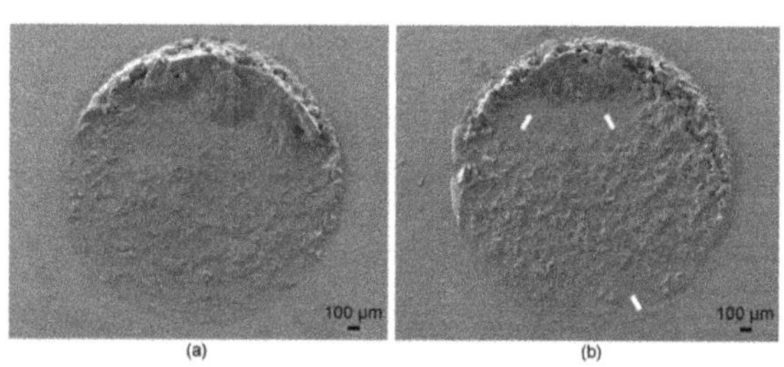

Figure 6. Representative SEM images of fractured zirconia sides (magnification: 37×) in the plasma-treated no-liner subgroups (**a**) and the liner subgroups (**b**). In Figure 6b, the white arrows represent complete delamination of porcelain or liner.

3. Discussion

Zirconia surface is rather hydrophobic and has very low surface concentrations of –OH groups [18,19]. It has been reported that non-thermal atmospheric argon plasma treatment of any dental substrate can clean the surface and produce a super-hydrophilic surface (contact angle < 10°) [13,20]. In this *in vitro* study, the results of the surface characterization (Figures 2–5 Table 1) indicate that the argon plasma treatment (group P-I) effectively cleaned the zirconia surfaces by reducing carbon-based contaminants and eventually created super-hydrophilic surfaces. Thus, the argon plasma treatment (group P-I) resulted in increased SBS between the zirconia and veneered porcelain in comparison with the control (untreated) group (Table 2), in accordance with the study by Canullo *et al.* [4]. Durability of surface properties after plasma treatment is also an important consideration in the practical application of this technology in a clinical setting [17]. When the veneering step was delayed for 24 h after plasma treatment, the water-storage condition (group P-W) maintained the properties of the cleaned, super-hydrophilic zirconia surfaces and did not significantly decrease the SBS. On the other hand, zirconia surfaces re-contaminated by being left in air (group P-A) underwent a significant decrease in SBS, although the value was still significantly higher than that of the control group. Thus, the null hypothesis that different storage conditions after plasma treatment do not affect the SBS between zirconia and veneered porcelain was rejected. In addition, liner application prior to veneering negatively affected the SBS values in groups P-I and P-W.

Previous studies showed that plasma surface treatment may alter the surface morphology [21,22]. In this study, a relatively long treatment time (600 s) and high power input (200 W) were employed during the plasma treatment. Nonetheless, the SEM and AFM examinations (Figure 2) clearly show that the plasma treatment used in this study did not induce much etching or destruction of the zirconia surfaces. The R_a values obtained in the AFM images also indicate no significant changes in surface topography. This finding may indicate great resistance of zirconia surface to plasma etching [23]. In addition, excited argon neutrals may have been less energetic than the high-energy electrons and ions, whose attack may cause surface damage [13].

Wettability and surface free energy of a surface are affected by both surface topography and physiochemistry [17]. Water CA reflects the wetting of the substrate by water and provides valuable information on the chemical composition of the substrate [13]. Despite only small changes in the surface morphology before and after the plasma treatment (Figure 2), the CA of zirconia (control: $71.0^\circ \pm 2.6^\circ$) was significantly reduced to $4.6^\circ \pm 1.0^\circ$ in group P-I (Figure 3), which is close to a value for a super-hydrophilic surface (CA $< 10^\circ$) [13,20]. A high CA indicates hydrophobic surface properties, and a dramatic drop in the CA may indicate the grafting of new polar functionalities onto the surface in addition to changes in chemical composition [13]. Thus, the reason for capability of the argon plasma treatment to reduce water CAs was examined using the XPS.

In the XPS analysis (Figures 4 and 5 Table 1), decreased carbon (C) content was noted in the plasma-treated groups compared to the control, suggesting that organic matter was removed via breaking of C–C and C–H bonds [13]. In contrast, the atomic percent of oxygen (O) increased after the plasma treatment. In general, plasma treatment generates reactive species through bond-splitting induced by the vigorous atom/ion bombardment [13]. These reactive species can then bind to other atoms or functional groups. Non-thermal plasma is driven by argon gas, the oxygen from ambient air being excited to form highly reactive oxygen species, then to react with the surfaces [13,24]. The increase in oxygen-containing polar moieties on the surface enhances surface hydrophilicity [11]. A high O/C ratio indicates good wettability of the surface due to an increase in oxygen-containing groups [13,21,25]. The XPS analysis showed that the O/C ratios increased noticeably after plasma treatment (Table 1). This finding suggests the reason for the dramatic drop in the zirconia CA after plasma treatment (Figure 3). Thus, although the argon plasma treatment did not induce any new reactive functionality on the zirconia surface, it seems that treatment with inert gas effectively removed the organic contaminant on the surface through ion bombardment [26].

Maintaining plasma-induced surface properties is another consideration of practical importance. Although the veneering step was delayed for 24 h, simple immersion of the specimens in distilled water (group P-W) effectively maintained the super-hydrophilicity induced by plasma (CA: $5.6^\circ \pm 1.2^\circ$) (Figure 3). In addition, group P-W showed only a slight decrease in the O content and an increase in the C content, resulting in a decrease in the O/C ratio compared with group P-I (Table 1), suggesting that such water storage of plasma-treated specimens minimized surface re-contamination [17]. For the specimens left in air for 24 h (group P-A), in contrast, substantial "hydrophobic" recovery (CA: $20.7^\circ \pm 2.9^\circ$) and a much lower O/C ratio than those of groups P-I and P-W were observed, probably due to significant surface re-contamination with moisture, contaminants, or carbon dioxide in ambient air. These findings indicate that appropriate storage conditions that can minimize re-contamination are important in maintaining plasma-induced surface properties.

A more detailed study of the O 1s high resolution spectrum was performed. The O 1s peak was deconvoluted into three components, as shown in Figure 5. The first component at binding energy 530.2 eV can be assigned to ZrO_2 (Zr–O–Zr), and the middle component at 531.8 eV to OH_{acidic} (acidic hydroxyl groups, Zr–O–H) [17,27]. The third component at 532.6 eV is assigned to OH_{basic} (basic hydroxyl groups, H–O–H) [17,27]. Among these, the third component can be ascribed to oxygen-containing impurities, such as water molecules, adsorbed on the surface [27,28]. The percentage area of the third component was greatest in the control group (36.07%) (Figure 5), indicating that cleaning only by ultrasonication (control) did not effectively remove the oxygen-containing contaminants from the specimens. As seen in the significant decrease in the third component in group P-I, oxygen-containing impurities were effectively removed from the zirconia surface by plasma cleaning. The third component of group P-W, similar to that of group P-I, suggests that most of the loosely bound water molecules were simply removed by a strong air blow after removing the specimen from the immersion water. The increased third component in group P-A compared to groups P-I and P-W indicates slight contamination (in particular the adsorption of water) in ambient air.

The SBS values for each group (Table 2) show that the bonding between zirconia and veneered porcelain was closely related to the surface properties of the zirconia specimens. Groups P-I and P-W showed superior zirconia-veneer bonding performance in comparison with the other groups, probably because the veneering porcelain wettability was effectively promoted on the clean, super-hydrophilic zirconia surfaces (Figure 3). In the failure pattern analysis, moreover, the control no-liner group showed primarily adhesive failure, whereas the three plasma-treated no-liner groups exhibited an increased area of cohesive failure (Figure 6a). However, a combination of plasma cleaning and liner application prior to veneering showed no positively synergistic effect on the bonding. The liner application was only positively effective for the untreated zirconia surfaces (control). When the liner was used after plasma cleaning, the SBS values significantly decreased in groups P-I and P-W (Table 2), which generally accords with the results of a recent previous study [4]. Moreover, complete delamination of the porcelain or liner was frequently observed in the liner-applied groups (Figure 6b), which is comparable to the finding of Kim *et al.* [5]. The substantially increased wettability in the two groups (Figure 3) may have caused a weakening in the liner bulk [4]. Moreover, it seems that great surface wettability of groups P-I and P-W put the veneering porcelain slurry in intimate contact with the zirconia surfaces without the intervention of a liner. Thus, simple plasma treatment seems to be a more effective and definite method than liner application for enhancing the bonding between zirconia and porcelain veneer. In addition, storage of plasma-treated specimens in water can be easily performed in dental laboratories and clinics without any additional equipment or facility. Long-term experimental and clinical investigations of plasma-treated zirconia-based restorations are still required to validate general clinical use of this relatively new technology. Moreover, this versatile technology may be effectively applied to various other dental bonding fields.

4. Materials and Methods

4.1. Zirconia Specimen Preparation

Rectangle-shaped (10 mm × 10 mm × 1 mm) sintered zirconia plates (KaVo Everest® ZS-Ronde, Kaltenbach and Voigt GmbH and Co., Bismarcking, Germany) were prepared according to the manufacturer's instructions. To produce a condition similar to that of milled zirconia, one surface (10 mm × 10 mm) of each specimen was ground with a polishing diamond disc (Allied High Tech Products, Rancho Domingues, CA, USA) under running water on a polishing machine (M-Prep3™, Allied High Tech Products) at a speed of 300 rpm [5]. Specimens were ultrasonically cleaned in acetone, alcohol, and distilled water for 10 min each, and finally air-dried. The average surface roughness R_a of each specimen was measured using a previously calibrated profilometer (Surftest SV-400, Mitutoyo Corp., Kawasaki, Japan) at a stylus speed of 0.1 mm/s, a cutoff of 0.8 mm, and a range of 600 mm and recorded as the average of the five readings [1]. The R_a of the zirconia specimens prepared using the polishing disc (0.233 ± 0.016 μm) was similar to that reported for milled zirconia (0.213 ± 0.002 μm) [5].

4.2. Plasma Treatment

Non-plasma-treated (untreated) zirconia specimens were used as control. Except for the control group, all specimens were subjected to plasma treatment. A dielectric barrier discharge atmospheric argon plasma system (CJ08-P180, Changjo Engineering Co., Ltd., Daegu, Korea) with a 13.56 MHz radio frequency power supply was used (Figure 7). The plasma treatment on each zirconia specimen was performed at a discharge power of 200 W with argon for 600 s. The flow rate of argon was 20 slm. In group P-I, the plasma-treated specimens were immediately subjected to the subsequent porcelain veneering procedures. In group P-W, the specimens were stored in distilled water in a plastic container for 24 h, and the surfaces then dried using a strong air blow to ensure the complete removal of residual water from the surfaces prior to veneering procedures. In group P-A, the specimens were left in air for 24 h and then subjected to veneering procedures. The storage was performed in a temperature controlled room at 23 °C ± 1 °C with relative humidity at 50% ± 5%.

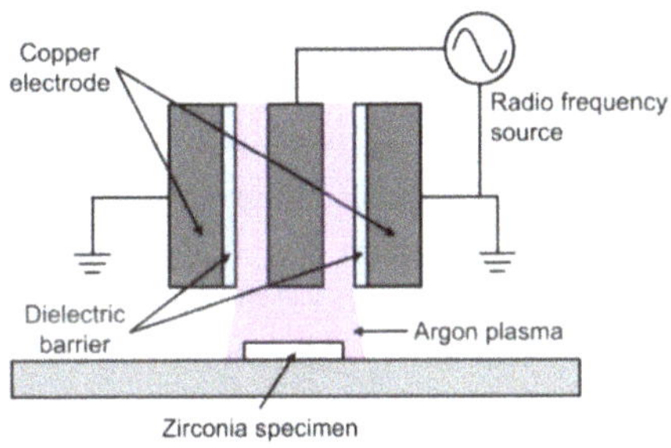

Figure 7. Schematic illustration of the atmospheric argon plasma system used in this study.

4.3. Surface Examinations

The untreated (control) and plasma-treated zirconia surfaces were observed by a field emission-scanning electron microscope (FE-SEM, JSM-6700F, Jeol, Tokyo, Japan). Specimens were sputter-coated with platinum and photographs of representative areas of the surfaces were taken at 2000× magnification.

In addition, the untreated and treated zirconia surfaces were examined by atomic force microscopy (AFM, XE-100, Park Systems Corp., Suwon, Korea). During analysis, the microscope was operated in non-contact mode with a Si_3N_4 V-shaped cantilever (k = 42 N/m). Images with 256 × 256 pixels were taken in air with a scan size of 5 μm × 5 μm and a scan rate of 0.5 Hz. Three measurements were performed for each specimen using a standardized rectangular spot (1.5 μm × 1.5 μm). The R_a of each specimen was recorded as the average of the three readings (n = 6).

4.4. Contact Angle Measurements

To compare the zirconia surface wettability, the contact angles (CAs) of water droplets on the surfaces were measured (using the static sessile drop method) by a surface goniometer (OCA 15 plus, Data-Physics Instrument GmbH, Filderstadt, Germany) (n = 6). CA measurements were performed in a temperature-controlled room at 23 °C ± 1 °C with relative humidity at 50% ± 5%.

4.5. X-ray Photoelectron Spectroscopy

The chemical compositions of the outermost surfaces of zirconia specimens were analyzed using X-ray photoelectron spectroscopy (XPS, K-Alpha, Thermo Fisher Scientific, East Grinstead, UK) with a monochromatic Al Kα X-ray source. Pass energies were 200 and 20 eV for survey scans and high resolution scans, respectively. The pressure within the XPS chamber was 10^{-7} to 10^{-8} Pa. Photo-emitted electrons were collected at a take-off angle of 90°. The intensities of Zr 3d, O 1s, Y 3d, and C ls were determined. Deconvolution of the O 1s core-level spectra was performed with a mixed Gaussian/Lorentzian model after a Shirley-type background subtraction [29]. Data acquisition and processing were performed using Thermo Advantage software (Thermo Fisher Scientific).

4.6. Veneering Procedures

Eighty rectangle-shaped (10 mm × 10 mm × 1 mm) sintered zirconia plates were prepared and embedded "loosely" in round silicone rubber molds using an acrylic resin. The embedded zirconia specimens were divided into four groups (control, groups P-I, P-W, and P-A), each group then being further separated into two subgroups according to the use of liner prior to veneering (no-liner and liner subgroups) ($n = 10$). For the plasma-treated groups, the zirconia plates were temporarily removed from the embedding resin during the plasma treatment.

The veneering porcelain onto the zirconia surfaces was performed using an Ultradent jig (Ultradent Products Inc., South Jordan, UT, USA), as shown in Figure 8. A loosely embedded zirconia specimen was placed into the jig and the surface to be veneered was isolated using a cylindrical-shaped plastic matrix (internal diameter: 2.38 mm). In the liner subgroups, a thin coat of a ceramic liner paste (Zr-Adhesive, Heraeus Kulzer GmbH, Hanau, Germany) was applied onto the zirconia surface through the matrix. The liner-coated zirconia plate was removed from the jig and the embedding resin and then subjected to the firing procedure in a porcelain furnace (Vacumat 40, Vita Zahnfabrik, Bad Säckingen, Germany). A mixed porcelain slurry (Heraceram Zirconia, Heraeus Kulzer GmbH) was applied onto the liner region through the matrix. After excess liquid was blotted with absorbent paper, the veneered zirconia plate was removed from the jig and the embedding resin, then subjected to the firing procedure in the Vacumat 40 porcelain furnace. In the no-liner subgroups, a mixed porcelain slurry was directly applied to the zirconia surface without the above liner application procedures. The firing temperature/holding time was 1050 °C/10 min and 870 °C/1 min for the liner and the veneering porcelain, respectively, following the manufacturer's instructions.

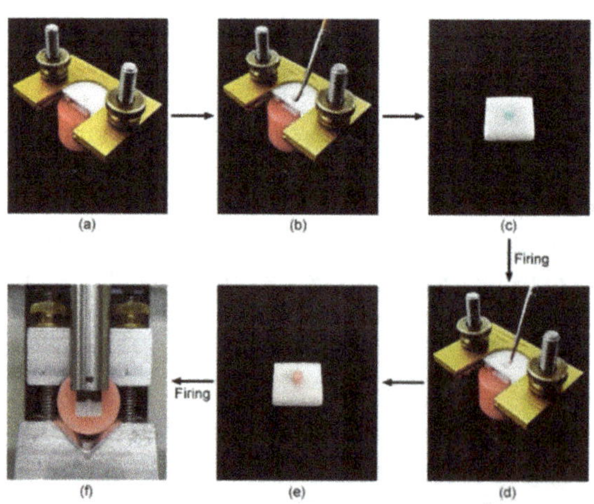

Figure 8. Porcelain veneering and subsequent shear bond strength testing procedures using Ultradent jig. (**a**) Zirconia specimen placed into the jig; (**b**) Liner application; (**c**) Liner applied to the zirconia surface; (**d**) Veneering porcelain on the liner; (**e**) Porcelain veneered on the surface; (**f**) Shear bond strength testing. In the no-liner subgroups, steps (b,c) were omitted.

4.7. Shear Bond Strength Testing

Prior to shear bond strength (SBS) testing, the veneered specimen was firmly fixed in the embedding resin by adding additional acrylic resin around the specimen (Figure 8f). Specimens were then perpendicularly engaged at their veneered porcelain cylinder bases with a round-notched custom shear blade (Ultradent Products Inc.) in a universal testing machine (Model 3343, Instron Inc., Canton, MA, USA) at a crosshead speed of 1.0 mm/min until bonding failure occurred (Figure 8f). Bond strengths (MPa) were calculated from the peak load of failure (N) divided by the veneered surface area, the diameter of which (~2.38 mm) for each specimen had been measured using a vernier caliper prior to SBS testing. Following debonding, all fractured interfaces were examined under an optical microscope (SZ61, Olympus, Tokyo, Japan) at 20× magnification to determine the mode of failure. These were classified into one of the three types: adhesive failure at the zirconia-veneer interface; cohesive failure within the veneered porcelain; and a combination of these two failure modes (mixed failure).

4.8. Statistical Analysis

All data were examined for the normality of their distribution (Shapiro-Wilk test) and for the equality of the variances (Levene test). The R_a of the zirconia surfaces before and after the plasma treatment was statistically compared by Student's t test. The CA values of the four groups were compared using one-way analysis of variance (ANOVA) followed by Tukey's *post hoc* test. For the SBS data, two-way ANOVA was used to test the main effects of the storage conditions (including control) and the use/non-use of the liner. Based on the results from the initial analysis, the data file was stratified and analyzed using one-way ANOVA followed by Tukey's *post hoc* test or Student's t test to determine whether the differences in SBS were statistically significant [30]. All statistical analyses were carried out using SPSS 17.0 for Windows (SPSS Inc., Chicago, IL, USA) at a level of significance of $\alpha = 0.05$.

5. Conclusions

The present *in vitro* study suggests that non-thermal atmospheric argon plasma treatment of zirconia effectively produced a clean, super-hydrophilic surface, which seems appropriate for the optimal wetting of veneering porcelain and, as a result, significantly enhanced the bonding between the zirconia and veneering porcelain. The plasma-induced super-hydrophilicity was effectively maintained by simple water immersion of the specimens until porcelain veneering.

Acknowledgments: This research was supported by Basic Science Research Program through the National Research Foundation of Korea (NRF) funded by the Ministry of Education (2013R1A1A2061732).

Author Contributions: Mun-Hwan Lee and Bong Ki Min prepared the specimens, performed the experiments and contributed to manuscript writing. Jun Sik Son performed data analysis and commented on various aspects during the manuscript preparation. Tae-Yub Kwon conceived and designed the overall research concept, analyzed the data, contributed to manuscript writing and supervised the work.

Conflicts of Interest: The authors declare no conflict of interest. The founding sponsors had no role in the design of the study; in the collection, analyses, or interpretation of data; in the writing of the manuscript, and in the decision to publish the results.

References

1. Kim, M.J.; Kim, Y.K.; Kim, K.H.; Kwon, T.Y. Shear bond strengths of various luting cements to zirconia ceramic: Surface chemical aspects. *J. Dent.* **2011**, *39*, 795–803. [CrossRef] [PubMed]
2. Janyavula, S.; Lawson, N.; Cakir, D.; Beck, P.; Ramp, L.C.; Burgess, J.O. The wear of polished and glazed zirconia against enamel. *J. Prosthet. Dent.* **2013**, *109*, 22–29. [CrossRef]
3. Chaiyabutr, Y.; McGowan, S.; Phillips, K.M.; Kois, J.C.; Giordano, R.A. The effect of hydrofluoric acid surface treatment and bond strength of a zirconia veneering ceramic. *J. Prosthet. Dent.* **2008**, *100*, 194–202. [CrossRef]

4. Canullo, L.; Micarelli, C.; Bettazzoni, L.; Magnelli, A.; Baldissara, P. Shear bond strength of veneering porcelain to zirconia after argon plasma treatment. *Int. J. Prosthodont.* **2014**, *27*, 137–139. [CrossRef] [PubMed]
5. Kim, H.J.; Lim, H.P.; Park, Y.J.; Vang, M.S. Effect of zirconia surface treatments on the shear bond strength of veneering ceramic. *J. Prosthet. Dent.* **2011**, *105*, 315–322. [CrossRef]
6. Sailer, I.; Feher, A.; Filser, F.; Gauckler, L.J.; Luthy, H.; Hammerle, C.H. Five-year clinical results of zirconia frameworks for posterior fixed partial dentures. *Int. J. Prosthodont.* **2007**, *20*, 383–388. [PubMed]
7. Wang, G.; Zhang, S.; Bian, C.; Kong, H. Interface toughness of a zirconia-veneer system and the effect of a liner application. *J. Prosthet. Dent.* **2014**, *112*, 576–583. [CrossRef] [PubMed]
8. Choi, Y.S.; Kim, S.H.; Lee, J.B.; Han, J.S.; Yeo, I.S. *In vitro* evaluation of fracture strength of zirconia restoration veneered with various ceramic materials. *J. Adv. Prosthodont.* **2012**, *4*, 162–169. [CrossRef] [PubMed]
9. Schmitt, J.; Goellner, M.; Lohbauer, U.; Wichmann, M.; Reich, S. Zirconia posterior fixed partial dentures: 5-year clinical results of a prospective clinical trial. *Int. J. Prosthodont.* **2012**, *25*, 585–589. [PubMed]
10. Tarumi, N.; Uo, M.; Yamaga, E.; Watari, F. SEM observation and wettability of variously processed and fractured surface of dental zirconia. *Appl. Surf. Sci.* **2012**, *262*, 253–257. [CrossRef]
11. Wang, G.; Zhang, S.; Bian, C.; Kong, H. Effect of zirconia surface treatment on zirconia/veneer interfacial toughness evaluated by fracture mechanics method. *J. Dent.* **2014**, *42*, 808–815. [CrossRef] [PubMed]
12. Ritts, A.C.; Li, H.; Yu, Q.; Xu, C.; Yao, X.; Hong, L.; Wang, Y. Dentin surface treatment using a non-thermal argon plasma brush for interfacial bonding improvement in composite restoration. *Eur. J. Oral Sci.* **2010**, *118*, 510–516. [CrossRef] [PubMed]
13. Chen, M.; Zhang, Y.; Sky Driver, M.; Caruso, A.N.; Yu, Q.; Wang, Y. Surface modification of several dental substrates by non-thermal, atmospheric plasma brush. *Dent. Mater.* **2013**, *29*, 871–880. [CrossRef] [PubMed]
14. Rodríguez-Villanueva, C.; Encinas, N.; Abenojar, J.; Martínez, M.A. Assessment of atmospheric plasma treatment cleaning effect on steel surfaces. *Surf. Coat. Technol.* **2013**, *236*, 450–456. [CrossRef]
15. Liebermann, A.; Keul, C.; Bahr, N.; Edelhoff, D.; Eichberger, M.; Roos, M.; Stawarczyk, B. Impact of plasma treatment of PMMA-based CAD/CAM blanks on surface properties as well as on adhesion to self-adhesive resin composite cements. *Dent. Mater.* **2013**, *29*, 935–944. [CrossRef] [PubMed]
16. Chan, Y.H.; Kim, J.K.; Liu, D.; Liu, P.C.K.; Cheung, Y.M.; Ng, M.W. Effect of plasma treatment of Au-Ni-Cu bond pads on process windows of Au wire bonding. *IEEE Trans. Adv. Packag.* **2005**, *28*, 674–684. [CrossRef]
17. Noro, A.; Kaneko, M.; Murata, I.; Yoshinari, M. Influence of surface topography and surface physicochemistry on wettability of zirconia (tetragonal zirconia polycrystal). *J. Biomed. Mater. Res. B* **2013**, *101*, 355–363. [CrossRef] [PubMed]
18. Lohbauer, U.; Zipperle, M.; Rischka, K.; Petschelt, A.; Müller, F.A. Hydroxylation of dental zirconia surfaces: Characterization and bonding potential. *J. Biomed. Mater. Res. B.* **2008**, *87*, 461–467. [CrossRef] [PubMed]
19. Piascik, J.R.; Swift, E.J.; Braswell, K.; Stoner, B.R. Surface fluorination of zirconia: Adhesive bond strength comparison to commercial primers. *Dent. Mater.* **2012**, *28*, 604–608. [CrossRef] [PubMed]
20. Nam, Y.; Sharratt, S.; Byon, C.; Kim, S.J.; Ju, Y.S. Fabrication and characterization of the capillary performance of superhydrophilic Cu micropost arrays. *J. Microelectromech. Syst.* **2010**, *19*, 581–588.
21. Yaman, N.; Özdoğan, E.; Seventekin, N. Atmospheric plasma treatment of polypropylene fabric for improved dyeability with insoluble textile dyestuff. *Fibers Polym.* **2011**, *12*, 35–41. [CrossRef]
22. Da Silva, M.A.M.; Guerra Neto, C.L.B.; Nunes Filho, A.; Freitas, D.O.; Braz, D.C.; Alves, C., Jr. Influence of topography on plasma treated titanium surface wettability. *Surf. Coat. Technol.* **2013**, *235*, 447–453. [CrossRef]
23. Choi, Y.R.; Kwon, J.S.; Song, D.H.; Choi, E.H.; Lee, Y.K.; Kim, K.N.; Kim, K.M. Surface modification of biphasic calcium phosphate scaffolds by non-thermal atmospheric pressure nitrogen and air plasma treatment for improving osteoblast attachment and proliferation. *Thin Solid Films* **2013**, *547*, 235–240. [CrossRef]
24. Fricke, K.; Steffen, H.; von Woedtke, T.; Schroder, K.; Weltmann, K.D. High rate etching of polymers by means of an atmospheric pressure plasma jet. *Plasma Process. Polym.* **2011**, *8*, 51–58. [CrossRef]
25. Sun, S.; Sun, J.; Yao, L.; Qiu, Y. Wettability and sizing property improvement of raw cotton yarns treated with He/O_2 atmospheric pressure plasma jet. *Appl. Surf. Sci.* **2011**, *257*, 2377–2382. [CrossRef]
26. Yavirach, P.; Chaijareenont, P.; Boonyawan, D.; Pattamapun, K.; Tunma, S.; Takahashi, H.; Arksornnukit, M. Effects of plasma treatment on the shear bond strength between fiber-reinforced composite posts and resin composite for core build-up. *Dent. Mater. J.* **2009**, *28*, 686–692. [CrossRef] [PubMed]

27. Mi, Y.; Wang, J.; Yang, Z.; Wang, Z.; Wang, H.; Yang, S. A simple one-step solution deposition process for constructing high-performance amorphous zirconium oxide thin film. *RSC Adv.* **2014**, *4*, 6060–6067. [CrossRef]
28. Homola, T.; Matoušek, J.; Medvecká, V.; Zahoranová, A.; Kormunda, M.; Kováčik, D.; Černák, M. Atmospheric pressure diffuse plasma in ambient air for ITO surface cleaning. *Appl. Surf. Sci.* **2012**, *258*, 7135–7139. [CrossRef]
29. Piascik, J.R.; Wolter, S.D.; Stoner, B.R. Development of a novel surface modification for improved bonding to zirconia. *Dent. Mater.* **2011**, *27*, e99–e105. [CrossRef] [PubMed]
30. Maryanchik, I.; Brendlinger, E.J.; Fallis, D.W.; Vandewalle, K.S. Shear bond strength of orthodontic brackets bonded to various esthetic pontic materials. *Am. J. Orthod. Dentofac. Orthop.* **2010**, *137*, 684–689. [CrossRef] [PubMed]

Permissions

All chapters in this book were first published in Materials, by MDPI; hereby published with permission under the Creative Commons Attribution License or equivalent. Every chapter published in this book has been scrutinized by our experts. Their significance has been extensively debated. The topics covered herein carry significant findings which will fuel the growth of the discipline. They may even be implemented as practical applications or may be referred to as a beginning point for another development.

The contributors of this book come from diverse backgrounds, making this book a truly international effort. This book will bring forth new frontiers with its revolutionizing research information and detailed analysis of the nascent developments around the world.

We would like to thank all the contributing authors for lending their expertise to make the book truly unique. They have played a crucial role in the development of this book. Without their invaluable contributions this book wouldn't have been possible. They have made vital efforts to compile up to date information on the varied aspects of this subject to make this book a valuable addition to the collection of many professionals and students.

This book was conceptualized with the vision of imparting up-to-date information and advanced data in this field. To ensure the same, a matchless editorial board was set up. Every individual on the board went through rigorous rounds of assessment to prove their worth. After which they invested a large part of their time researching and compiling the most relevant data for our readers.

The editorial board has been involved in producing this book since its inception. They have spent rigorous hours researching and exploring the diverse topics which have resulted in the successful publishing of this book. They have passed on their knowledge of decades through this book. To expedite this challenging task, the publisher supported the team at every step. A small team of assistant editors was also appointed to further simplify the editing procedure and attain best results for the readers.

Apart from the editorial board, the designing team has also invested a significant amount of their time in understanding the subject and creating the most relevant covers. They scrutinized every image to scout for the most suitable representation of the subject and create an appropriate cover for the book.

The publishing team has been an ardent support to the editorial, designing and production team. Their endless efforts to recruit the best for this project, has resulted in the accomplishment of this book. They are a veteran in the field of academics and their pool of knowledge is as vast as their experience in printing. Their expertise and guidance has proved useful at every step. Their uncompromising quality standards have made this book an exceptional effort. Their encouragement from time to time has been an inspiration for everyone.

The publisher and the editorial board hope that this book will prove to be a valuable piece of knowledge for researchers, students, practitioners and scholars across the globe.

List of Contributors

Aitziber Adrados , Isabel De Marco, Alexander López-Urionabarrenechea, Jon Solar, Blanca M. Caballero and Naia Gastelu
Department of Environmental and Chemical Engineering, University of the Basque Country UPV/EHU,Bilbao 48013, Spain

Piotr Rychter and Wojciech Prochwicz
Faculty of Mathematics and Natural Sciences, Jan Dlugosz University in Czestochowa, Armii Krajowej 13/15 Av., Czestochowa 42-218, Poland

Natalia Śmigiel-Gac and Piotr Dobrzyński
Faculty of Mathematics and Natural Sciences, Jan Dlugosz University in Czestochowa, Armii Krajowej 13/15 Av., Czestochowa 42-218, Poland
Centre of Polymer and Carbon Materials, Polish Academy of Sciences, Curie-Sklodowska 34 St., Zabrze 41-800, Poland

Elżbieta Pamuła
Department of Biomaterials, Faculty of Materials Science and Ceramics, AGH University of Science and Technology, Al Mickiewicza 30, Kraków 30-059, Poland

Anna Smola-Dmochowska and Henryk Janeczek
Centre of Polymer and Carbon Materials, Polish Academy of Sciences, Curie-Sklodowska 34 St., Zabrze 41-800, Poland

Luiza R. Brancher
Laboratory of Materials Chemistry Research, University of Caxias do Sul, Rua
Francisco Getúlio Vargas, 1130, Caxias do Sul RS 95070-560, Brazil
Laboratory of Constructive Technology, University of Caxias do Sul, Avenida Frederico Segala, 3099, Caxias do Sul RS 95010-550, Brazil

Ana Maria C. Grisa and Mára Zeni
Laboratory of Materials Chemistry Research, University of Caxias do Sul, Rua
Francisco Getúlio Vargas, 1130, Caxias do Sul RS 95070-560, Brazil

Maria Fernanda de O. Nunes and Daniel T. Pagnussat
Laboratory of Constructive Technology, University of Caxias do Sul, Avenida Frederico Segala, 3099, Caxias do Sul RS 95010-550, Brazil

Philip Van den Heede, Niels Ringoot, Arno Beirnaert, Geert De Schutter and Nele De Belie
Magnel Laboratory for Concrete Research, Ghent University, Technologiepark Zwijnaarde 904, Ghent B-9052, Belgium

Andres Van Brecht and Erwin Van den Brande
Indaver nv, Dijle 17a, Mechelen B-2800, Belgium

Eva Jesenská
Department of Materials Science and Technology, Nagaoka University of Technology, Nagaoka,Niigata 1603-1, Japan
Institute of Physics, Charles University, Prague 12116, Czech Republic

Takahiro Hashinaka and Takayuki Ishibashi
Department of Materials Science and Technology, Nagaoka University of Technology, Nagaoka,Niigata 1603-1, Japan

Kiyoshi Kuga, Ken-ichi Aoshima, Kenji Machida and Hidekazu Kinjo
Science and Technology Research Laboratories, Japan Broadcasting Corporation (NHK),Tokyo 157-8510, Japan

Martin Veis, Lukáš Beran, Ján Dušek and Roman Antoš
Institute of Physics, Charles University, Prague 12116, Czech Republic

Qingsong Mei
Department of Materials Engineering, School of Power and Mechanical Engineering, Wuhan University, Wuhan 430072, China

Juying Li
School of Mechanical Engineering, Wuhan Polytechnic University, Wuhan 430023, China

Csaba Hegedűs
Department of Biomaterials and Prosthetic Dentistry, University of Debrecen, Nagyerdei krt. 98, H-4032 Debrecen, Hungary

Chia-Che Ho
Institute of Oral Science, Chung Shan Medical University, Taichung City 402, Taiwan

Attila Csik and Sándor Biri
Institute for Nuclear Research (ATOMKI), Hungarian Academy of Sciences, Bem tér 18/c, H-4026 Debrecen, Hungary

Shinn-Jyh Ding
Institute of Oral Science, Chung Shan Medical University, Taichung City 402, Taiwan
Department of Dentistry, Chung Shan Medical University Hospital, Taichung City 402, Taiwan

Jessica Giro-Paloma and Ana Inés Fernández
Department of Materials Science and Metallurgical Engineering, Faculty of Chemistry,Universitat de Barcelona, C/ Martí i Franquès, Barcelona 1. 08028, Spain

Refat Al-Shannaq and Mohammed M. Farid
Department of Chemical and Materials Engineering, University of Auckland, Private Bag 92019, 20 Symonds Street, Auckland 1142, New Zealand

Kuijian Yang, Yuli Chen, Fei Pan, Shengtao Wang and Yong Ma
Institute of Solid Mechanics, International Research Institute for Multidisciplinary Science, Beihang University, Beijing 100191, China

Qijun Liu
Department of Aerospace Engineering, University of Illinois, Champaign, IL 61801, USA

Arash Ghaedizadeh, Jianhu Shen and Yi Min Xie
Centre for Innovative Structures and Materials, School of Engineering, RMIT University, GPO Box 2476, Melbourne 3001, Australia

Xin Ren
Centre for Innovative Structures and Materials, School of Engineering, RMIT University, GPO Box 2476, Melbourne 3001, Australia
Key Laboratory of Traffic Safety on Track, School of Traffic & Transportation Engineering, Central South University, Changsha 410075, China

Yanlin Yu and JianweiWang
General Research Institute for Nonferrous Metals, Beijing 100088, China

Wei Xiao
General Research Institute for Nonferrous Metals, Beijing 100088, China
School of Materials Science and Engineering, University of Science and Technology Beijing, Beijing 100083, China

Ligen Wang
General Research Institute for Nonferrous Metals, Beijing 100088, China
Power Environmental Energy Research Institute, Covina, CA 91722, USA

Mónika Sándor 1 and József Fazakas
Faculty of Chemistry and Chemical Engineering, Babes-Bolyai University, Arany János Street, No.11, Cluj-Napoca 400028, Romania

Cristina Lavinia Nistor, Rusandica Stoica, Cristian Andi Nicolae, Florin Oancea and Dan Donescu
National Research & Development Institute for Chemistry and Petrochemistry—ICECHIM, Splaiul Independentei No.202, 6th district, PO Box 35-174, Bucharest 060021, Romania

Elvira Alexandrescu
Faculty of Chemistry and Chemical Engineering, Babes-Bolyai University, Arany János Street, No.11, Cluj-Napoca 400028, Romania
National Research & Development Institute for Chemistry and Petrochemistry—ICECHIM, Splaiul Independentei No.202, 6th district, PO Box 35-174, Bucharest 060021, Romania

Gábor Szalontai
NMR Laboratory, Institute of Materials Engineering, University of Pannonia, Egyetem Street, No. 10, Veszprém 8200, Hungary

Song Wang, Miao Jiang, Lihong Gao, Zhuang Ma and Fuchi Wang
School of Materials Science and Engineering, Beijing Institute of Technology, Beijing 100081, China

Helena Pereira and António José Alves Santos
Centro de Estudos Florestais, Instituto Superior de Agronomia, Universidade de Lisboa, Tapada da Ajuda,1349-017 Lisboa, Portugal

Ofélia Anjos
Centro de Estudos Florestais, Instituto Superior de Agronomia, Universidade de Lisboa, Tapada da Ajuda,1349-017 Lisboa, Portugal
Instituto Politécnico de Castelo Branco, Escola Superior Agrária, Apartado 119, 6001-909 Castelo Branco, Portugal

Rodrigo García-Rodríguez and Gerko Oskam
Departamento de Física Aplicada, Centro de Investigación y de Estudios Avanzados del Instituto Politécnico Nacional (CINVESTAV-IPN), Mérida, Yucatán 97310, Mexico

Julio Villanueva-Cab
Instituto de Física, Benemérita Universidad Autónoma de Puebla (BUAP), Puebla, Puebla 72570, Mexico

Juan A. Anta
Nanostructured Solar Cells Group, Department of Physical, Chemical and Natural Systems, Universidad Pablo de Olavide, Seville ES-41013, Spain

B. Deepa
Department of Chemistry, Bishop Moore College, Mavelikara-690101, Kerala, India
Department of Chemistry, Church Missionary Society(C.M.S) College, Kottayam-686001, Kerala, India

Eldho Abraham
Robert H. Smith Faculty of Agriculture, Food and Environment, Hebrew University, Jerusalem, POB 12, Rehovot 76100, Israel

Laly A. Pothan
Department of Chemistry, Bishop Moore College, Mavelikara-690101, Kerala, India

Nereida Cordeiro and Marisa Faria
Faculty of Exact Science and Engineering, University of Madeira, 9000-390 Funchal, Portugal

Sabu Thomas
School of Chemical Sciences, Mahatma Gandhi University, Kottayam-686560, Kerala, India

Mun-Hwan
Department of Medical & Biological Engineering, Graduate School, Kyungpook National University, 2-188-1 Samduk-dong, Jung-gu, Daegu 700-412, Korea

Bong Ki Min
Center for Research Facilities, Yeungnam University, 214-1 Dae-dong, Gyeongsan 712-749, Korea

LeeJun Sik Son
Korea Textile Development Institute, 1083 Jungri-dong, Seo-gu, Daegu 703-712, Korea

Tae-Yub Kwon
Department of Dental Biomaterials, School of Dentistry, Kyungpook National University, 2-188-1 Samduk-dong, Jung-gu, Daegu 700-412, Korea

Index

www.ingramcontent.com/pod-product-compliance
Lightning Source LLC
LaVergne TN
LVHW062235200926
848484LV00006B/1545
* 9 7 8 1 6 8 2 8 5 3 9 3 1 *